Includes a trial copy of ArcGIS® Desktop 9.2 software on DVD

GIS Tutorial

Updated for ArcGIS 9.2

WORKBOOK FOR ARCVIEW® 9, SECOND EDITION

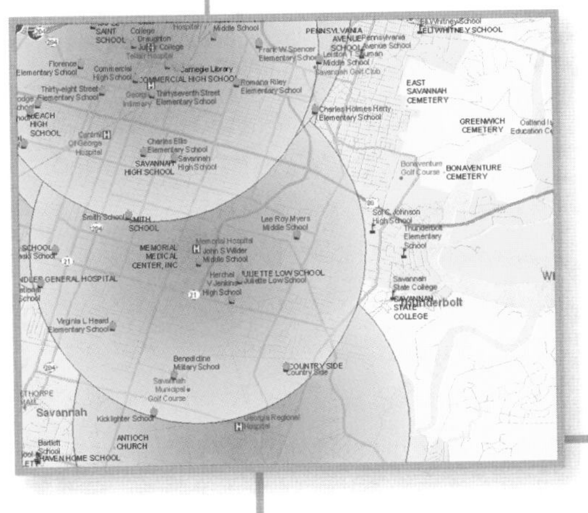

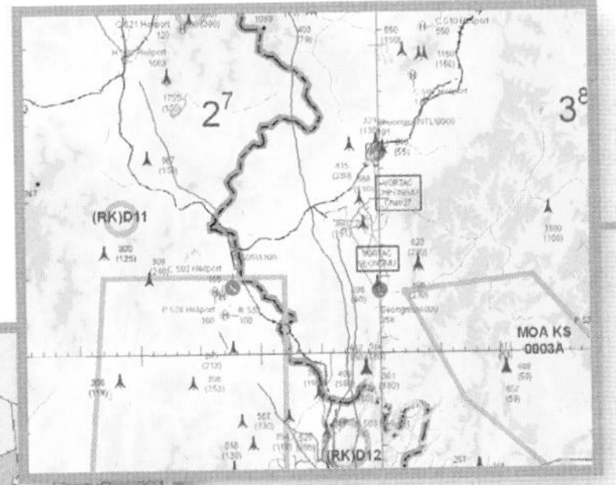

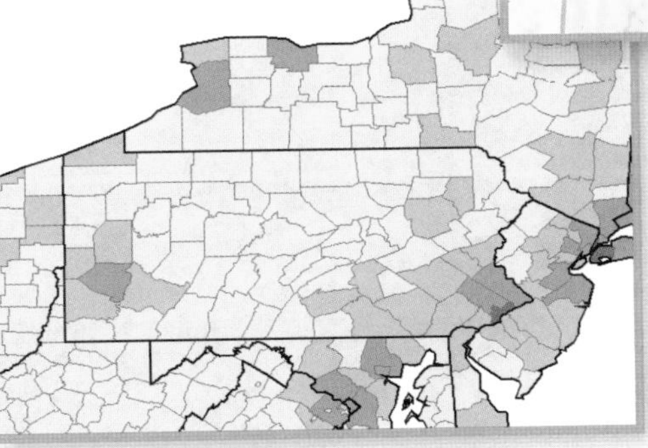

WILPEN L. GORR
KRISTEN S. KURLAND

ESRI PRESS
REDLANDS, CALIFORNIA

ESRI Press, 380 New York Street, Redlands, California 92373-8100

 First edition 2005. Second edition 2007
10 09 08   4 5 6 7 8 9 10

Printed in the United States of America

Library of Congress Cataloging-in-Publication Data
Gorr, Wilpen L.
GIS tutorial : workbook for ArcView 9 / Wilpen L. Gorr, Kristen S. Kurland.—2nd ed.
p. cm.
ISBN 978-1-58948-178-7
1. ArcView. 2. Geographic information systems. I. Kurland, Kristen Seamens, 1966– II. Title.
G70.212.G74 2007
910.285'53—dc22
2007015138

Ask for ESRI Press titles at your local bookstore or order by calling 1-800-447-9778. You can also shop online at www.esri.com/esripress. Outside the United States, contact your local ESRI distributor.

ESRI Press titles are distributed to the trade by the following:

In North America:
Ingram Publisher Services
Toll-free telephone: (800) 648-3104
Toll-free fax: (800) 838-1149
E-mail: customerservice@ingrampublisherservices.com

In the United Kingdom, Europe, and the Middle East:
Transatlantic Publishers Group Ltd.
Telephone: 44 20 7373 2515
Fax: 44 20 7244 1018
E-mail: richard@tpgltd.co.uk

Cover design by Suzanne Davis

Contents

Contents

GIS Tutorial 1 Introduction (continued)

GIS Tutorial 2 Map Design 41

Acknowledgments

We would like to thank all who made this book possible.

GIS Tutorial was used for several years in GIS courses at Carnegie Mellon University before it went to ESRI Press for publication. During this time, the students and teaching assistants at the university who used the book in their classes provided us with significant feedback. Their thoughtful comments guided our revisions and helped improve the content and overall quality of this book.

Faculty at other universities who have taught GIS using drafts of *GIS Tutorial* have also provided valuable feedback. They include Don Dixon of California State University at Sacramento, Mike Rock of Columbus State Community College, Piyusha Singh of State University of New York at Albany, An Lewis of the University of Pittsburgh, and George Tita at the University of California at Irvine.

We are very grateful to the many public servants and vendors who have generously supplied us with interesting GIS applications and data, including Kevin Ford of Facilities Management Services, Carnegie Mellon University; Barb Kviz of the Green Practices Program, Carnegie Mellon University; Susan Golomb and Mike Homa of the City Planning Department, City of Pittsburgh; Richard Chapin of infoUSA, Inc.; Pat Clark and Traci Jackson of Jackson Clark Partners, Pennsylvania Resources Council; Commander Kathleen McNeely, Sgt. Mona Wallace, and John Shultie of the Pittsburgh Bureau of Police; Chief Robert Duffy, Lt. Todd Baxter, Lt. Michael Wood; and Jeff Cheal of the Rochester, New York Police Department; Kirk Brethauer of Southwestern Pennsylvania Commission *(www.spcregion.org)*; and Tele Atlas for use of their USA datasets contained within the ESRI Data & Maps 2004 Media Kit.

Finally, thanks to the team at ESRI who tested, designed, and produced this book.

Preface

GIS Tutorial is the direct result of the authors' experiences teaching GIS to students at high schools and universities, as well as working professionals. *GIS Tutorial* is a hands-on workbook that takes the reader from the basics of using ArcGIS® Desktop interfaces to performing advanced spatial analysis. Instructors can use this book in the classroom, or the individual can use it for self-study.

If you are new to ArcGIS Desktop and are using this book as a self-study guide, we recommend you work through the exercises in order. However, the tutorials are independent of each other, and you can use them in the order that best fits you or your class's needs.

In tutorial 1, readers will learn the basics of working with existing GIS data and maps. In tutorials 2 and 3, readers will learn how to build maps from GIS data and how to create, then add, graphs and reports to their maps. The exercises in tutorial 4 teach the reader how to create geodatabases and import data into them. Tutorial 5 has the reader explore the basic data types used within GIS and use the Internet to gather GIS data. Editing spatial data is a large part of GIS work, and in tutorial 6 readers learn how to manage an edit session, digitize vector data, and transform data to match real-world coordinates. In tutorial 7, the reader will learn how to map address data as points through a process called Address Geocoding. In tutorials 8 and 9, the reader performs spatial analysis using geoprocessing tools and analysis workflow models.

To reinforce the skills learned in the step-based content and to provoke critical problem-solving skills, there are *Your Turn* tasks and advanced exercise assignments within each tutorial. The quickest way to increase your GIS skills is to follow step-based content with independent work. The Your Turn tasks and Exercise Assignments provide this important follow-up by having the reader perform tasks and solve problems without the aid of step-by-step direction.

This book comes with a CD with exercise data and a DVD with a 180-day trial of ArcView® 9.2. You will need to install the software and the data in order to perform the exercises in this book. (If you already have ArcView 9, ArcEditor™ 9, or ArcInfo™ 9 installed, you will need to uninstall it.) Instructions for installing the data and the software that come with this book are included in the appendix.

We sincerely hope you enjoy *GIS Tutorial*. For teacher resources and updates related to this book, go to *www.esri.com/esripress/GISTutorial.*

OBJECTIVES

Work with map layers
Zoom and pan
Magnifier and Overview windows
Spatial bookmarks
Measure distances
Identify features
Select features on a map
Find features
Work with feature attribute tables
Label features

GIS Tutorial 1

Introduction

This first tutorial familiarizes you with some of the basic functionality of ArcMap and illustrates the fundamentals of GIS. You will work with map layers and underlying attribute data tables for U.S. states, cities, counties, and streets. All layers you will use are made up of geographic or spatial features consisting of points, lines, and polygons. Each geographic feature has a corresponding data record, and you will work with both features and their data records.

Launch ArcMap

ArcMap is the primary mapping component of ArcGIS Desktop software from ESRI. ESRI offers three licensing levels of ArcGIS Desktop, each with increasing capabilities; namely, ArcView, ArcEditor, and ArcInfo. Together, ArcMap and two other components that you will use later in this workbook (ArcCatalog and ArcToolbox) make up ArcView, the world's most popular GIS software.

1 From the Windows taskbar, click Start, All Programs, ArcGIS, ArcMap.

Depending on your operating system and how ArcGIS and ArcMap have been installed, you may have a different navigation menu.

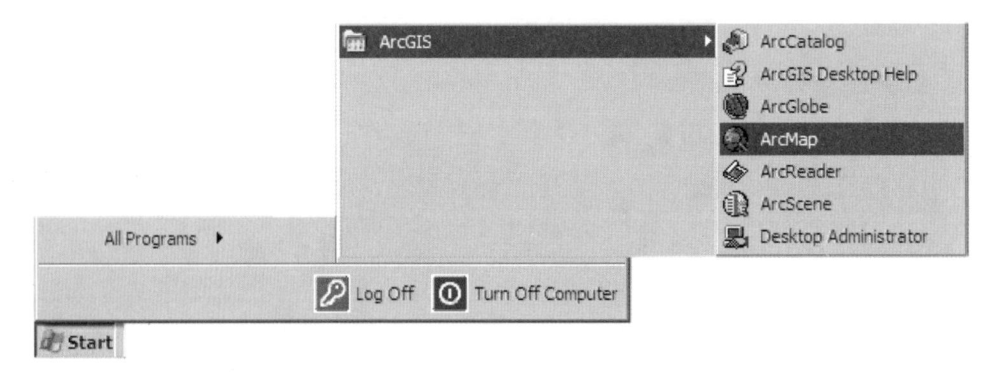

2 In the resulting ArcMap window, click the An existing map radio button and click OK.

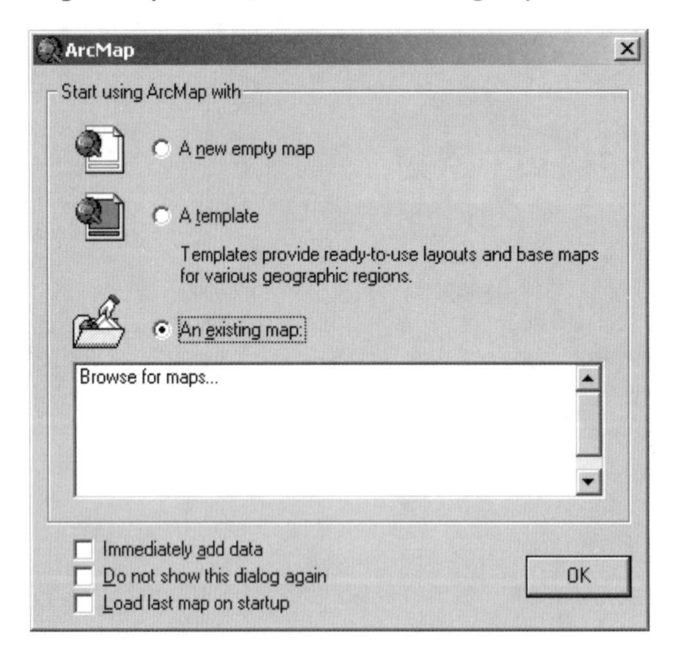

Open an existing map

1 **Browse to the drive on which the Gistutorial folder has been installed (e.g., C:\Gistutorial), click the Tutorial1-1.mxd (or Tutorial1-1) icon and click Open.**

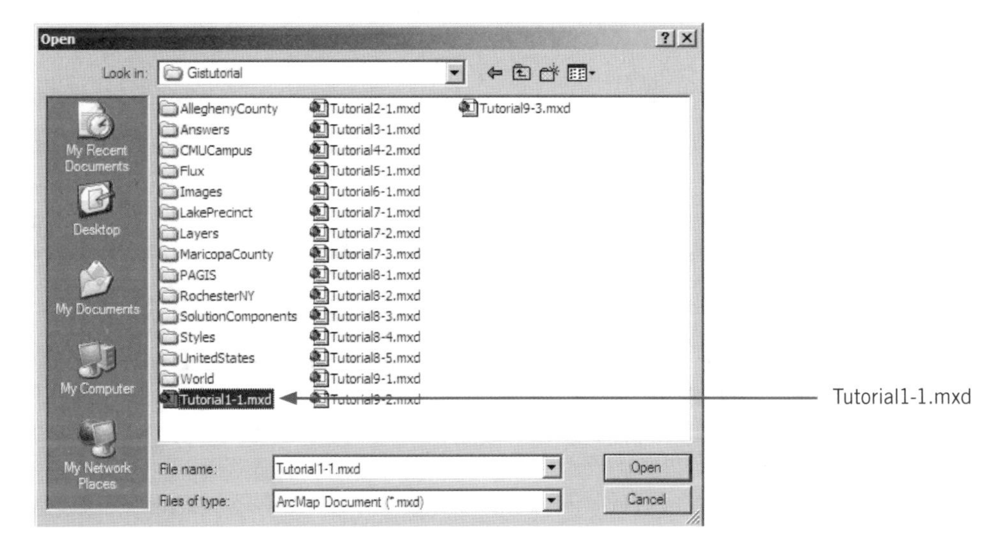

The Tutorial1-1.mxd map document opens in ArcMap showing a map consisting of the US States layer (boundaries of the lower forty-eight contiguous states). The US Cities layer (not yet turned on) is the subset of cities with population greater than 300,000. Note that your Tools menu, which is anchored on the right side of the screen below, may be free-floating on your screen or docked somewhere else on the interface. If you wish, you can anchor it by clicking in its top area, dragging it to a side of the map display window and releasing when you see a thin rectangle materialize. If you do not see the Tools menu at all, click View, Toolbars, Tools to make it visible.

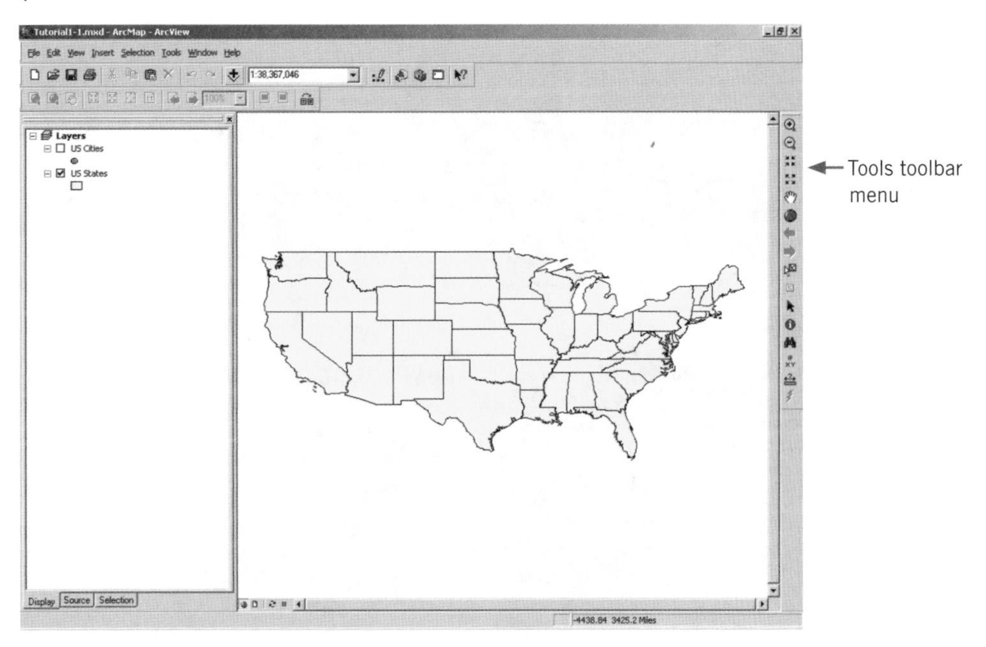

Map layers

Map layers are references to data sources such as point, line, and polygon shapefiles; geodatabase feature classes; raster images; and so forth, representing spatial features that can be displayed on a map.

Turn a layer on

1 Click the small check box to the left of the US Cities layer in the table of contents to turn that layer on.

The table of contents is the panel on the left side of the view window. If the table of contents accidentally closes, click Window, Table of Contents to reopen it. A check mark appears if the layer is turned on. Nothing appears if it is turned off.

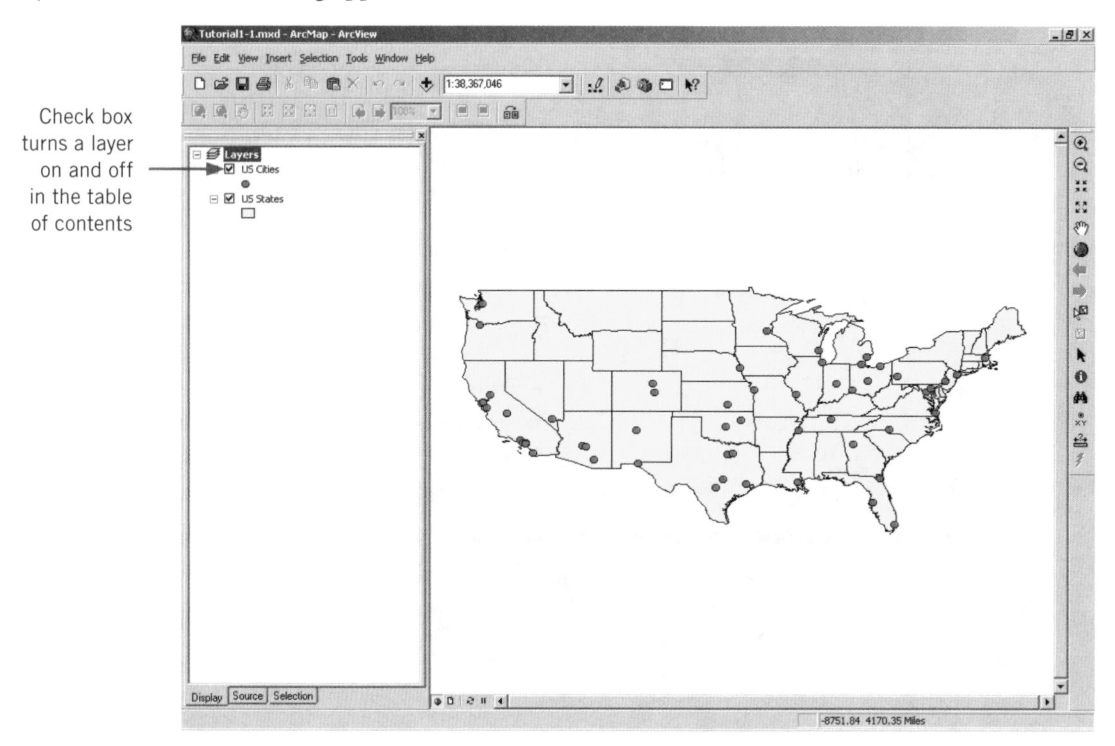

Turn a layer off

1 Click the small check box to the left of the US Cities layer in the table of contents again to turn the layer off.

2 Click the check box again to turn the layer on.

Add a layer

1 **Click the Add Data button.**

2 **In the Add Data browser, browse to \Gistutorial\UnitedStates\Colorado.**

3 **Click Counties.shp.**

4 **Click Add.**

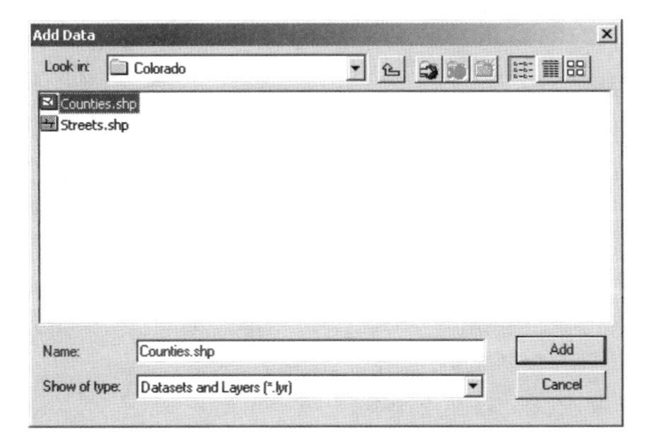

5 **In the Geographic Coordinate Systems Warning dialog box, click the Don't warn me again ever check box and click Close.**

ArcMap randomly picks a color for the Colorado counties layer. The color can be changed later.

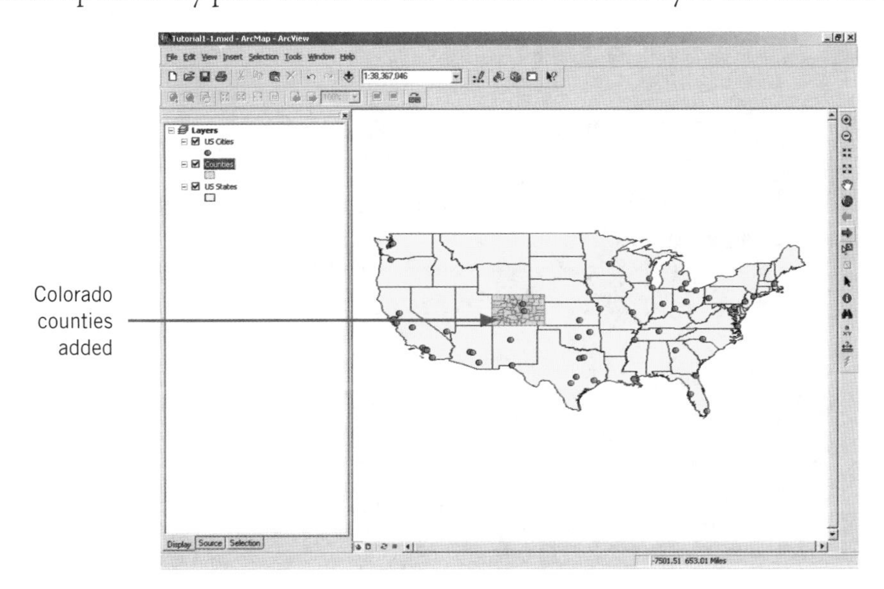

YOUR TURN

Use the Add Data button to add Streets.shp, also found in the \Gistutorial\UnitedStates\Colorado folder. These are street centerlines for Jefferson County, Colorado.

Note: You may not be able to see the streets because they occupy a small area of the map and are partially obscured by the top city point marker in Colorado.

Change a layer's display order

1 **In the table of contents, click and hold down the left mouse button on the name of the US Cities layer.**

2 **Drag the US Cities layer down to the bottom of the table of contents.**

ArcMap draws layers from the bottom up. Because the US Cities layer is now drawn first, its points are covered by the US States and Counties layers and cannot be seen.

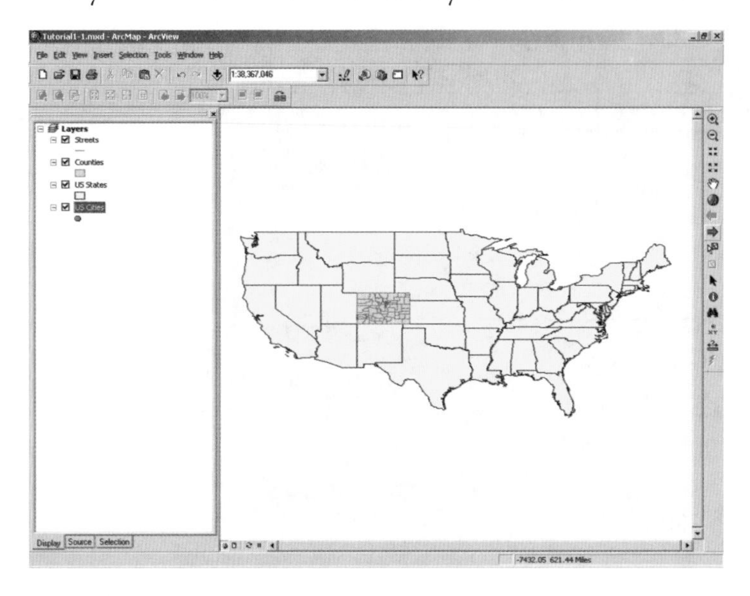

3 **Click and hold down the left mouse button on the US Cities layer.**

4 **Drag the US Cities layer back to the top of the table of contents.**

Because US Cities is now drawn last, its points can be seen again.

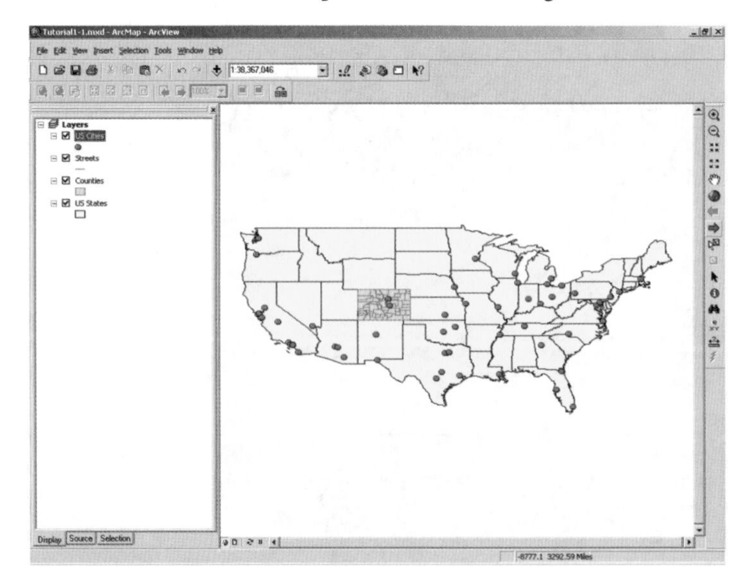

Change a layer's color

1 **Click the Counties layer's legend symbol.**

The legend symbol is the rectangle below the layer name in the table of contents.

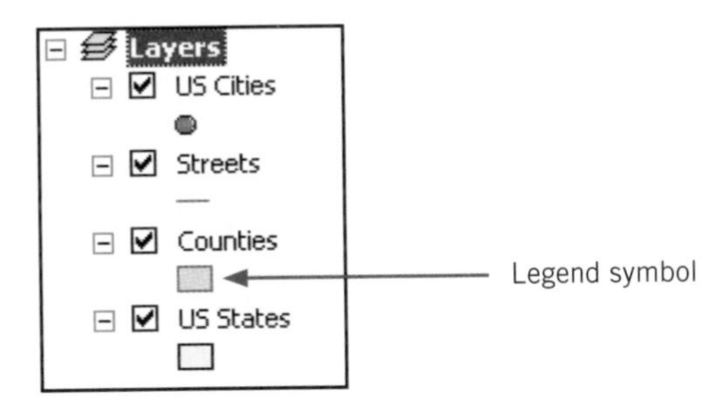

2 **In the resulting Symbol Selector window, click the Fill Color button in the Options section.**

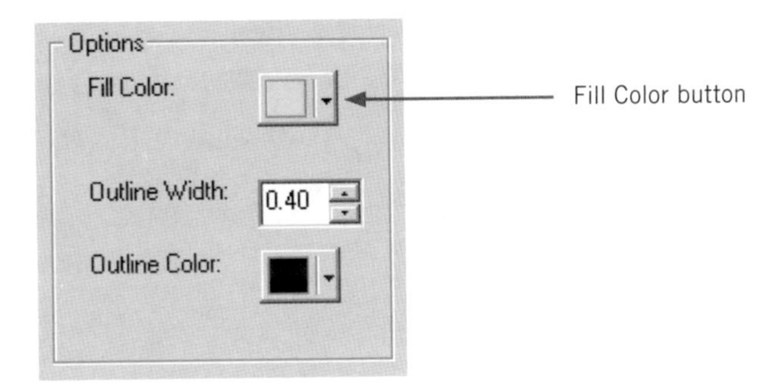

3 **Click the Tarragon Green tile in the Color Palette.**

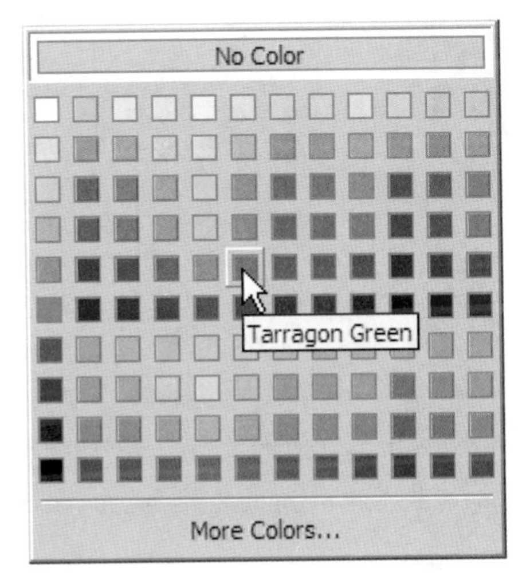

4 **Click OK.**

The layer's color changes to Tarragon Green on the map.

Change a layer's outline color

1 **Click the Counties layer's legend symbol.**

2 **Click the Outline Color button in the Options section of the Symbol Selector dialog box.**

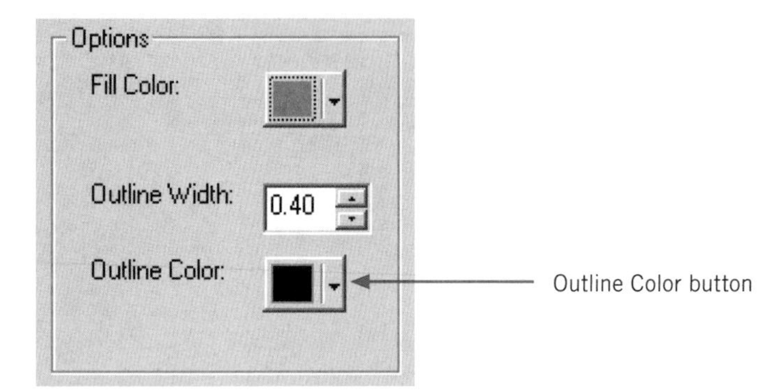

3 **Click the Black tile in the Color Palette.**

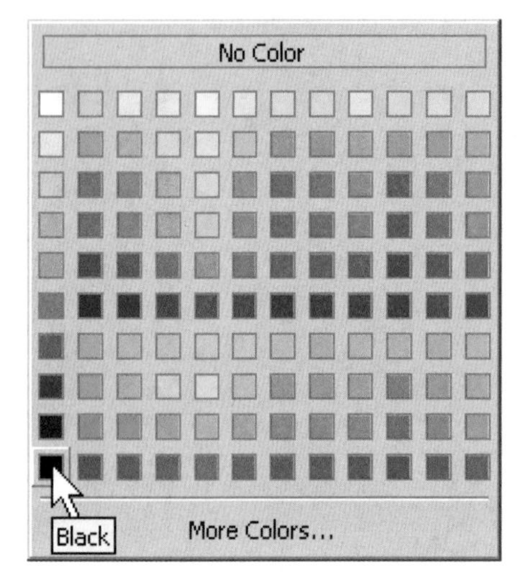

4 **Click OK.**

YOUR TURN

Repeat the steps outlined above, this time changing the colors of the other layers to your liking.

Zoom and pan

Zooming and panning enlarges or reduces the display and shifts the display to reveal different areas of the map. The zoom and pan buttons are found on the Tools toolbar.

Zoom In

1 **Click the Zoom In button.** 

2 **Click and hold down the mouse button on a point above and to the left of the state of Florida.**

3 **Drag the mouse below and to the right of the state of Florida and release the mouse button.**

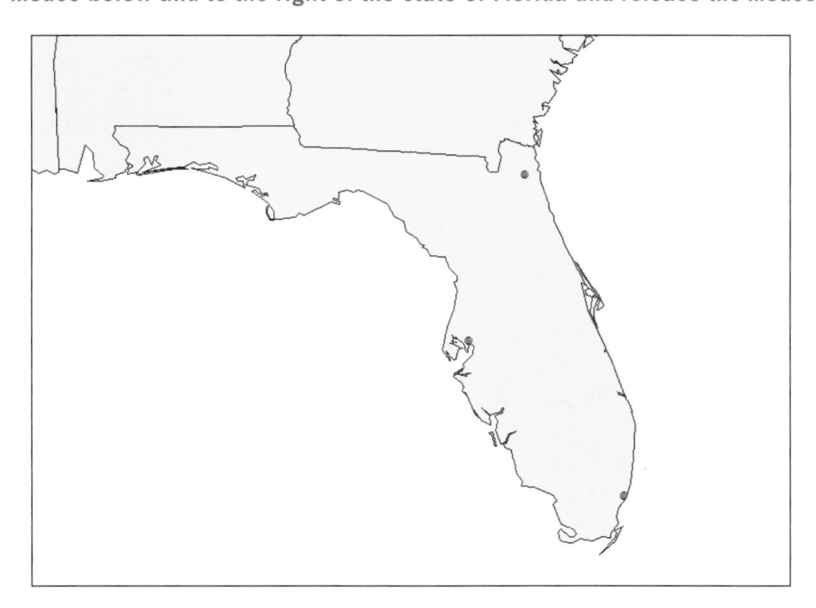

4 **Click once on the map to zoom in.**

The display recenters on the point you clicked. This is an alternative to dragging a rectangle for zooming in.

Fixed Zoom In

1 **Click the Fixed Zoom In button.** 

This zooms in a fixed distance on the center of the current zoomed display.

Zoom Out

1 **Click the Zoom Out button.** 

2 **Click once on the map to zoom out from the point you pick.**

Fixed Zoom Out

1 **Click the Fixed Zoom Out button.** 

This zooms out a fixed distance from the center of the current zoomed display.

Pan

Panning shifts the current display to the left, right, up, or down without changing the current scale.

1 **Click the Pan button.** 

2 **Move the cursor anywhere onto the map display.**

3 **Hold down the left mouse button and drag the mouse in any direction.**

4 **Release the mouse button.**

Full Extent

1 **Click the Full Extent button.** 

This zooms to a full display of all layers, regardless of whether they are turned on or turned off.

Go Back to Previous Extent

1 Click the Go Back to Previous Extent button. 

This returns the map display to its previous extent.

2 Continue to click this button to step back through all of the views.

Go to Next Extent

1 Click the Go to Next Extent button. 

This moves forward through the sequence of zoomed extents you have viewed.

2 Continue to click this button until you reach the most recently viewed extent.

YOUR TURN

Zoom to the county polygons in Colorado and then to the streets in Jefferson County, Colorado. Leave your map zoomed into the streets.

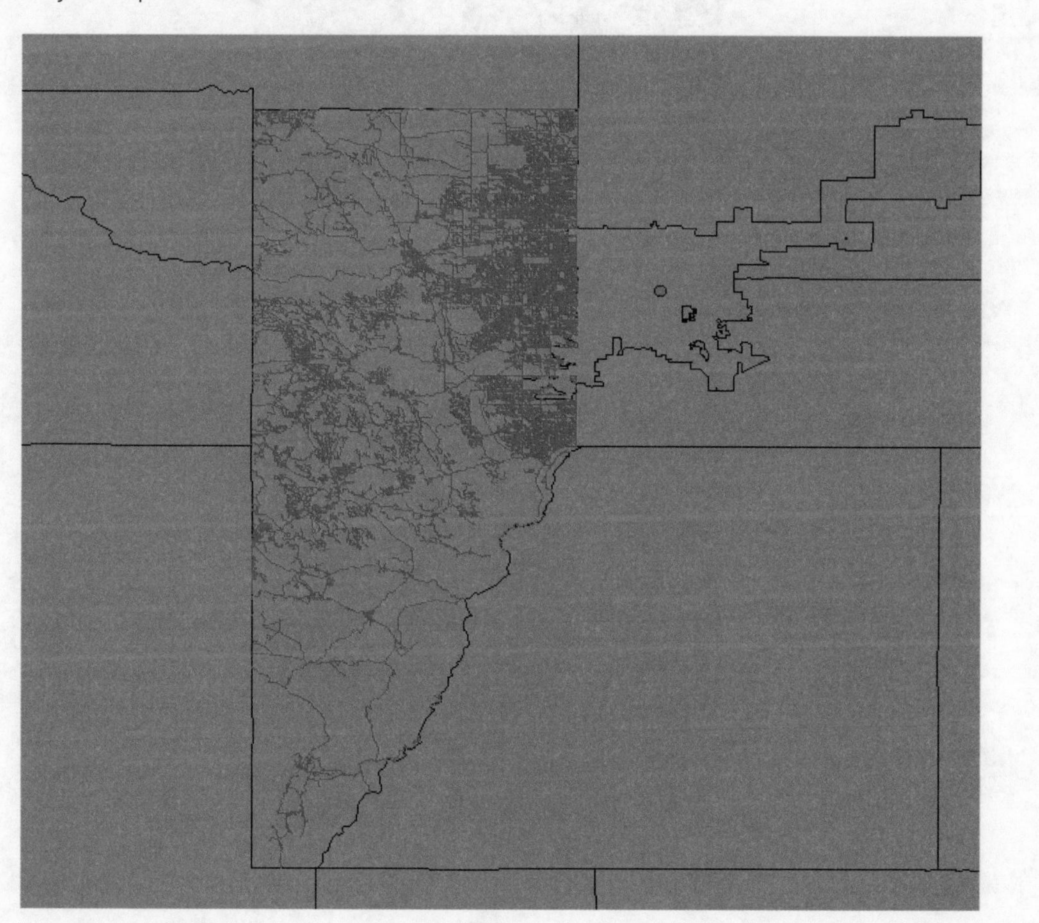

Magnifier window

The magnifier window adjusts the map display to see more detail or get an overview of an area. This window works like a magnifying glass: as you pass the window over the map display, you see a magnified view of the location under the window. Moving the window does not affect the current map display.

Opening the magnifier

1 **Click Window, Magnifier.**

2 **Drag the Magnifier over an area of the map to see crosshairs for area selection and then release to see the zoomed details.**

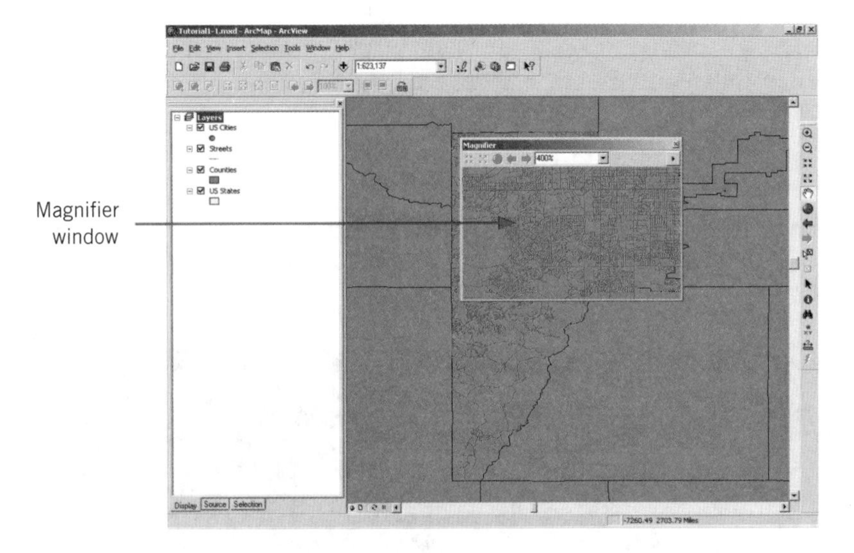

3 **Drag the Magnifier to a new area to see another detail on the map.**

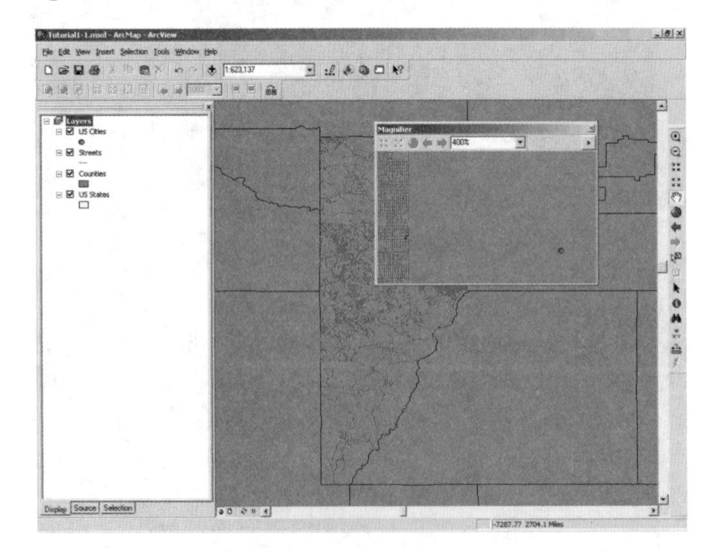

Magnifier properties

1 **Right-click the title bar of the Magnifier window.**

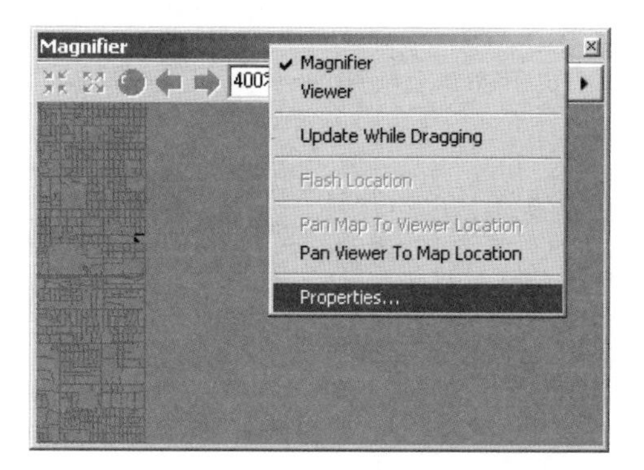

2 **Click Properties.**

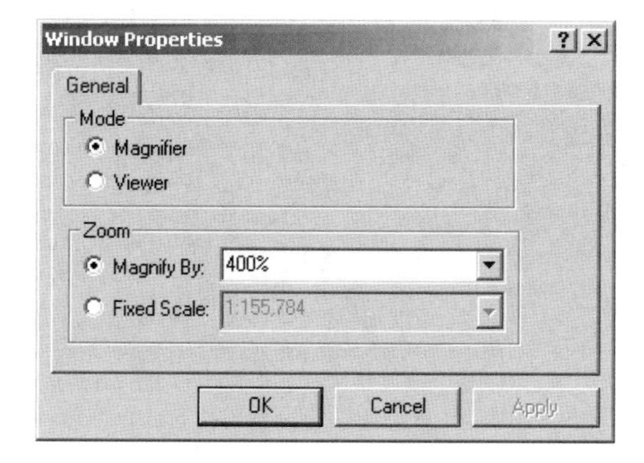

3 **Change the Zoom percentage to 200%.**

4 **Click OK.**

5 **Drag the Magnifier to a different location and see the resulting view.**

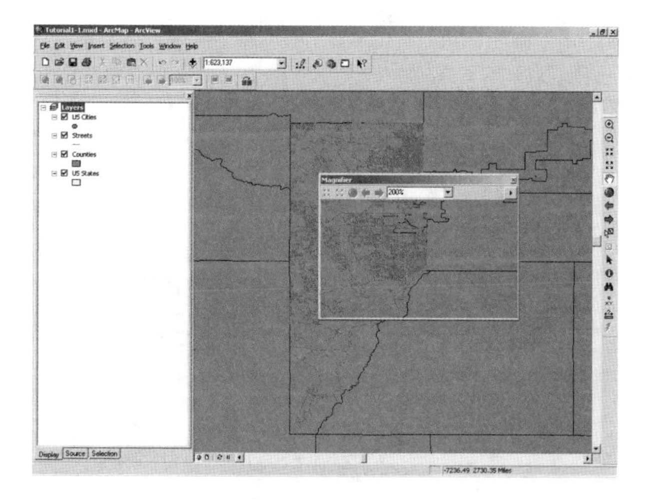

6 **Close the Magnifier window.**

Overview window

The Overview window shows the full extent of the layers in a map. A box shows the area currently zoomed to. You can move the box to pan the map display. You can also make the box smaller or larger to zoom the map display in or out.

Opening the Overview window

1 **Click View, Zoom Data, Full Extent.**

2 **Zoom to a small area of the map (one or two states).**

3 **Click Window, Overview.**

The current extent of the map display is highlighted with a box and red hatch pattern.

4 **Move the cursor to the center of the hatch pattern, click and drag to move it to a new location, and release.**

The extent of the map display updates to reflect the changes made in the Layers Overview window. Note: If you right-click the top of the Layers Overview window and then click Properties, you can modify the display.

Spatial bookmarks

Spatial bookmarks save the extent of a map display or geographic location so you can return to it later without having to use the Pan and Zoom tools.

1 **Close the Layers Overview window.**

2 **Click View, Zoom Data, Full Extent.**

3 **Zoom to the state of Florida.**

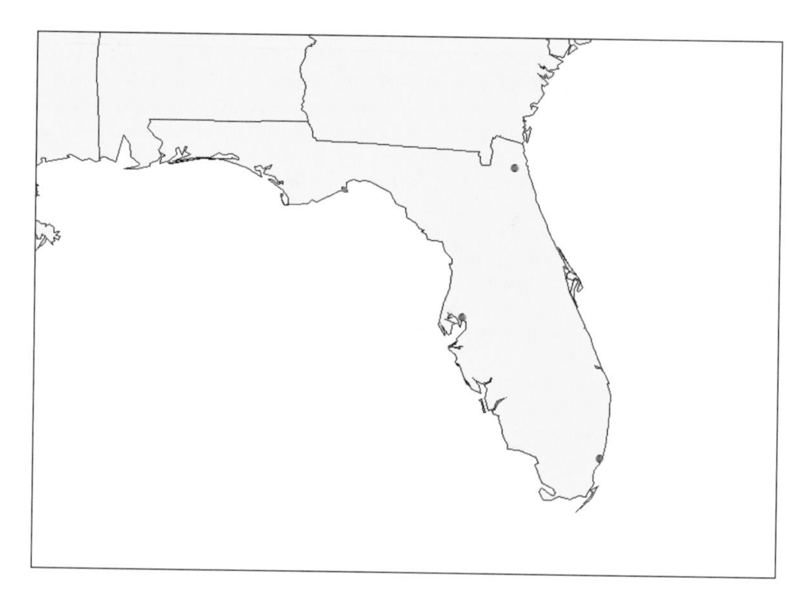

4 **Click View, Bookmarks, Create.**

5 **Type Florida in the Bookmark Name field.**

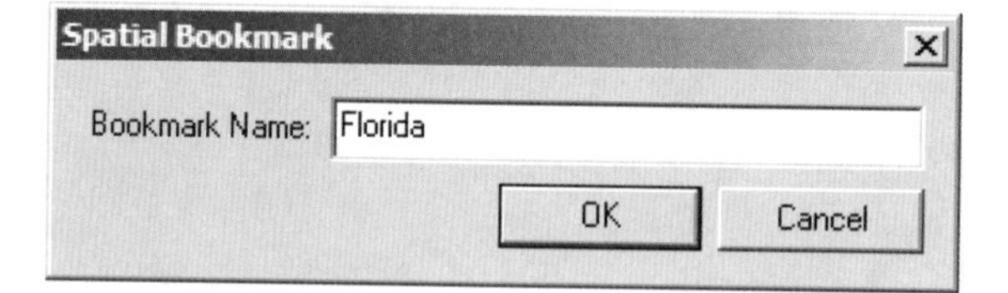

6 **Click OK.**

7 **Click View, Zoom Data, Full Extent.**

8 **Click View, Bookmarks, Florida.**

ArcMap zooms to the saved bookmark of Florida.

YOUR TURN

Create spatial bookmarks for the states of California, New York, and Colorado. Try out your bookmarks. Use View, Bookmarks, Manage to remove the California bookmark.

Measure distances

Measure the horizontal distance of Colorado

1 **Use your bookmark (or use the Full Extent and Zoom In tools) to zoom to Colorado.**

2 **From the Tools toolbar, click the Measure button.** 

The Measure dialog box opens with the Measure Line tool enabled.

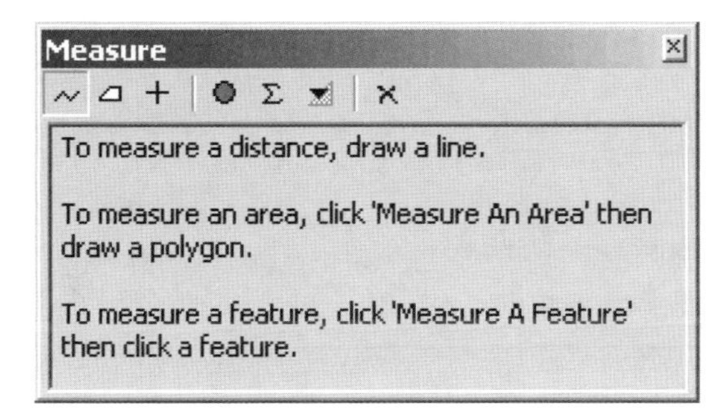

3 **In the Measure dialog box, click the Choose Units button**  **and click Distance, Miles. Leave the Measure dialog box open.**

4 **Click once on the western boundary of the state of Colorado.**

5 **Drag the mouse to the eastern boundary of Colorado and double-click.**

The distance should be about 377 miles.

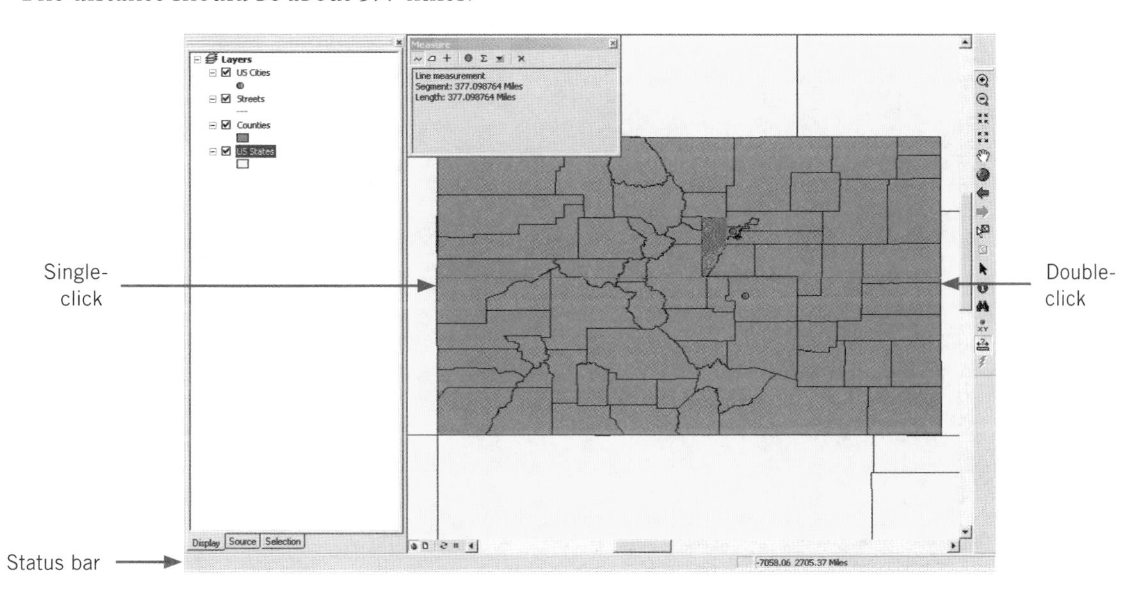

Change distance measurements

1 In the Measure dialog box, click the Choose Units tool and click Distance, Kilometers.

The distance is approximately 607 kilometers.

2 Close the Measure dialog box and from the Tools toolbar click the Select Elements button.

YOUR TURN

Measure the vertical distance (top to bottom) of Colorado. The distance should be about 275 miles or 445 kilometers.

Measure the length and height of the continental United States. The length is approximately 2,700 miles and the height 1,600 miles, but these measurements are difficult to make precisely.

Identify features

To display the data attributes of a map feature, click the feature with the Identify tool. This tool is the easiest way to learn something about a location on a map.

Identify various U.S. states

1 **If necessary, zoom to the full extent.**

2 **From the Tools toolbar, click the Identify button.**

3 **Click anywhere on the map.**

4 **From the Identify window, click the Identify from drop-down list and click US States.**

5 **Click inside the state of Texas.**

The state temporarily flashes and its attributes appear in the Identify dialog box.

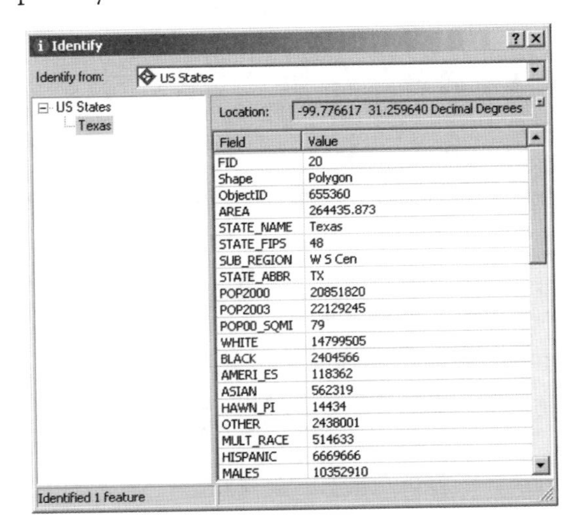

Leave this window open.

Identify various U.S. cities

1 **From the Identify window, click the Identify from drop-down list and click US Cities.**

2 **Click the red circular point marker for Miami (at the southern tip of Florida).**

Make sure the point of the arrow is inside the circle when you click the mouse button. Notice which feature flashes—that is the feature for which you get information.

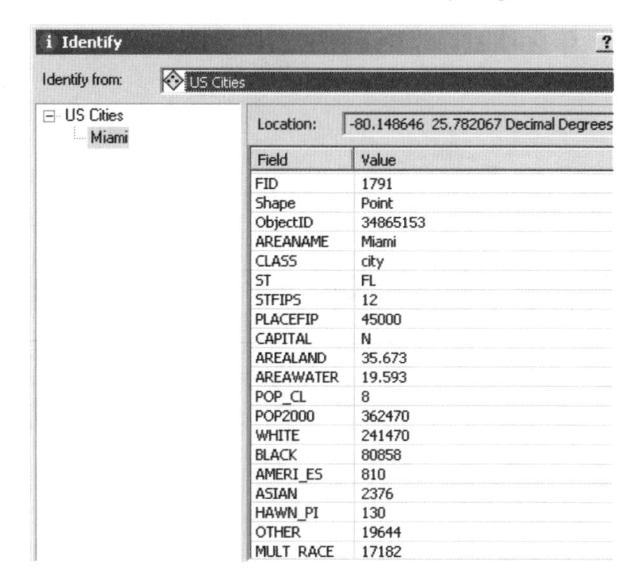

3 Continue clicking other cities to see the identify results.

Hold down the Shift key to retain information for more than one city. Then click the name of a city in the left panel of the Identify window to view that city's information.

Use more Identify tool capabilities

1 Without holding down the Shift key, click Miami with the Identify tool.

2 Right-click the name Miami in the Identify window and click Flash. This flashes Miami's point marker.

3 Right-click the name Miami again and click Zoom To. The map display zooms to Miami.

ArcMap will identify the US States only because its layer is set in the dialog box.

4 Right-click the name Miami once again and click Create Bookmark.

5 Close the Identify window.

6 Click View, Zoom Data, Full Extent.

7 Click View, Bookmarks, Miami.

YOUR TURN

Restrict the Identify results to the Counties layer and identify Colorado counties. Practice making bookmarks for various counties using the Identify tool.

Selecting features

Selecting features identifies the features on which you want to perform certain operations. For example, before you move, delete, or copy a feature, you must select it. Selected features appear highlighted in the layer's attribute table and in the map.

Select button

1 **Click View, Zoom Data, Full Extent.**

2 **Turn off the Streets and Counties layers.**

3 **From the Tools toolbar, click the Select Features tool.** 

4 **Click inside Texas.**

Texas is selected and highlighted with a thick, bright blue line.

Selecting multiple features

1 **Hold down the Shift key and click inside the three states adjacent to Texas.**

All of the selected states are selected and highlighted.

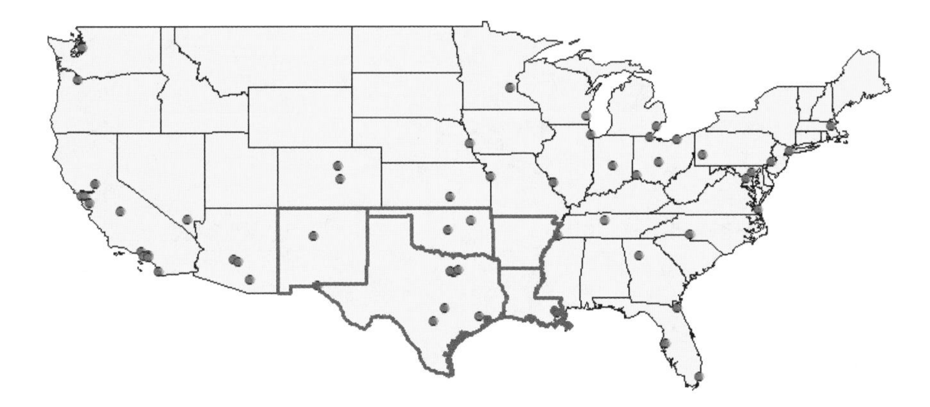

Clearing selected features

1 **Click Selection, Clear Selected Features.**

Selection Color

1 **Click Selection, Options.**

2 **Click the color box in the Selection Color frame.**

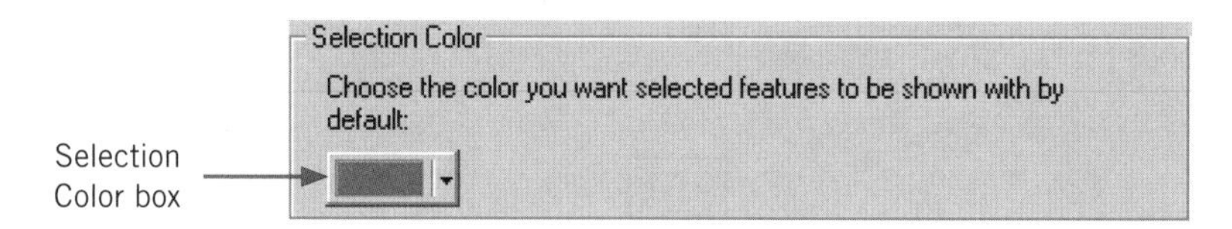

3 **Pick a new color.**

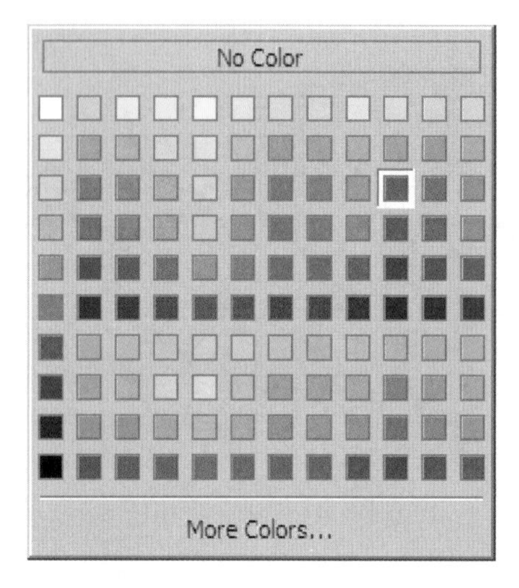

4 **Click OK.**

Changing selection symbol

1 **Right-click the US Cities layer in the table of contents.**

2 **Click Properties.**

The resulting Layer Properties dialog box is one that you will use often. It allows you to modify many properties of a map layer.

3 **Click the Selection tab.**

4 **Pick a new symbol and/or color for the point features.**

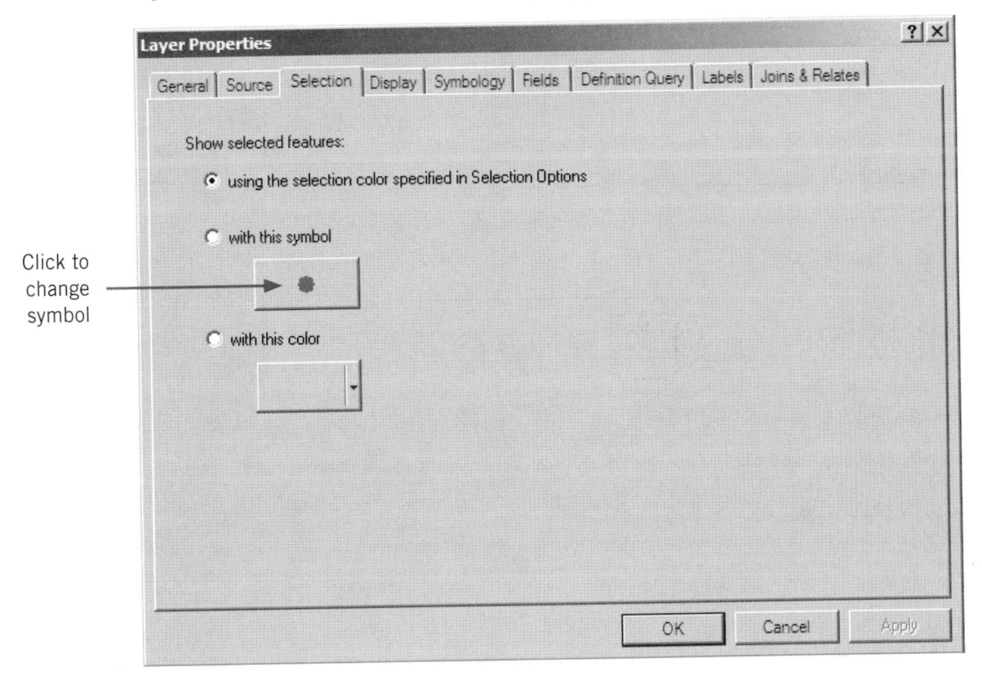

5 **Click OK and click OK again.**

Selectable layers

Making a layer selectable allows its features to be selected using the Select Features tool, Select by Graphics tool, Find tool, and so forth.

1 Click Selection, Set Selectable Layers.

2 Click off the check boxes for Streets, Counties, and US States to make only US Cities selectable.

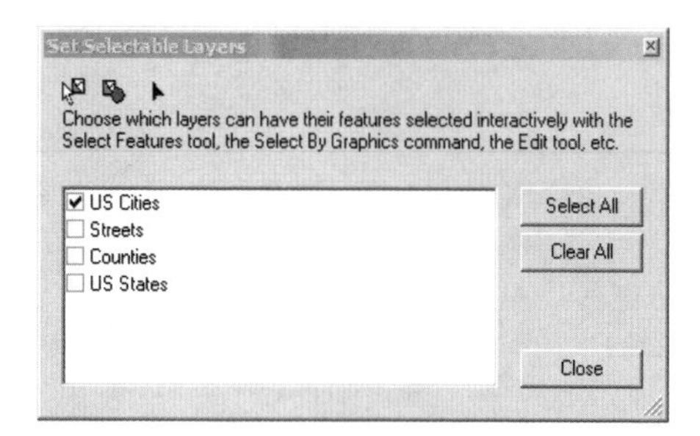

3 Click Close.

4 Click the Select Features button  **and click a city.**

The selected city gets the selection symbol and color that you chose on the previous page.

5 Clear the selected features.

YOUR TURN

Create a new layer from selected features by selecting multiple cities in one state. After the cities are selected, right-click the US Cities layer, click Selection, and click Create Layer from Selected Features.

Find features

The Find tool is used to locate features in a layer or layers based on their attribute values. You can also use this tool to select, flash, zoom, bookmark, identify, or unselect the feature in question.

1 **From the Tools toolbar, click the Find button.**

2 **Click the Features tab.**

3 **Type Dallas as a city name to find.**

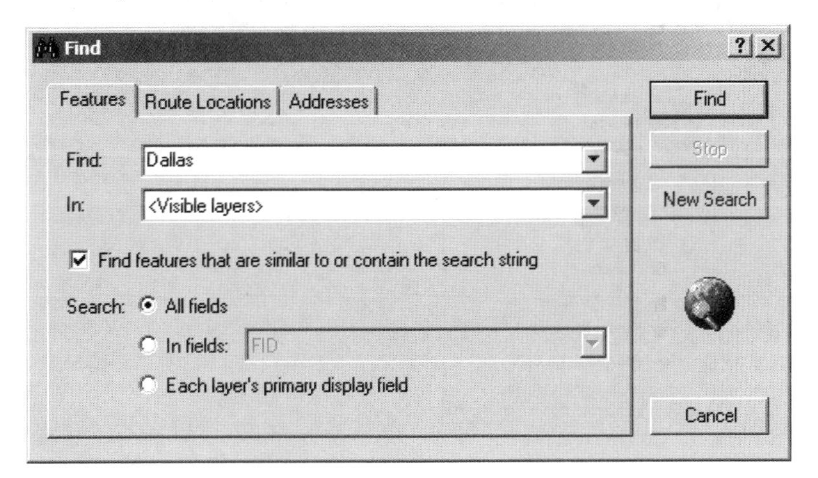

4 **Click Find.**

The results will appear in the following section of the Find dialog box.

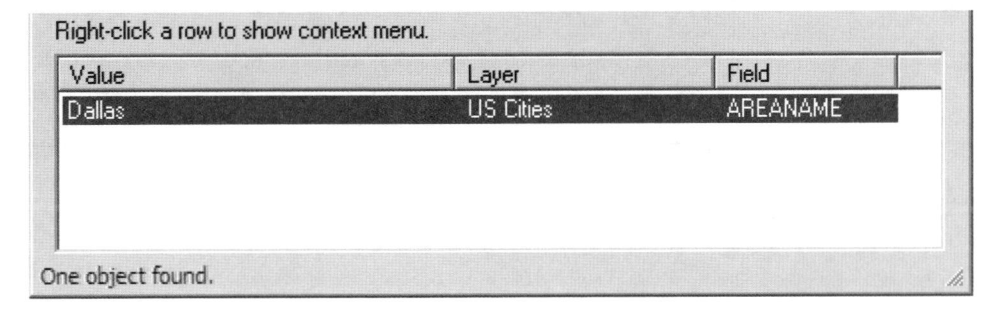

5 **Right-click the city name and click Zoom To.**

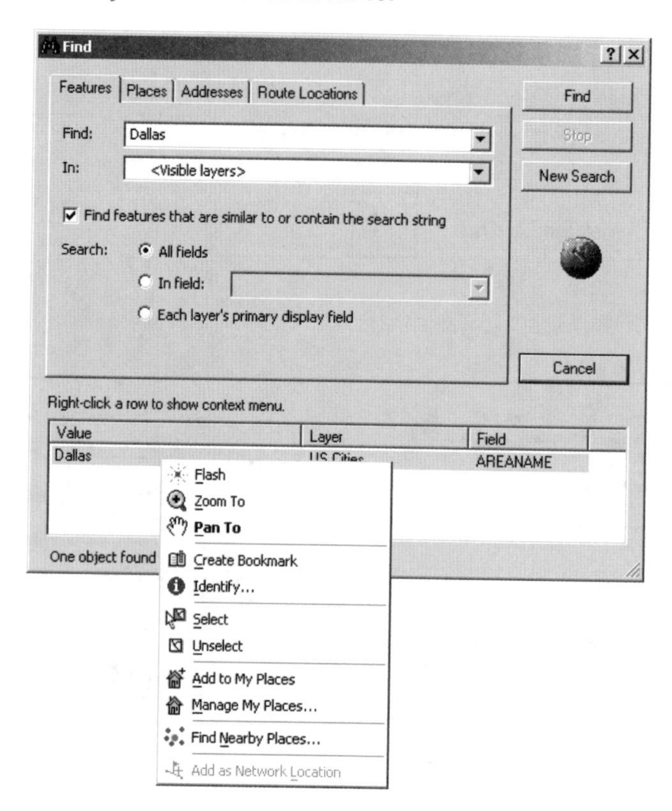

6 **Click Cancel in the Find dialog box.**

YOUR TURN

Find other cities and practice showing them using other find options such as Flash Features, Identify Feature(s), and Set Bookmark. When finished, clear any selected features and zoom to the full extent.

Work with feature attribute tables

Tabular attribute data associated with map features can be viewed using the layer's attribute table. To explore the attributes of a layer on a map, open its attribute table to select and find features with particular attribute values.

Open the table of the US Cities layer

1 **Right-click the US Cities layer in the table of contents.**

2 **Click Open Attribute Table.**

The table opens, containing one record for each US City point feature. Every layer has an attribute table with one record per feature.

3 **Scroll down in the table until you find Chicago (the order of records may be different on your computer) and click the record selector for Chicago to select that record.**

If a feature is selected in the attribute table, it will also be selected on the map. You will see this on the next page.

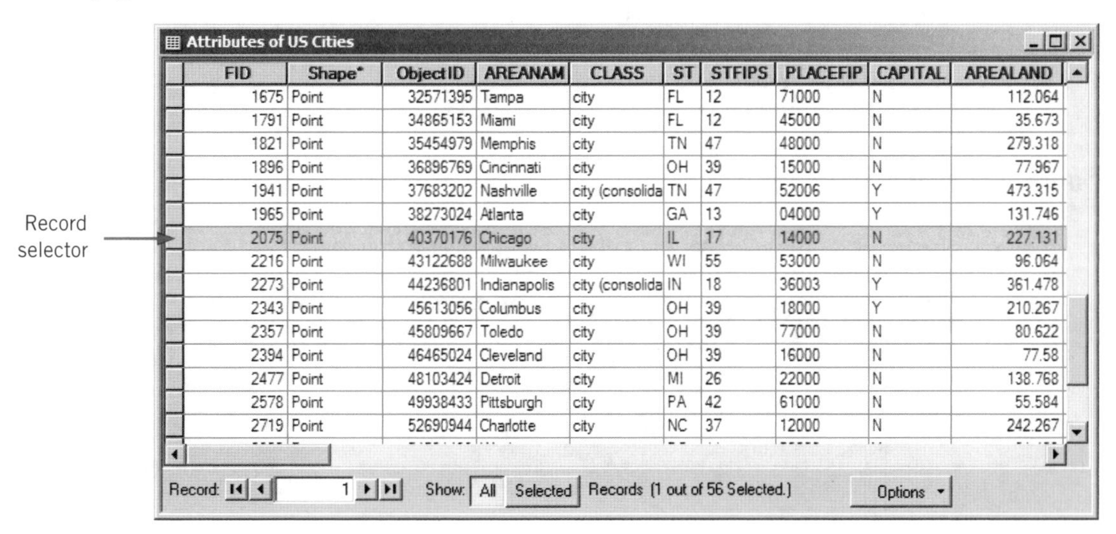

FID	Shape*	ObjectID	AREANAM	CLASS	ST	STFIPS	PLACEFIP	CAPITAL	AREALAND
1675	Point	32571395	Tampa	city	FL	12	71000	N	112.064
1791	Point	34865153	Miami	city	FL	12	45000	N	35.673
1821	Point	35454979	Memphis	city	TN	47	48000	N	279.318
1896	Point	36896769	Cincinnati	city	OH	39	15000	N	77.967
1941	Point	37683202	Nashville	city (consolida	TN	47	52006	Y	473.315
1965	Point	38273024	Atlanta	city	GA	13	04000	Y	131.746
2075	Point	40370176	Chicago	city	IL	17	14000	N	227.131
2216	Point	43122688	Milwaukee	city	WI	55	53000	N	96.064
2273	Point	44236801	Indianapolis	city (consolida	IN	18	36003	Y	361.478
2343	Point	45613056	Columbus	city	OH	39	18000	Y	210.267
2357	Point	45809667	Toledo	city	OH	39	77000	N	80.622
2394	Point	46465024	Cleveland	city	OH	39	16000	N	77.58
2477	Point	48103424	Detroit	city	MI	26	22000	N	138.768
2578	Point	49938433	Pittsburgh	city	PA	42	61000	N	55.584
2719	Point	52690944	Charlotte	city	NC	37	12000	N	242.267

Show the connection between layers and tables

1 **Resize the Attributes of US Cities table to see both the map and table on the screen.**

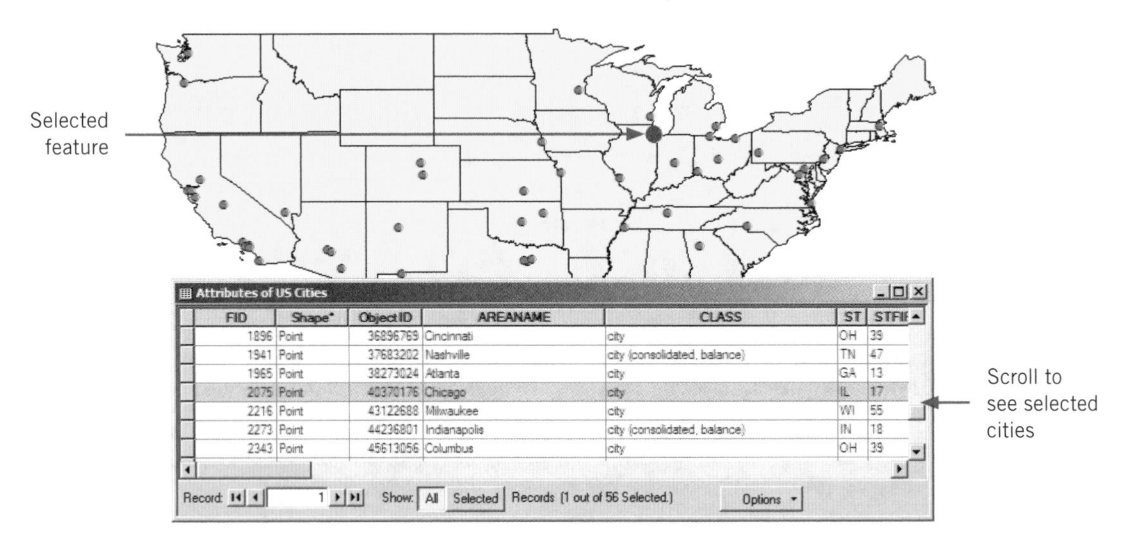

2 **Click the Select Features button** [button icon] **and click various cities on the map.**

3 **Scroll through the Attributes of US Cities to see the selected cities.**

Show only selected records

1 **In the Attributes of US Cities table, click the Selected Records button.**

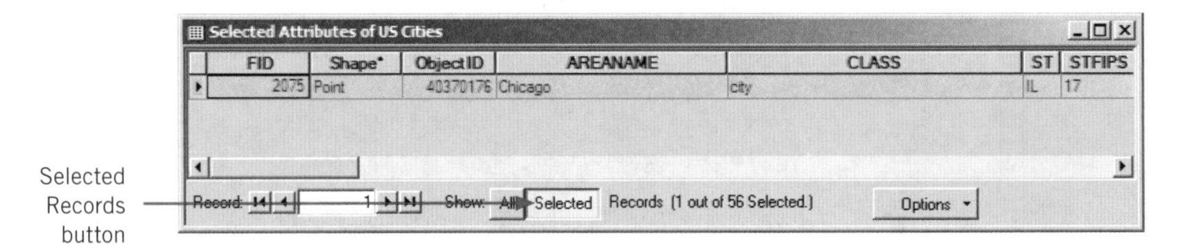

This will show only the records for the features selected in the map.

2 **Click the All Records button to show all records again.**

Clear selections

1 In the Attributes of US Cities table, click the Options button. (If you cannot see the Options button, widen the table to the right until it appears.)

2 Click Clear Selection.

Select more than one record from the table

1 In the Attributes of US Cities table, click the record selector for Atlanta.

2 Hold down the Ctrl key and select the following records: Jacksonville, Miami, and Tampa.

This will highlight these selected features both in the table and the map. Scroll up or down to find these records.

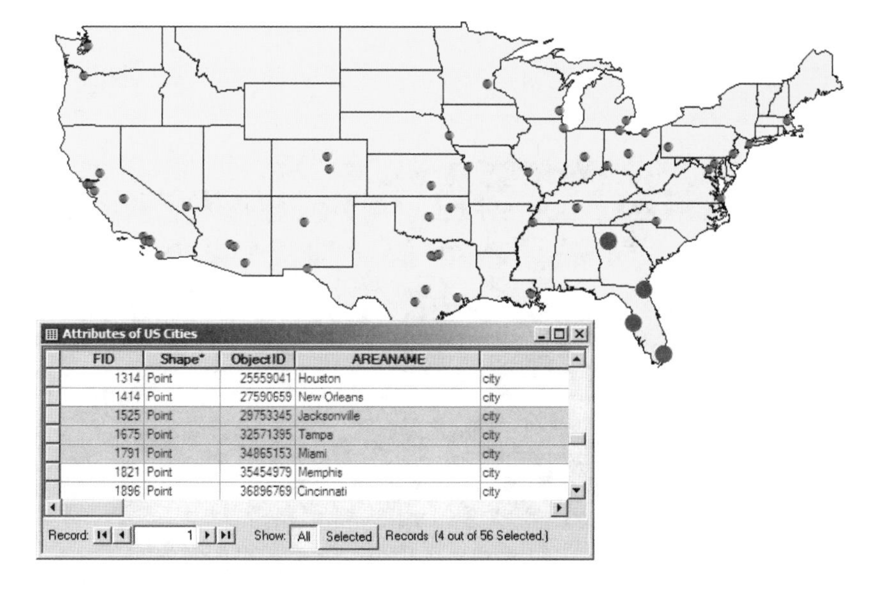

FID	Shape*	ObjectID	AREANAME	
1314	Point	25559041	Houston	city
1414	Point	27590659	New Orleans	city
1525	Point	29753345	Jacksonville	city
1675	Point	32571395	Tampa	city
1791	Point	34865153	Miami	city
1821	Point	35454979	Memphis	city
1896	Point	36896769	Cincinnati	city

3 Hold down the Ctrl key and click the Atlanta record again to deselect it.

Zoom to selected feature

1 Click View, Zoom Data, Zoom to Selected Features.

This will zoom to the three selected cities in Florida.

Switch selections

1 In the Attributes of US Cities table, click the Options button.

2 Click Switch Selection.

This reverses the selection: it selects all of those that were not selected and deselects those that were selected.

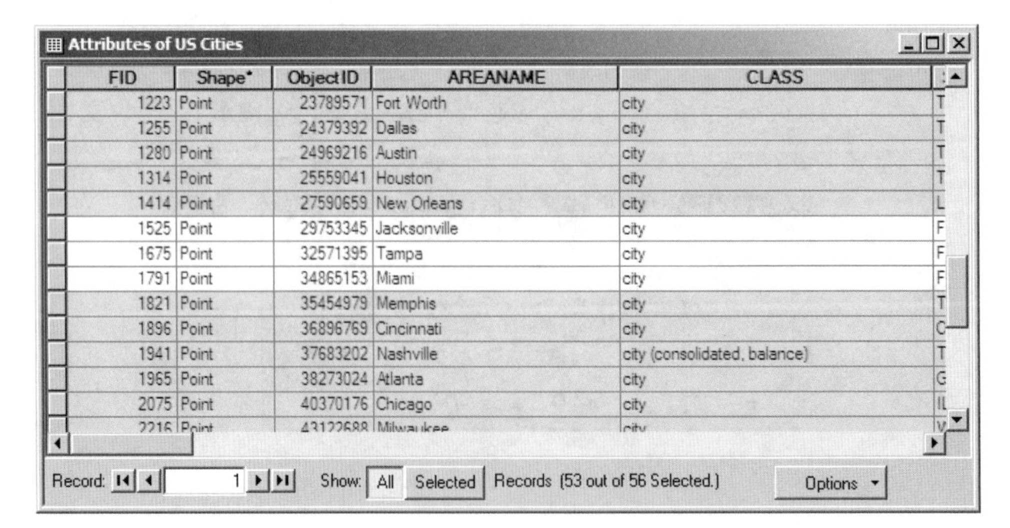

FID	Shape*	Object ID	AREANAME	CLASS
1223	Point	23789571	Fort Worth	city
1255	Point	24379392	Dallas	city
1280	Point	24969216	Austin	city
1314	Point	25559041	Houston	city
1414	Point	27590659	New Orleans	city
1525	Point	29753345	Jacksonville	city
1675	Point	32571395	Tampa	city
1791	Point	34865153	Miami	city
1821	Point	35454979	Memphis	city
1896	Point	36896769	Cincinnati	city
1941	Point	37683202	Nashville	city (consolidated, balance)
1965	Point	38273024	Atlanta	city
2075	Point	40370176	Chicago	city

Clear selections

1 In the Attributes of US Cities table, click Options and Clear Selection. Clear Selection

Sort a field

1 In the Attributes of US Cities table, right-click the AREANAME field name.

2 Click Sort Ascending. Sort Ascending

This will sort the table from A to Z by the name of each U.S. city.

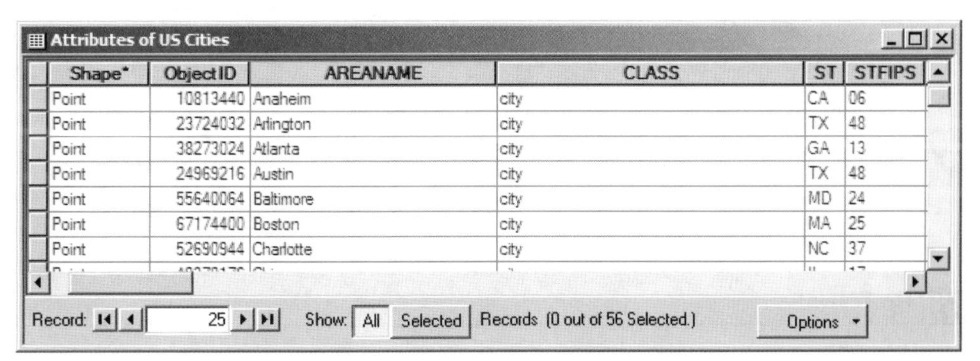
Attributes of US Cities

Shape*	ObjectID	AREANAME	CLASS	ST	STFIPS
Point	10813440	Anaheim	city	CA	06
Point	23724032	Arlington	city	TX	48
Point	38273024	Atlanta	city	GA	13
Point	24969216	Austin	city	TX	48
Point	55640064	Baltimore	city	MD	24
Point	67174400	Boston	city	MA	25
Point	52690944	Charlotte	city	NC	37

Record: 25 Show: All Selected Records (0 out of 56 Selected.) Options

3 Scroll to the right in the table and right-click the POP2000 field name.

4 Click Sort Descending. Sort Descending

This will sort the field from the highest populated city to the lowest populated city.

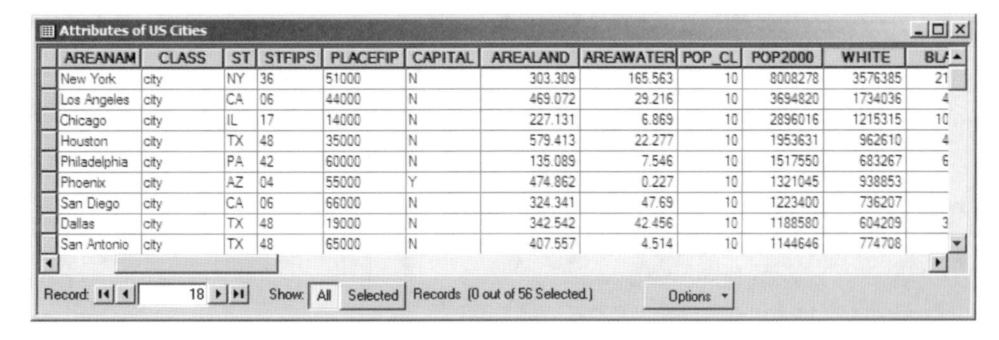
Attributes of US Cities

AREANAM	CLASS	ST	STFIPS	PLACEFIP	CAPITAL	AREALAND	AREAWATER	POP_CL	POP2000	WHITE	BL
New York	city	NY	36	51000	N	303.309	165.563	10	8008278	3576385	21
Los Angeles	city	CA	06	44000	N	469.072	29.216	10	3694820	1734036	4
Chicago	city	IL	17	14000	N	227.131	6.869	10	2896016	1215315	10
Houston	city	TX	48	35000	N	579.413	22.277	10	1953631	962610	4
Philadelphia	city	PA	42	60000	N	135.089	7.546	10	1517550	683267	6
Phoenix	city	AZ	04	55000	Y	474.862	0.227	10	1321045	938853	
San Diego	city	CA	06	66000	N	324.341	47.69	10	1223400	736207	
Dallas	city	TX	48	19000	N	342.542	42.456	10	1188580	604209	3
San Antonio	city	TX	48	65000	N	407.557	4.514	10	1144646	774708	

Record: 18 Show: All Selected Records (0 out of 56 Selected.) Options

Move a field

1 **Click the gray title of the POP2000 field in the Attributes of US Cities table.**

2 **Click, drag, and release the POP2000 field to the left of another field.**

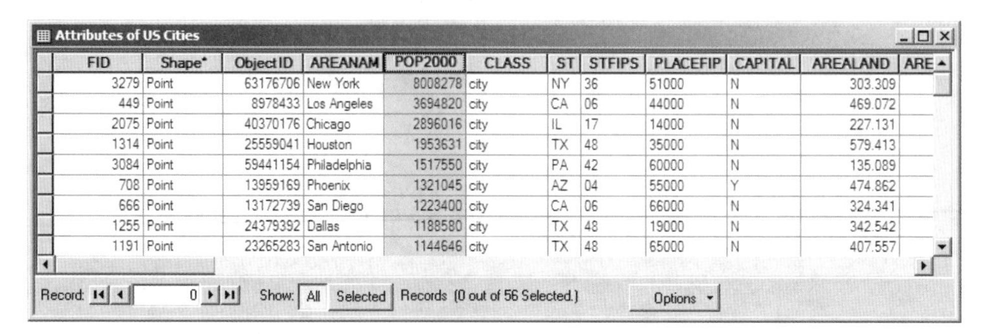
Attributes of US Cities

FID	Shape*	ObjectID	AREANAM	POP2000	CLASS	ST	STFIPS	PLACEFIP	CAPITAL	AREALAND	ARE
3279	Point	63176706	New York	8008278	city	NY	36	51000	N	303.309	
449	Point	8978433	Los Angeles	3694820	city	CA	06	44000	N	469.072	
2075	Point	40370176	Chicago	2896016	city	IL	17	14000	N	227.131	
1314	Point	25559041	Houston	1953631	city	TX	48	35000	N	579.413	
3084	Point	59441154	Philadelphia	1517550	city	PA	42	60000	N	135.089	
708	Point	13959169	Phoenix	1321045	city	AZ	04	55000	Y	474.862	
666	Point	13172739	San Diego	1223400	city	CA	06	66000	N	324.341	
1255	Point	24379392	Dallas	1188580	city	TX	48	19000	N	342.542	
1191	Point	23265283	San Antonio	1144646	city	TX	48	65000	N	407.557	

Record: 0 Show: All Selected Records (0 out of 56 Selected.) Options

3 **Close the Attributes of US Cities.**

YOUR TURN

Move and sort by other field names. Try sorting by multiple fields. For example, you could sort US Cities alphabetically or by whether or not they are state capitals.

To sort by multiple fields, rearrange the table's fields so the field whose values will be sorted first appears directly to the left of the field whose values will be sorted second. While holding down the Ctrl key, click the heading of the two fields you want to use to sort the records. Right-click the name of one of the selected fields and choose a sort order. When you sort, the selected fields will be in the sort order you chose.

Label features on the map

Labels are text items on the map that are dynamically placed and whose text values are derived from one or more feature attributes.

Set label properties

1 **Zoom to Florida if the map is not already zoomed to that state.**

2 **Right-click the US Cities layer in the table of contents.**

3 **Click Properties.**

4 **Click the Labels tab.**

5 **Click the Label Field drop-down arrow and click AREANAME if not already selected.**

6 **Click OK.**

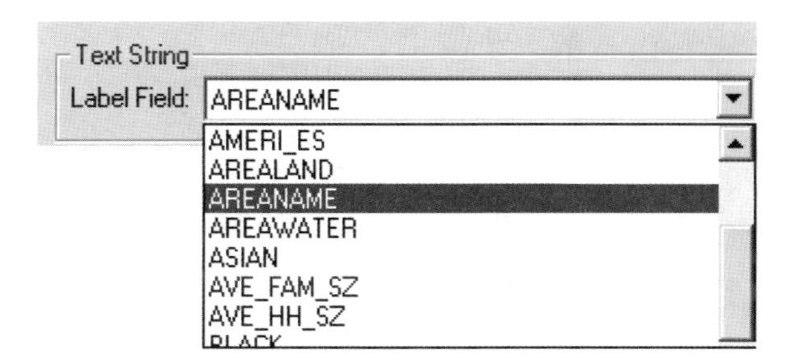

Label features

1 **Right-click the US Cities layer in the table of contents.**

2 **Click Label Features.**

All of the features on the map are labeled.

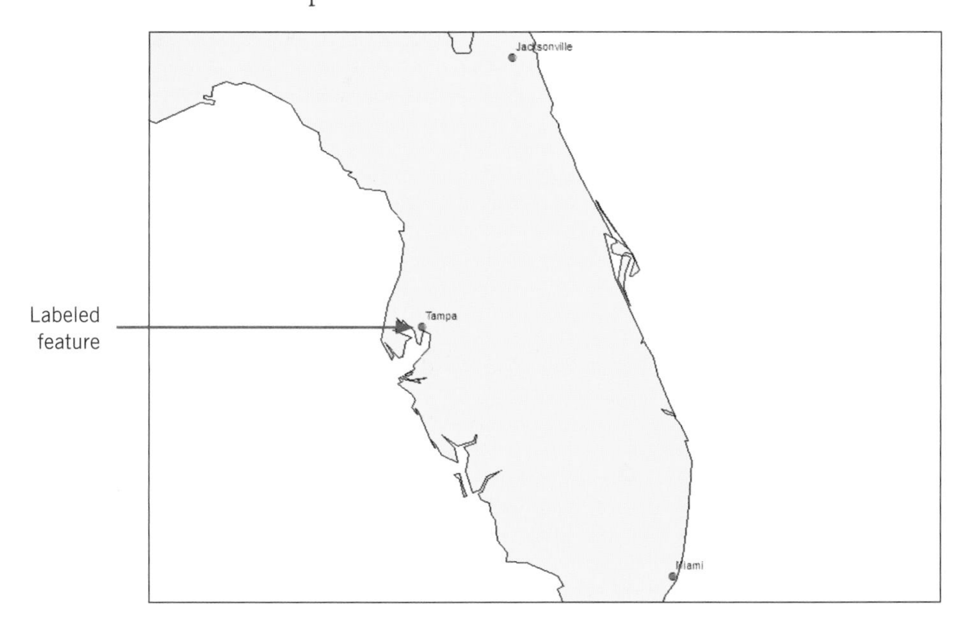

Remove labels

1 Right-click the US Cities layer in the table of contents.

2 Click Label Features.

Labels in the map are turned off.

3 Click Label Features again to turn them back on.

Get statistics

You can get descriptive statistics, such as the mean and maximum value of an attribute, in ArcMap by opening a map layer's attribute table and applying a tool.

1 Zoom to the full extent.

2 At the bottom of the table of contents, click the Selection tab.

3 Check US States so that it is selectable.

4 At the bottom of the table of contents, click the Display tab.

5 Hold down the Shift key and use the Select Features tool to select the state of Texas and the four states adjacent to it.

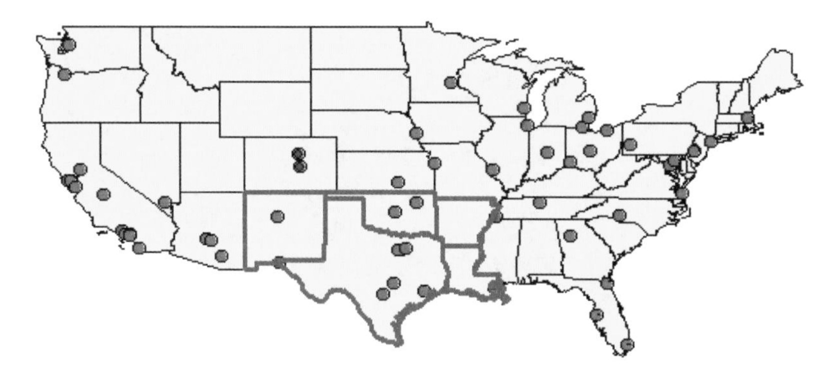

6 In the table of contents, right-click US States.

7 Click Open Attribute Table.

8 Right-click the gray column heading for the POP2000 attribute.

9 Click Statistics.

The resulting window has statistics for the five selected states; for example, the mean 2000 population is 6,652,779.

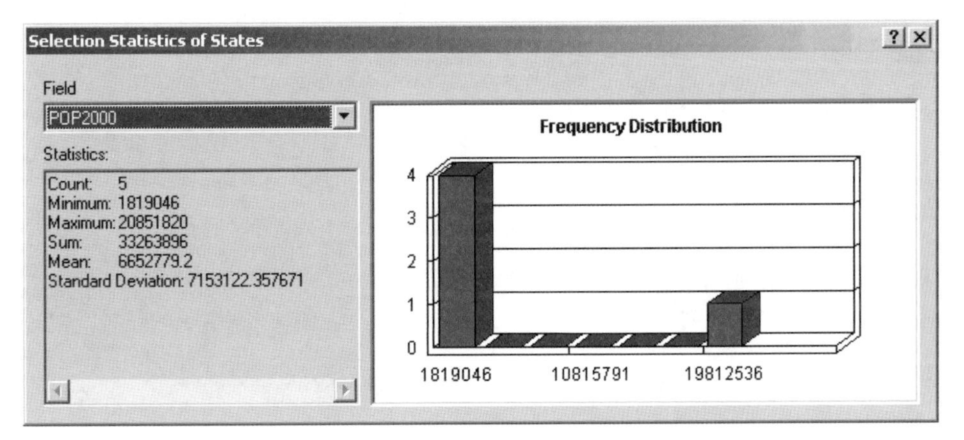

YOUR TURN

Get statistics for a new selection of states and attributes of your choice.

Relative paths and saving maps

When a layer is added to a map, the path name to the data is stored in the map, but the layer is not copied from its original location. When a map is opened, ArcMap locates the layer data it needs using these stored path names. If ArcMap cannot find the data for a layer, the layer will appear in the ArcMap table of contents but it won't be drawn. Instead, a red exclamation mark (!) will appear next to the layer to indicate that it needs to be repaired.

Absolute path names

An example of an absolute full path is C:\Gistutorial\Tutorial1.mxd. To share maps saved with absolute paths, everyone who uses the map must either do so on the same computer or have the data on their computer in exactly the same folder structure (e.g., C:\Gistutorial). This is not conducive for a computer lab environment because instructors, teaching assistants, and students all work on different machines. Instead, the relative path option is favored.

You can view information about the data source for a layer by clicking the Source tab in the Layers Properties box.

Relative path names

An example of a relative path is \Gistutorial\Tutorial1.mxd. Relative paths in a map specify the location of the layers relative to the current location on disk of the map document (.mxd file). Because relative paths do not contain drive letter names, they enable the map and its associated data to point to the same directory structure regardless of the drive that the map resides on. If a project is moved to a new drive, ArcMap will still be able to find the maps and their data by traversing the relative paths.

This option, for example, allows you to share maps that you made with data on your local F:\ drive with people who only have a C:\ drive. This also allows you to easily move the map and its data to a different hard drive on your computer, or give the map and its data to another person to copy to their computer.

Saving layers as relative path names

1 **Click File, Document Properties.**

2 **Click Data Source Options.**

3 **Click the Store relative path names to data sources radio button.**

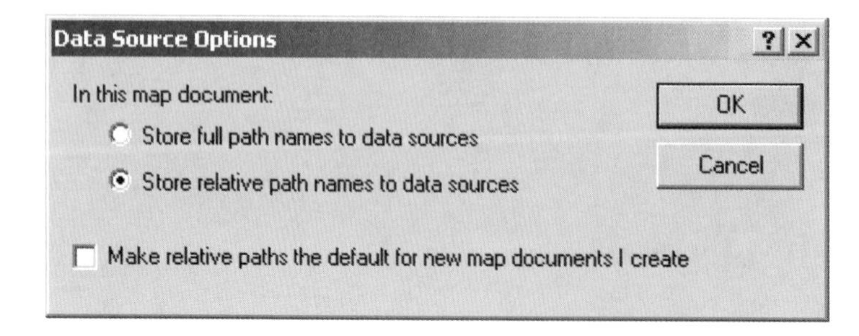

4 **Click OK and OK again to close the Data Source Options and Tutorial1-1.mxd Properties windows.**

Save the project and exit ArcMap

1 Click File, Save As.

2 Navigate to the \Gistutorial folder and save the map as Tutorial1-2.mxd.

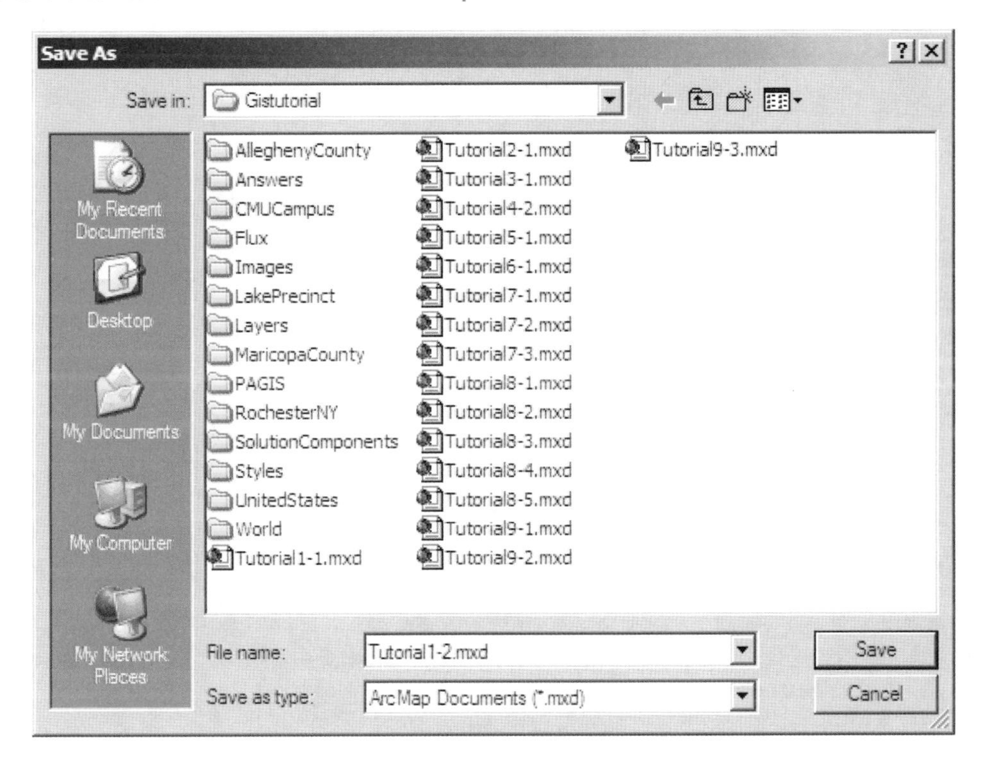

When you save a map, you save it to a map document file, which has an .mxd file extension. When working with ArcMap you usually spend time setting properties that affect the look and functionality of your map and its layers. For example, in this map you added layers, changed their symbology, and created bookmarks. By saving this map document, all of the work is preserved in the Tutorial1-2.mxd file, which you can reopen anytime or share with others.

3 Click File, Exit.

Assignment 1-1

Statistics on U.S. housing

In this assignment, you will compare statistics for U.S. states on the number of housing units, number of renter- and owner-occupied units, and highest number of vacant units.

Start with the following:

- **\Gistutorial\UnitedStates\States**—polygon layer of U.S. states with census 2000 data.
  Attributes of States table—attribute table for U.S. states that includes the following fields needed for the assignment:
  STATE_ABBR two-letter state abbreviation
  HSE_UNITS number of housing units per state
  RENTER_OCC number of renter-occupied units per state
  OWNER_OCC number of owner-occupied units per state
  VACANT number of vacant units per state

Change the map and get statistics

Create a new map document called C:\Gistutorial\Answers\Assignment1\Assignment1-1.mxd with relative paths and with the above layer of the US States added and symbolized with a hollow-filled color and a medium gray outline. Using the States attribute table and a bright red selection color, select the five states having the highest number of vacant units. Label each state with its abbreviation.

Create data document

- Create a Microsoft Word document called **C:\Gistutorial\Answers\Assignment1\Assignment1-1.doc.**
- In the Word file, create a table with statistics as follows for the five states with the highest number of vacant units only.
- With your finished map document, in ArcMap click File, Export Map and browse to your **Gistutorial\Answers\Assignment1** folder to save **Assignment1-1.jpg** there. In your Word document, place the insertion point after your table and click Insert, Picture, From File. Then browse to **Gistutorial\Answers\Assignment1\Assignment1-1.jpg** and insert the map image.

Attribute	Mean	Minimum	Maximum
HSE_Units			
Renter_Occ			
Owner_Occ			
Vacant			

Hint: Copying and pasting statistics

- Select a statistic in the Statistics output table, press Ctrl + C to copy the statistic, click in the appropriate cell of your Word table, and press Ctrl + V to paste it.

Assignment 1-2

Erin Street Crime Watch

Crime prevention depends very much on what the criminology literature calls "informal guardianship": residents and their neighbors keep an eye on suspicious behavior and intervene in some fashion. Police departments therefore actively promote and support crime watch or block watch citizen groups and keep them informed on crime trends. Suppose that the police commander of a precinct has a notebook computer, ArcMap, and a portable color projector for use at crime watch meetings. Your job is to get the commander ready for a meeting with the 100 block Erin Street crime watch group.

Start with the following:

- **C:\Gistutorial\PAGIS\Midhill\Street**—arc layer for street centerlines in the Middle Hill neighborhood of Pittsburgh. Note: This is a TIGER street centerline map from the U.S. Census Bureau. You will study and use TIGER maps extensively in GIS.
  - **Attributes of Street table**—attribute table for streets in the Middle Hill neighborhood that includes the following fields needed for the assignment:
    - **Fname** = street name
    - ***Address Ranges***
    - **LeftAdd1** = beginning house number on the left side of the street
    - **LeftAdd2** = ending house number on the left side of the street
    - **RgtAdd1** = beginning house number on the right side of the street
    - **RgtAdd2** = ending house number on the right side of the street
- **C:\Gistutorial\PAGIS\Midhill\Curbs**—arc layer for curbs in the Middle Hill neighborhood of Pittsburgh. Note that this and the remaining layers are in a map format called a coverage. After you browse to the Midhill folder, you have to double-click the Curbs coverage icon and then double-click Arc to add the curbs to your map document.
- **C:\Gistutorial\PAGIS\Midhill\Building**—polygon layer for buildings in the Middle Hill neighborhood of Pittsburgh.
- **C:\Gistutorial\PAGIS\Midhill\Mid911**—point layer for 911 emergency calls for service in the Middle Hill neighborhood of Pittsburgh.
  - **Attributes of Mid911 table**—attribute table for mid911 points that includes the following attributes needed for the assignment:
    - **Nature_Code** = call type
    - **Date** = date of crime
    - **Address** = addresses of crime locations

Change the map and get statistics

Create a new map document with the above layers saved as C:\Gistutorial\Anwers\Assignment1\Assignment1-2.mxd with relative paths that includes a zoomed view of the Erin Street block selected and labeled with street names. Display streets, curbs, and buildings as medium-light gray lines, and 911 calls as bright red circles. Create a spatial bookmark of the zoomed area called Erin Street.

Create data document

- Create a table of addresses, dates of calls, and call types for crimes in the 100 block of Erin Street (see hints). The street names will include Davenport, Erin, and Trent. Create **C:\Gistutorial\Answers\Assignment1\Assignment1-2.doc** and paste the table into it as directed in the hints below.

Address	Date	Call Type

Hints

- The 100 block of Erin Street is the segment of Erin Street whose address range is from 100 to 199 and perpendicular to streets Webster and Wylie. The crime reports are prepared for the two blocks on either side of Erin Street in this range. Use both the attribute table and Identify tool to find and label these streets.
- Although it appears that there are only six points, there are actually thirteen total because multiple calls are at the same location. Use the Select Features button and information in the attribute table to get the data on all relevant calls.
- Data can be exported from the attribute table. In the table, choose "Options" and "Export." Save the selected records to a .dbf file. Open the .dbf file in Microsoft Excel, edit the records, and paste from there into Assignment1-2.doc.

What to turn in

If you are working in a classroom setting with an instructor, you may be required to submit the exercises you created in tutorial 1. Below are the files you are required to turn in. Be sure to use a compression program such as PKZIP or WinZip to include all required files as one .zip document for review and grading. Include your name and assignment number in the .zip document (YourNameAssn1.zip).

ArcMap documents

C:\Gistutorial\Answers\Assignment1\Assignment1-1.mxd
C:\Gistutorial\Answers\Assignment1\Assignment1-2.mxd

Word documents

C:\Gistutorial\Answers\Assignment1\Assignment1-1.doc
C:\Gistutorial\Answers\Assignment1\Assignment1-2.doc

OBJECTIVES

Create choropleth maps
Create group layers
Create threshold scales for dynamic display
Create choropleth maps using custom attribute scales
Create pin (point) maps
Create hyperlinks
Create Tool Tips

GIS Tutorial 2

Map Design

In this tutorial you will learn all steps necessary to compose common maps from available map layers. One type of map that you will create is a choropleth map that color codes polygons to convey information about areas. The second is a "pin map" that uses point markers to display spatial patterns in point data. You will continue to use U.S. states and counties, plus census tracts and detailed census data for the Commonwealth of Pennsylvania. All maps that you will produce are of interest to demographers and policy makers.

Launch ArcMap

1 From the Windows taskbar, click Start, All Programs, ArcGIS, ArcMap.

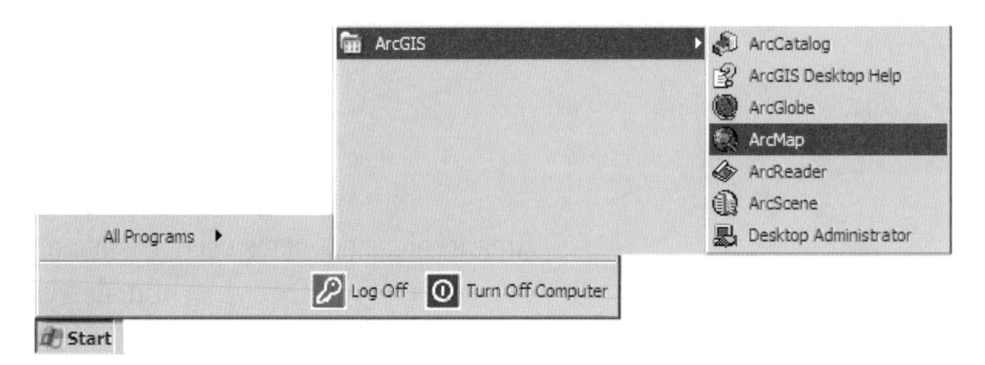

2 Click the An existing map radio button in the ArcMap dialog box.

3 Click OK.

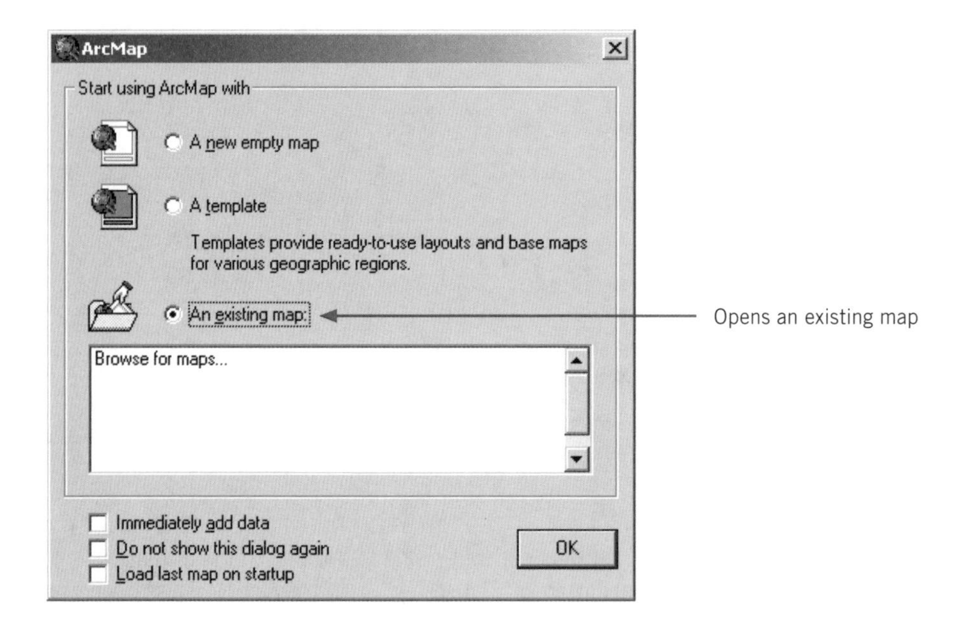

4 Browse to the drive on which the Gistutorial folder has been installed (e.g., C:\Gistutorial), select the Tutorial2-1.mxd project, and click the Open button.

ArcMap opens a map with no layers added. You will add the layers needed for the tutorial next.

Create choropleth maps

A choropleth map is a map in which polygon areas are colored or shaded to represent attribute values. You will use population to create choropleth maps for states, counties, and census tracts.

Add a layer to the view

1 **Click the Add Data button.** 

2 **Navigate to the folder where you have the Gistutorial data installed and click \Gistutorial\UnitedStates\.**

3 **Click States.shp, Add.**

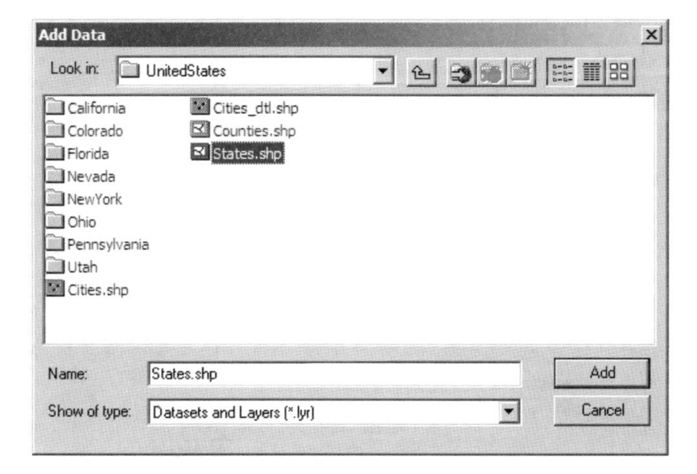

The forty-eight states of the continental United States are drawn in the map display in a color randomly picked by ArcMap. You will change the color of the states later in the tutorial.

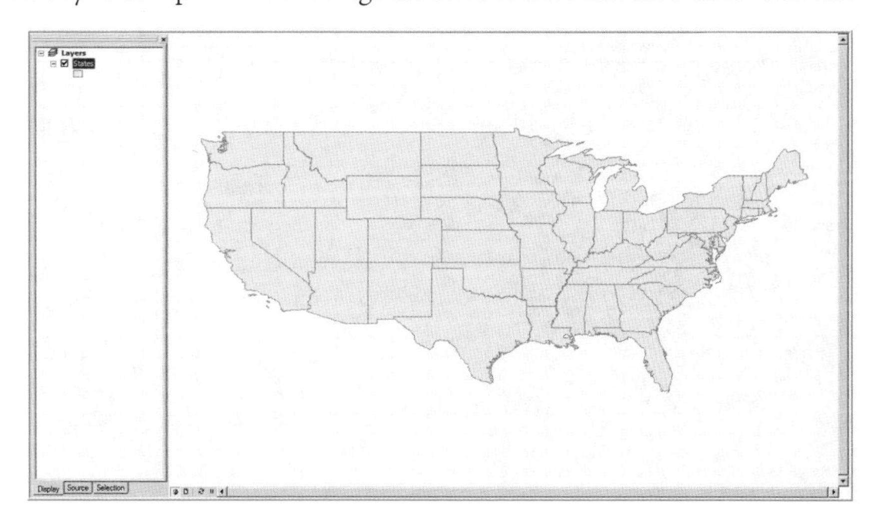

Change a layer's name

1 **Right-click the States layer in the table of contents.**

2 **Click Properties.**

3 **Click the General tab.**

Notice that the current layer name is "States."

4 **Type Population by State as the new layer name.**

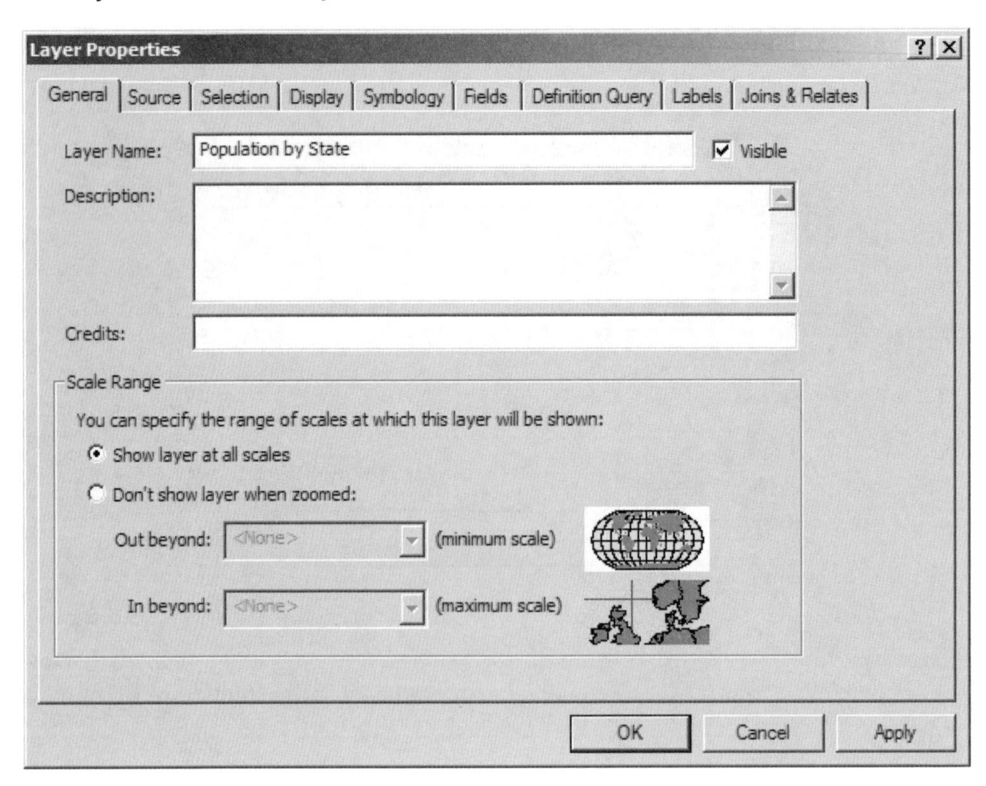

5 **Click OK.**

Select a census attribute to display state population

1. **Right-click the Population by State layer in the table of contents.**
2. **Click Properties.**
3. **Click the Symbology tab.**
4. **In the Show box, click Quantities, Graduated colors.**
5. **In the Fields box, click the Value drop-down list and click POP2003.**

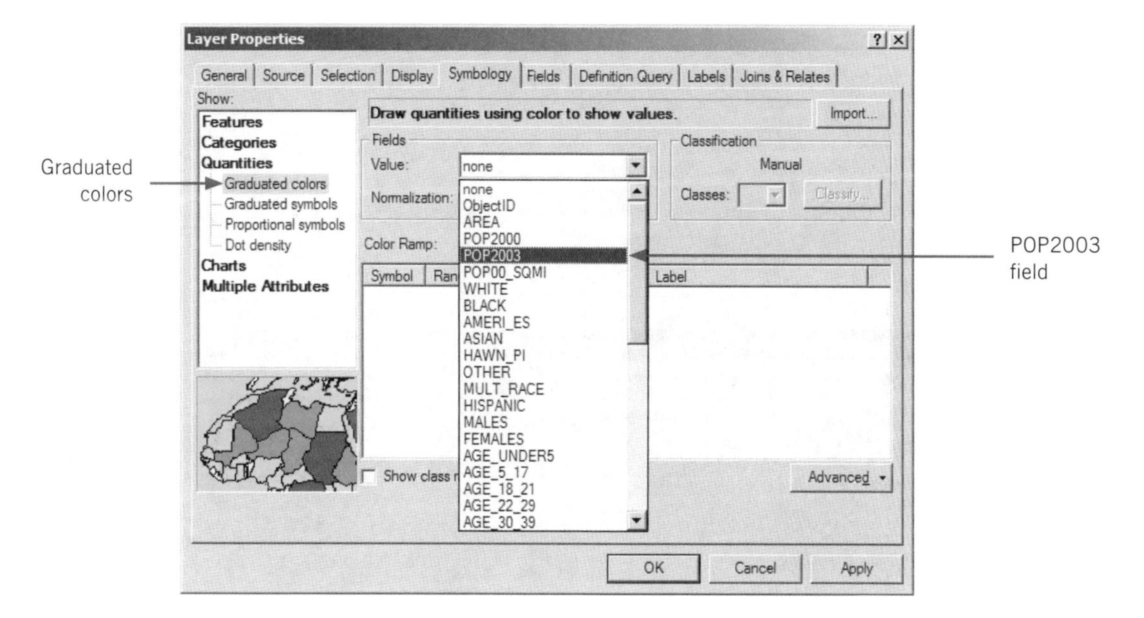

6. **Click the Color Ramp drop-down list and scroll down to and click the yellow-to-brown color ramp.**
7. **Click OK.**

The result is a classification consisting of five value intervals of 2003 population ranging from lowest to highest population with darker colors for higher population. By default, ArcMap uses a method called Natural Breaks to construct the classification intervals. You will learn how to change classifications later.

Create group layers

Group layers contain other layers, allowing for better organization of the layers in your map. Group layers have behavior similar to other layers in the table of contents. Turning off the visibility of a group layer turns off the visibility of all its component layers.

Add a group layer to the map

1 **Right-click Layers in the table of contents.**

2 **Click New Group Layer.**

3 **Right-click the resulting New Group Layer and click Properties.**

4 **Click the General tab.**

5 **Type Population by County as the group layer name.**

Do not click OK yet.

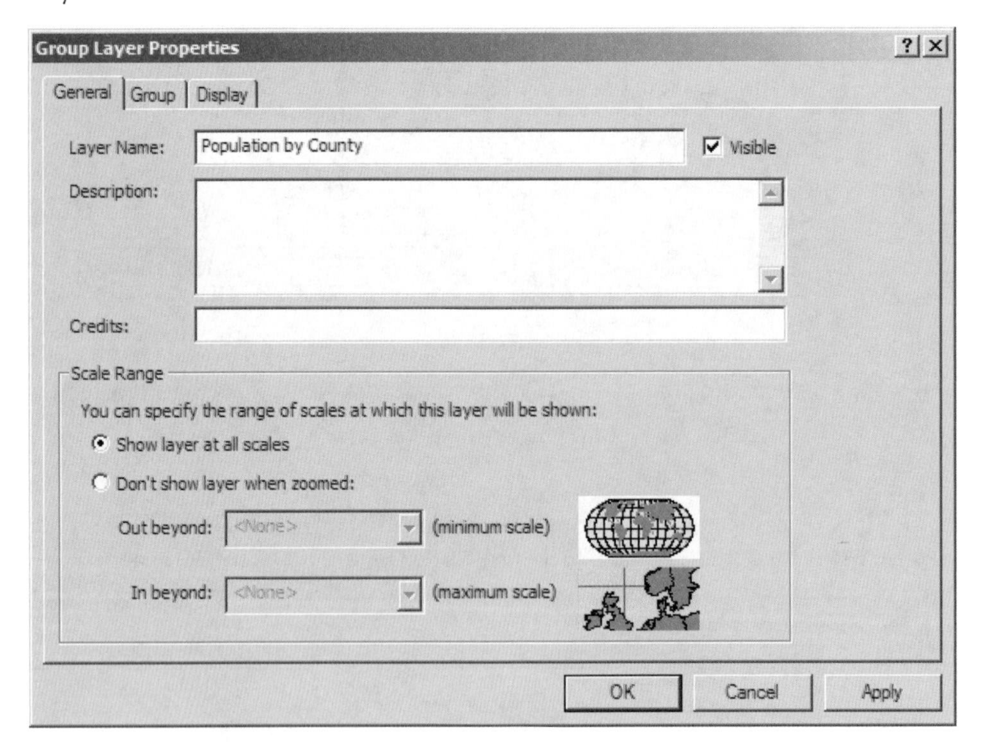

Add a layer to the group

1 **Click the Group tab in the Group Layer Properties window.**

2 **Click the Add button.**

3 **Navigate to your \Gistutorial\UnitedStates folder.**

4 **Press and hold down the Ctrl key.**

5 **Click States.shp, Counties.shp, Add.**

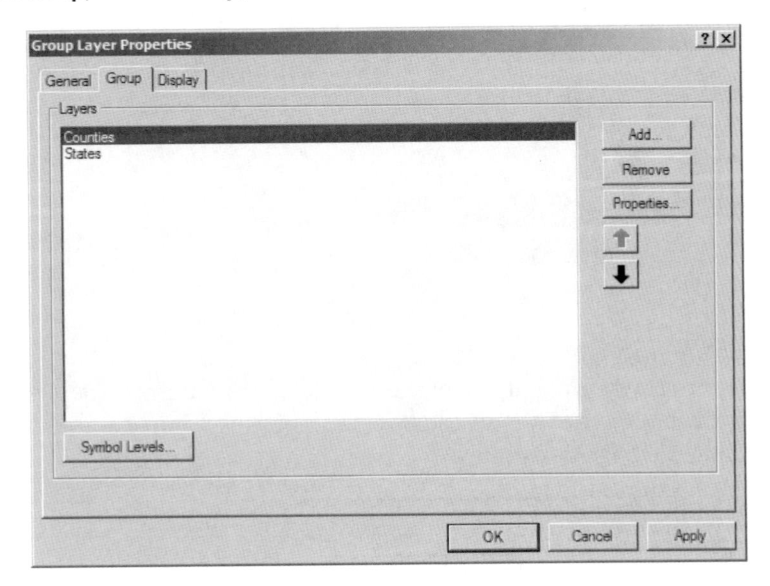

6 **Click OK.**

The U.S. counties are displayed in the map with a color randomly selected by ArcMap.

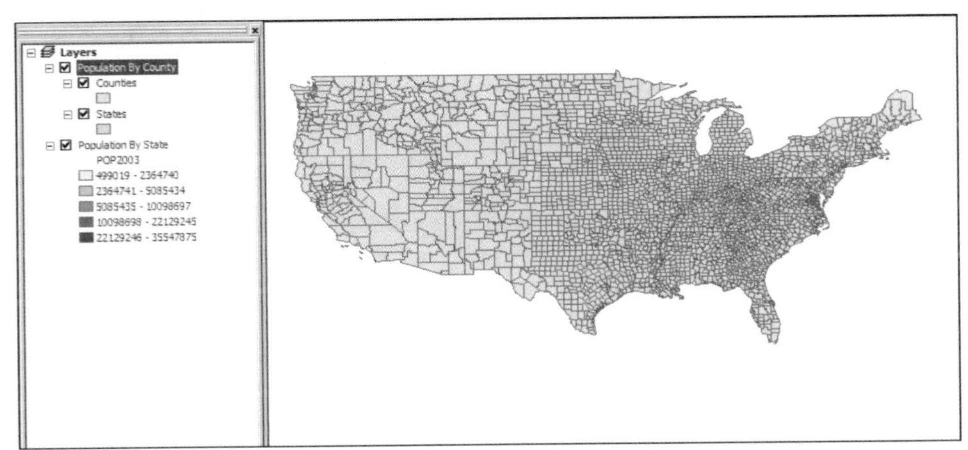

Change the symbology for states

1 **Within the Population By County group layer, click the States layer and drag it above the Counties layer.**

2 **Click the legend symbol below the States layer name in the group layer.**

3 **In the Symbol Selector's Options panel, change the Fill Color to No Color, type an Outline Width of 1.5, and change Outline Color to Black.**

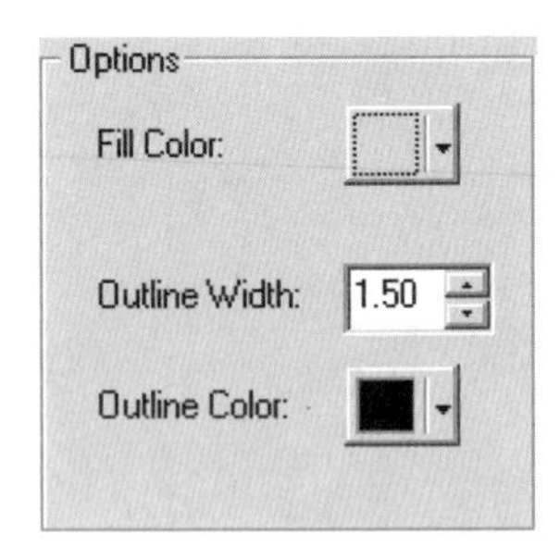

4 **Click OK.**

If by mistake you drag the States layer outside of the Population by County group, just drag it back inside the group. This is another way to add layers to a group—simply add them to the table of contents and then drag them inside a group.

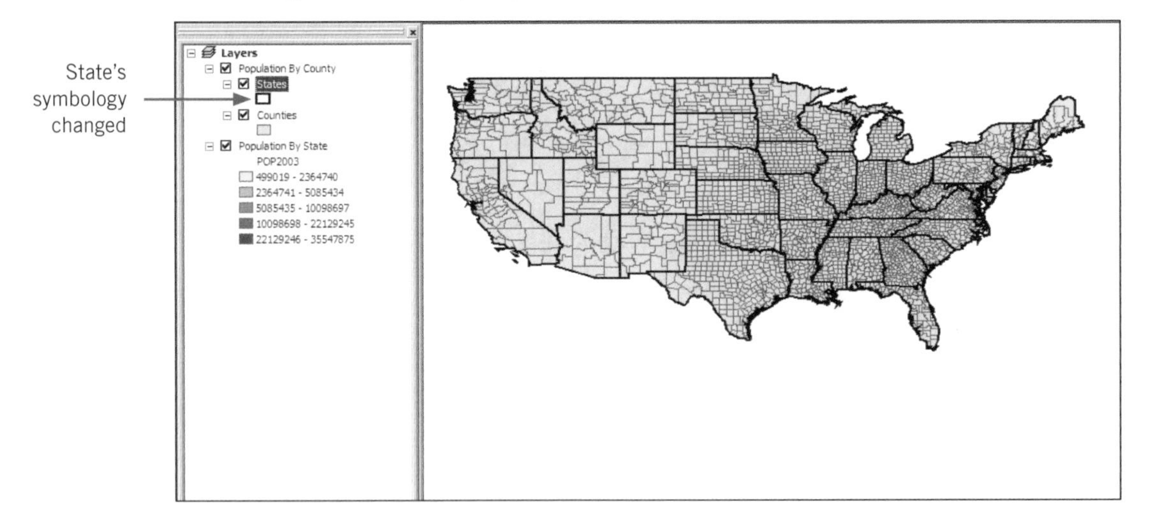

Select a census attribute to display county population

1 **Right-click the Counties layer in the group layer.**

2 **Click Properties.**

3 **Click the Symbology tab.**

The current symbol for the counties layer is Single symbol.

4 **In the Show box, click Quantities, Graduated colors.**

5 **In the Fields panel, click the Value drop-down list and click POP2003.**

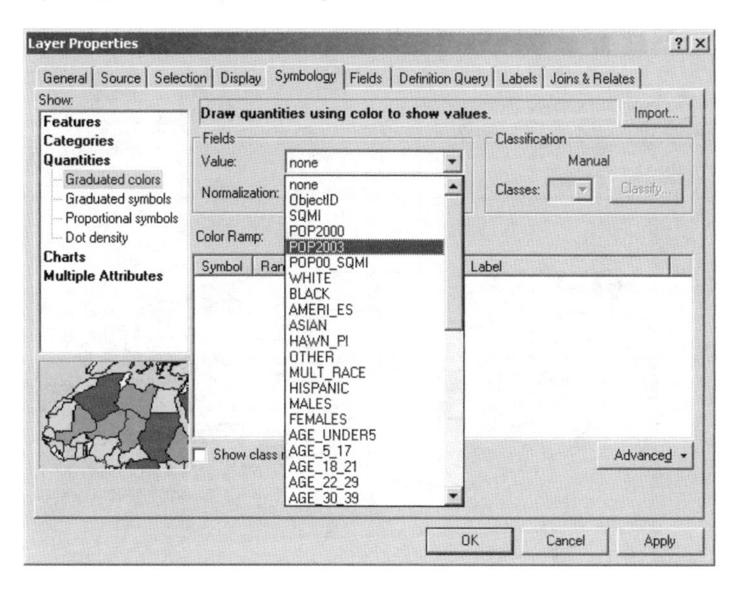

6 **Click OK.**

The result is a classification of the U.S. counties into five value ranges of 2003 population.

7 **Collapse the tree structures in the table of contents by clicking the boxes that have minus signs (–) for Population by County and Population by State.**

You can reverse this process by clicking the boxes again, which now have plus signs (+) indicating that they can be expanded.

YOUR TURN

Note: You must complete this Your Turn *exercise before continuing. You will use this group layer later in the tutorial.*

Turn off the Population By County group layer and the Population By State layer.

Create a new group layer called Population By Census Tract.

Add the census tract layers for Utah and Nevada and the States layer to the Population By Census Tract group layer. (The census tract layers are located at \Gistutorial\UnitedStates\Utah\UtahTracts.shp and \Gistutorial\UnitedStates\Nevada\NevadaTracts.)

Classify the census tracts using graduated colors based on the POP2003 field. Choose No Color for Fill Color and a black 1.5-width line for the States layer.

Notice that the resulting scales for the two states differ. Later in this tutorial you will learn how to build custom numerical scales. Then you could create a single scale for both states.

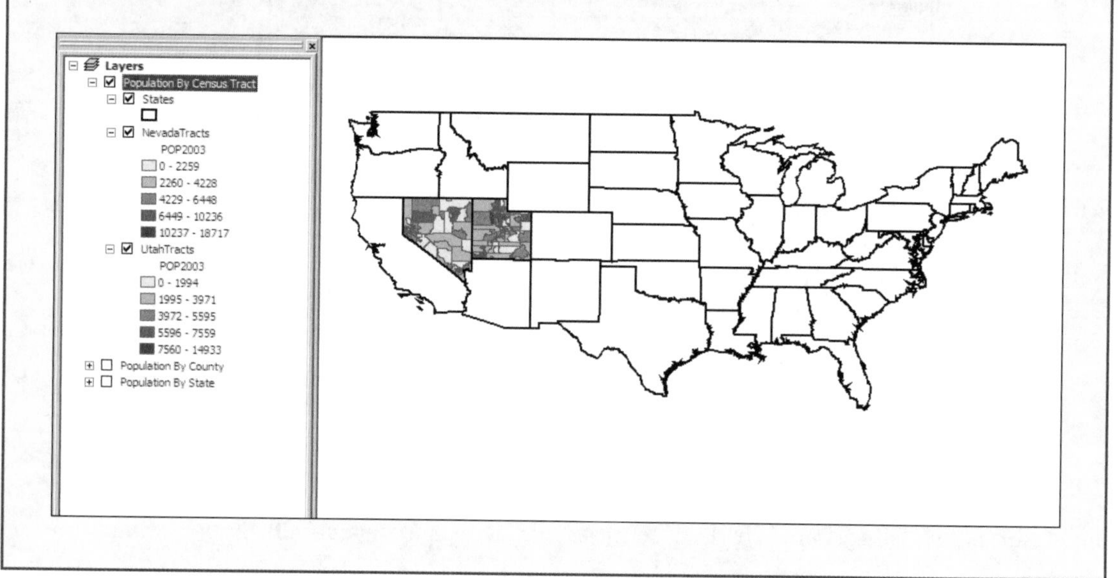

Saving layer files

1 Right-click the layer group named Population by County.

2 Click Save As Layer File.

3 Navigate to the **\Gistutorial\Layers** folder.

4 Type **PopulationByCounty.lyr** in the Name field.

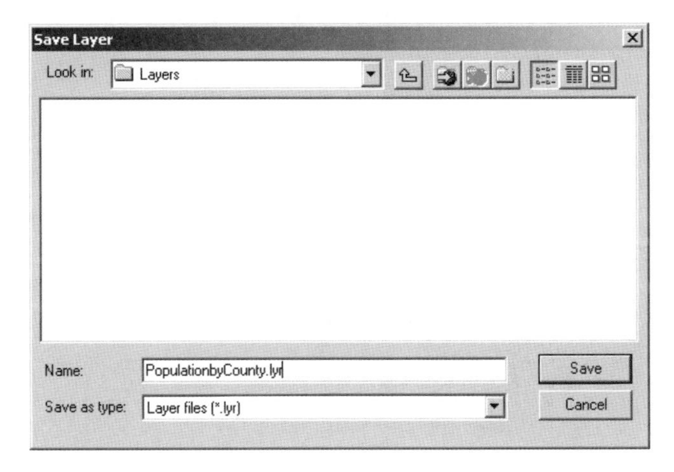

5 Click Save.

The layer file is now stored in the Layers folder. It can be added to any map that you create. Note that you also can save ungrouped layers, such as Population By State, as a layer file for reuse.

Adding group layers

1 Click the Add Data button. 

2 Navigate to **\Gistutorial\Layers**.

3 Click **PopulationByCounty.lyr**.

4 Click the Add button.

You now have a second copy of the group layer in the table of contents.

Removing group layers

1 Right-click the duplicate Population By County layer group that was just added.

2 Click Remove.

Create threshold scales for dynamic display

If a layer is turned on in the table of contents, ArcMap will draw it, regardless of the map scale (that is, how far you are zoomed in or out). To help you automatically display layers at an appropriate map scale, you can set a layer's visible scale range and define the range of scales at which ArcMap draws the layer.

Set a visible scale based on the current scale

1 **Turn off the Population By Census Tract group layer and turn on Population By County and Population By State.**

2 **Zoom to a few states in the northeastern part of the country.**

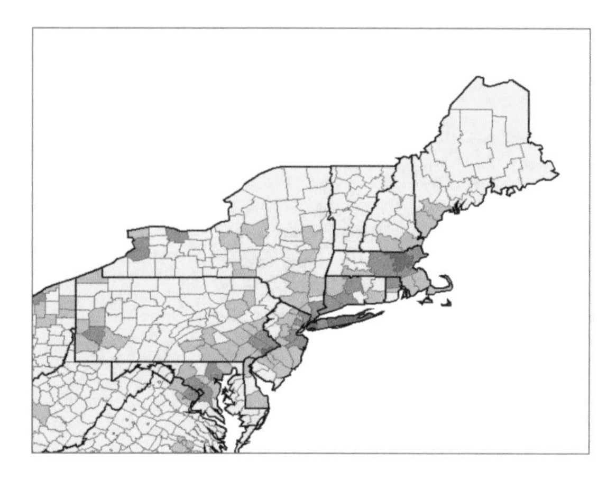

3 **If necessary, expand the Population By County group layer, then right-click the Counties layer in the Population By County group layer.**

4 **Click Visible Scale Range, Set Minimum Scale.**

ArcMap sets the scale to display this layer when zoomed in this close or closer. Zooming out any further will turn off the polygons for this layer.

5 **Click the Full Extent button.** 

Now the county polygons will not display.

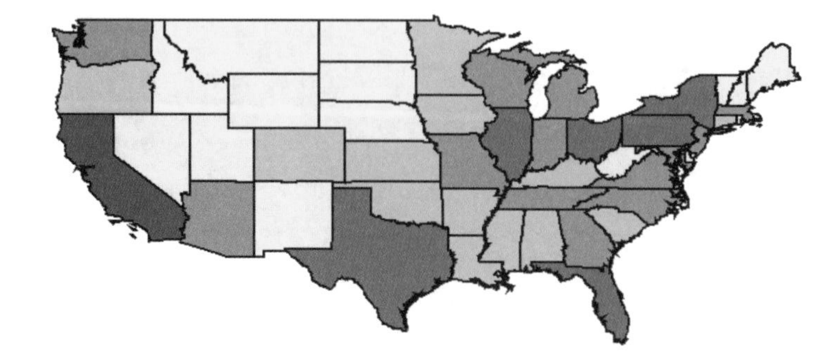

Set a maximum scale based on the current scale

1 **Zoom in until the county polygons are displayed again.**

2 **Right-click the States layer in the Population By County group layer.**

3 **Click Visible Scale Range, Set Maximum Scale.**

4 **Zoom in a little closer.**

The black outline polygons for the states are not displayed when zoomed in beyond the maximum scale just set. Zooming out enough will turn on the state polygons again.

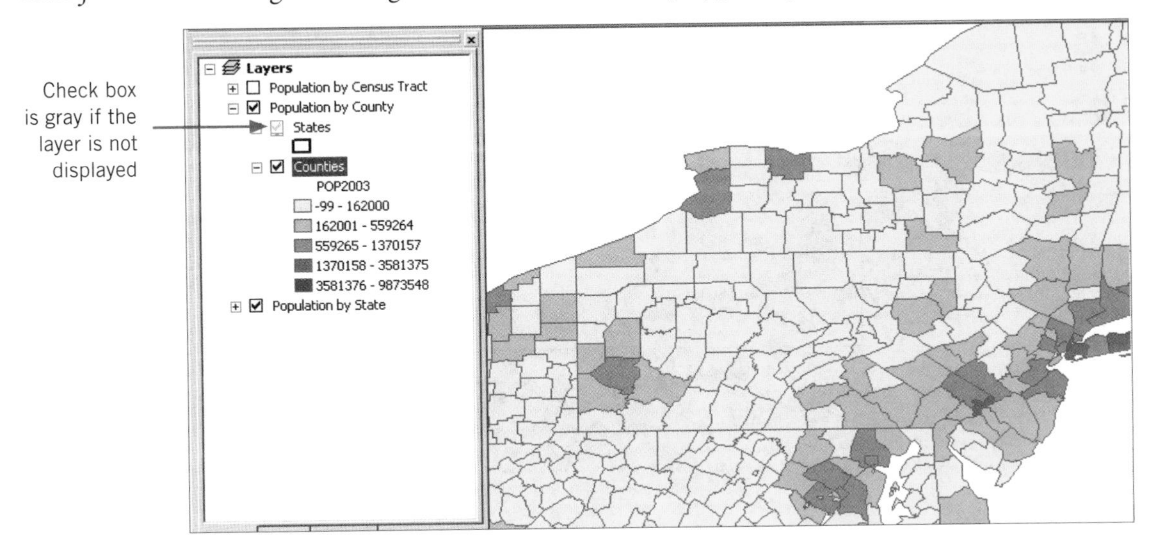

Clear a layer's visible scale

1 **Right-click the States layer in the Population By County layer group.**

2 **Click Visible Scale Range, Clear Scale Range.**

The outline polygons for the states are displayed again when zoomed at this scale.

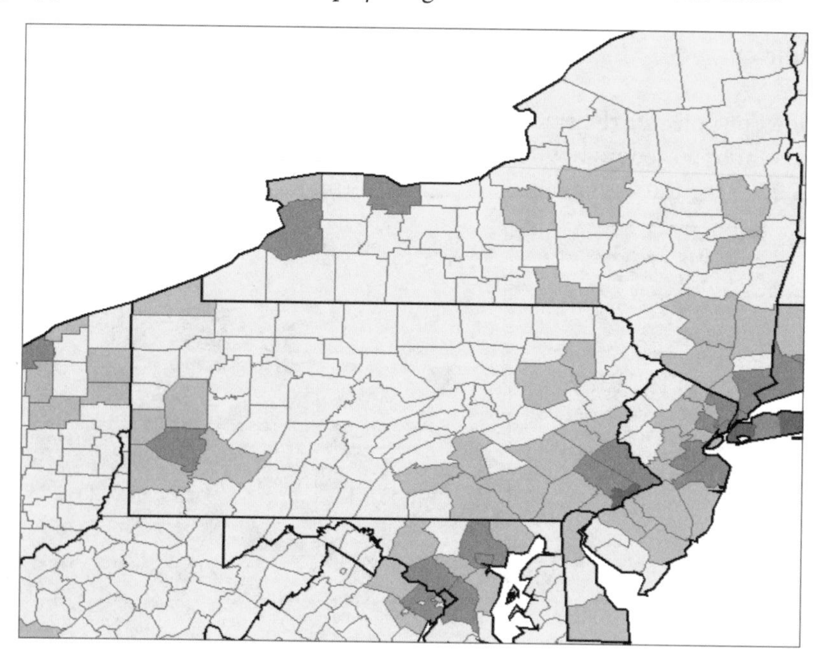

Set a minimum visible scale for a specific layer

1 Zoom to the full extent of the map. 

2 Turn the Population By Census Tract group layer back on.

3 In the table of contents, right-click the NevadaTracts layer and click Properties.

4 Click the General tab.

5 Click the Don't show layer when zoomed radio button.

6 Type **8,000,000** in the Out beyond field.

If you zoom out beyond this scale, the layer will not be visible.

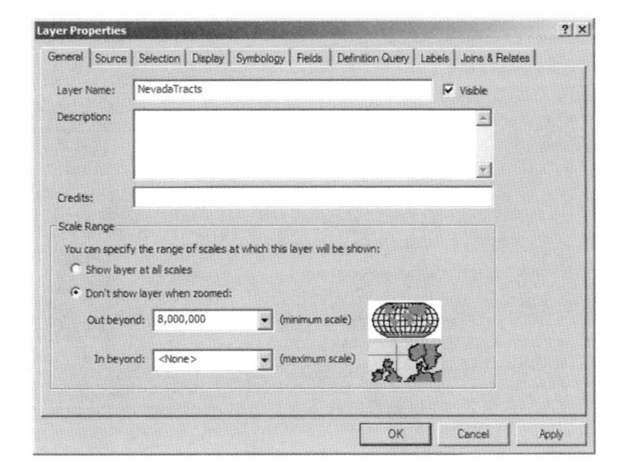

7 Click OK.

ArcMap does not show the Nevada census tract polygons when zoomed out past a scale of 1:8,000,000.

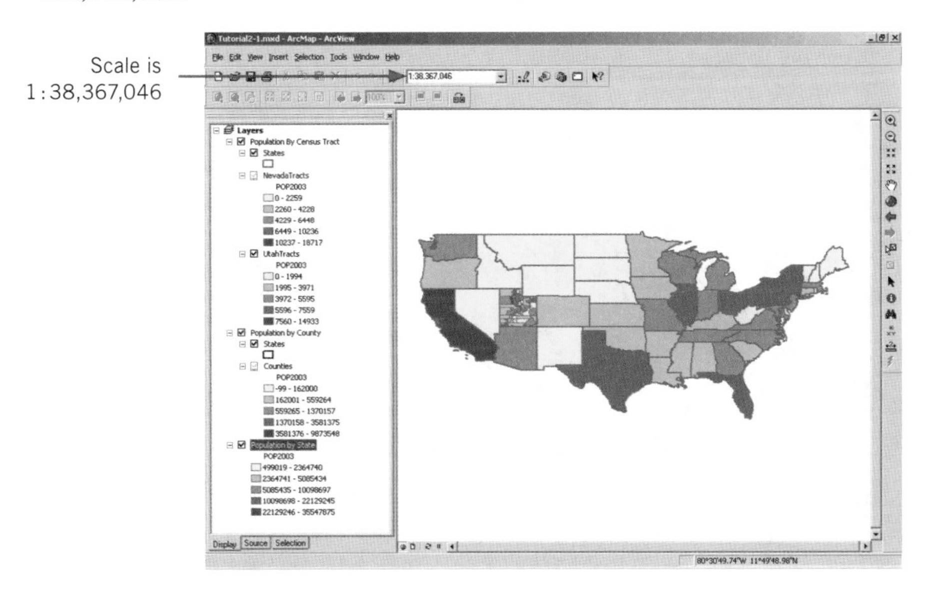

YOUR TURN

Set the Utah Tracts minimum scale to the same as Nevada Tracts. In the Out Beyond field of the General tab of the Properties window for this map layer, click the drop-down list to find your previously set scale and click it.

Set a maximum visible scale for a specific layer

1 **Right-click the Population By State layer and click Properties.**

2 **Click the General tab.**

3 **Click the Don't show layer when zoomed radio button.**

4 Type **10,000,000** in the In beyond field.

5 **Click OK.**

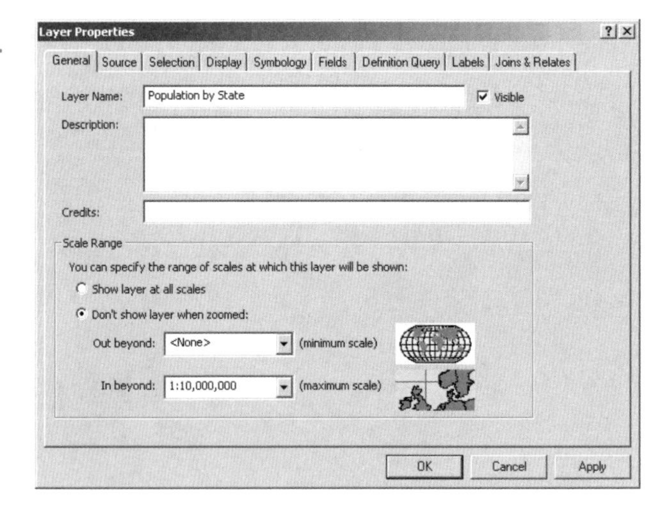

6 Type **1:9,500,000** in the scale field.

The map is now zoomed in too far for Population By State to display (<1:10,000,000) and not far enough for Nevada or Utah tracts to display (>1:8,000,000).

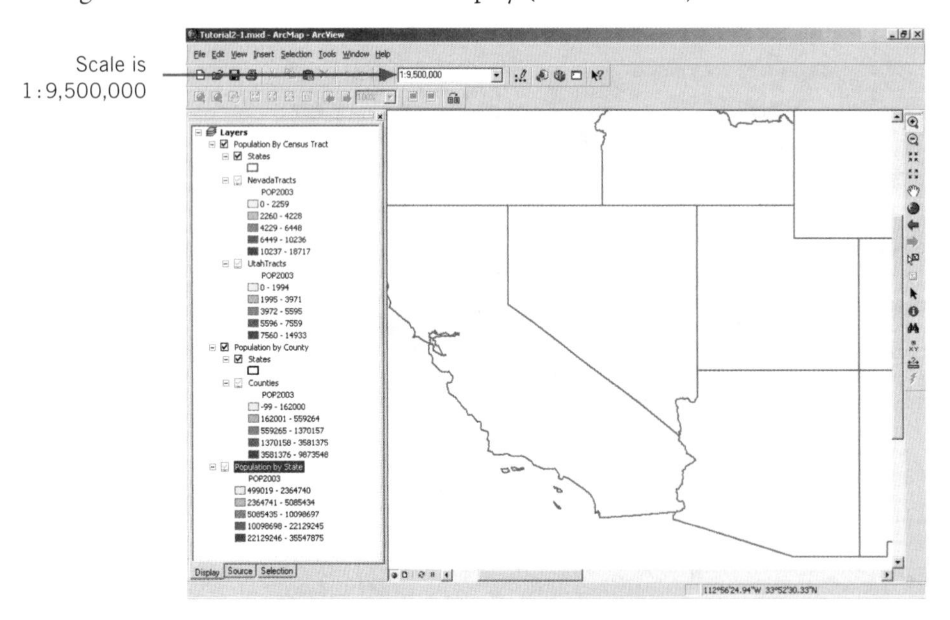

Create choropleth maps using custom attribute scales

Earlier in this tutorial, you created a choropleth map from the Population By States layer using a classification method called Natural Breaks to divide the features in the map into five value classes. Although Natural Breaks is the default method, ArcMap allows you to choose other methods for classifying your data. Here you will learn how to create your own custom classification (or numerical scale).

Create custom classes in a legend

1 **Zoom to the full extent.** 

2 **Turn off all layers except Population By State, and expand that layer in the table of contents so that you can see its classes.**

3 **Right-click the Population by State layer and click Properties.**

4 **In the Layer Properties dialog box, click the Symbology tab.**

5 **In the Classification panel, click the Classes drop-down list and select 6.**

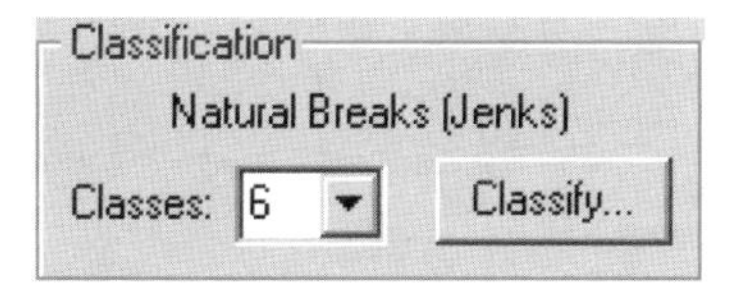

6 **In the Classification panel, click Classify.**

The Classification dialog box shows the current classifications, statistics, and break values.

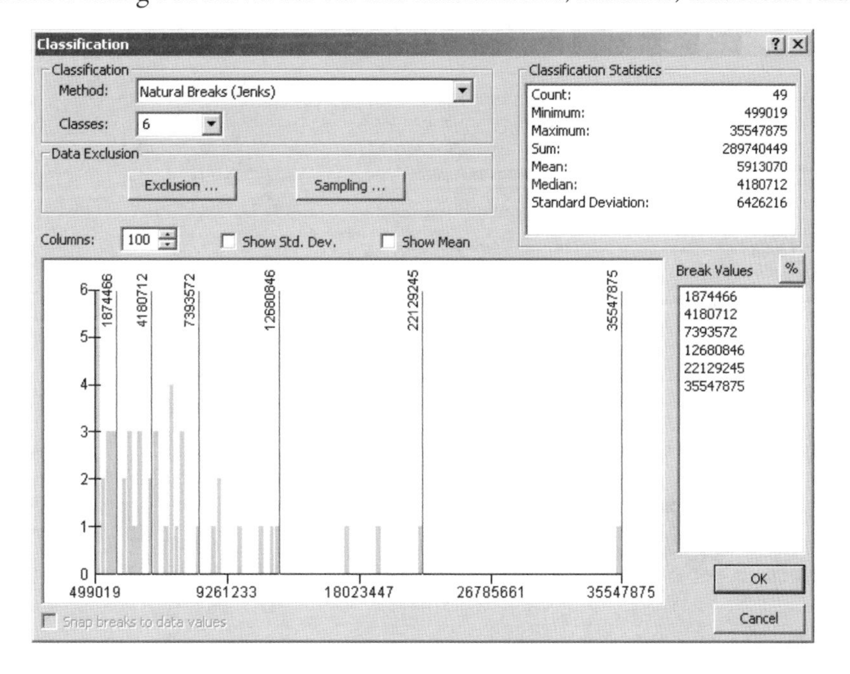

Manually change class values

1 Click the drop-down list for the Classification Method and click Manual.

2 In the Break Values panel, click the first value, 1874466, to highlight it.

Notice that the blue graph line corresponding to that value turns red.

3 Type 2000000 and press Enter to move to the next break value.

4 Continue by entering the following break values: 4000000, 8000000, 16000000, and 32000000 but let the last (maximum) value remain 35547875.

These break values create increasing-width intervals that double in each successive class.

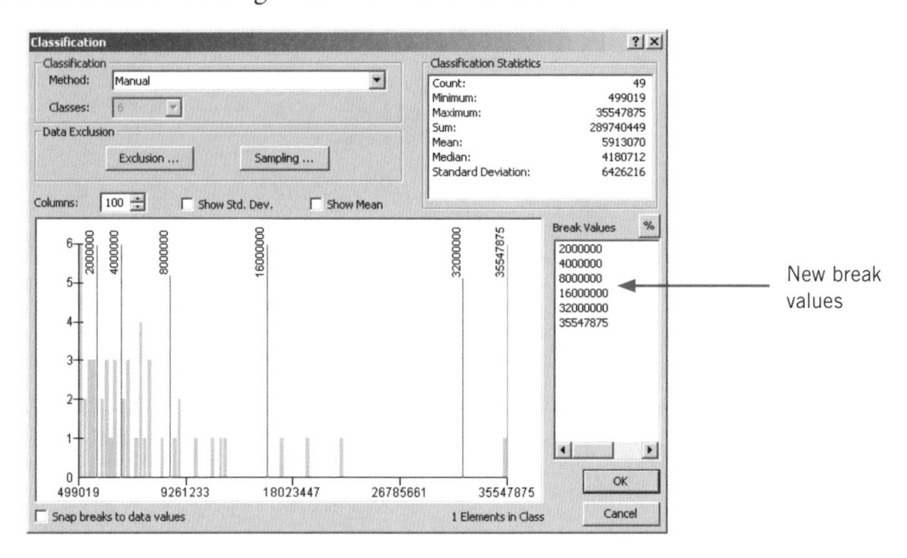

5 Click OK.

6 Click the gray Label heading to the right of the gray Range heading and click Label format. In the Number Format dialog box, select Show thousands separators, OK.

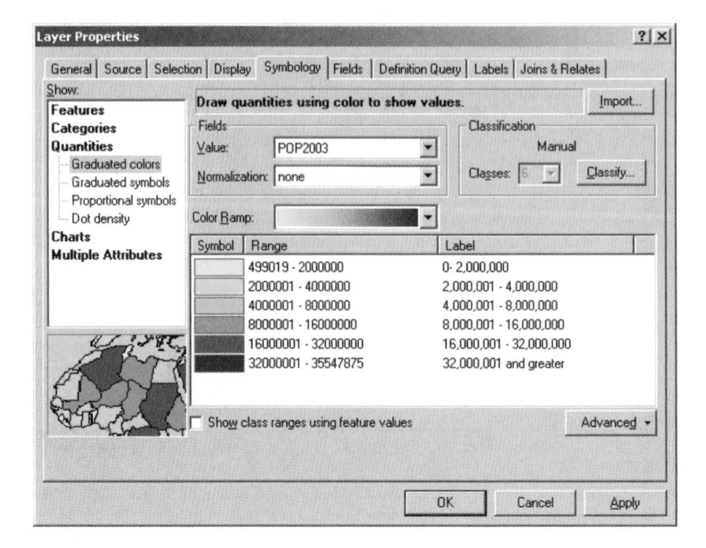

7 **In the Label field of the Symbology tab, change the first value to 0–2,000,000 and the last label to 32,000,001 and greater.**

8 **Click OK.**

The Population By State layer changes to reflect the new break values and labels. Besides being easier to read and interpret, the classification is appropriate for the long-tailed state population distribution, with its increasing-width intervals.

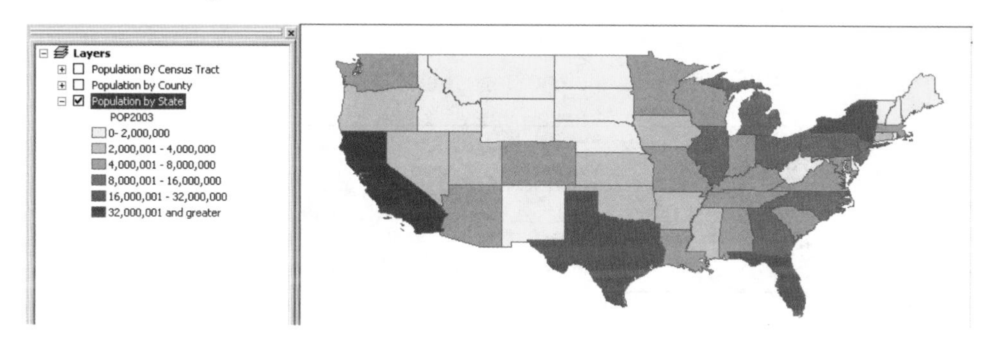

9 **Right-click the Population By State layer in the table of contents and click Save As Layer File.**

10 **Browse to the \Gistutorial\Layers folder and type PopulationByState in the name field, and click Save.**

YOUR TURN

Change the classification break values for the Counties layer based on the population (POP2003) field. Use the same method as above to manually change the values. Be sure to change the labels in the legend.

Hints: Use five classes and start with the Quantile Classification method. With five classes, each resulting interval has 20 percent of the counties. Using the quantile break values as a guideline, design a doubling scale that has multiples of 1,000.

Zoom into the Counties layer to see how your classification scheme looks on the choropleth map.

Manually change class colors and hues

Colors for classes can be changed manually. Generally, it is best to have more classes with light colors and a few with dark colors (the human eye can differentiate light colors more easily). So here you will create a custom color ramp that starts with white and ends with a bright blue.

1 **Right-click the Population By State layer and click Properties.**

2 **In the Layer Properties dialog box, click the Symbology tab.**

3 **Right-click the Color Ramp and click Properties.**

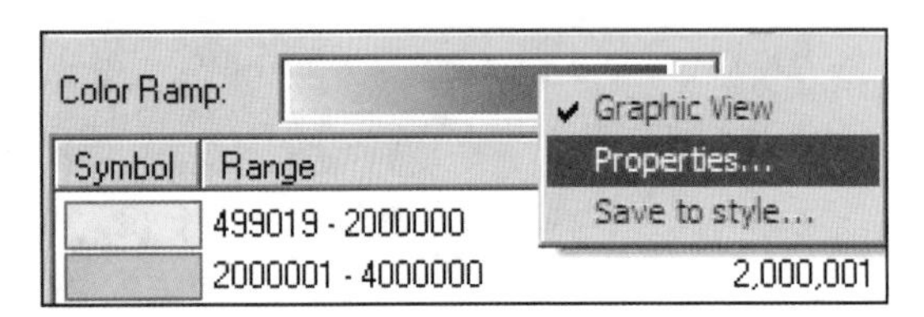

4 **Click the color box beside Color 1 and click the Arctic White paint chip.**

5 **Click the color box beside Color 2, click the Ultra Blue paint chip, and click OK twice.**

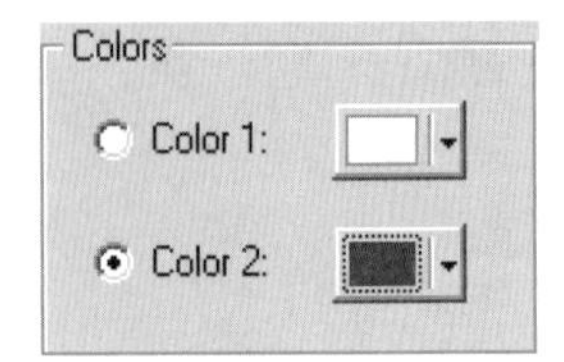

The Population By State map changes to reflect the new color ramp.

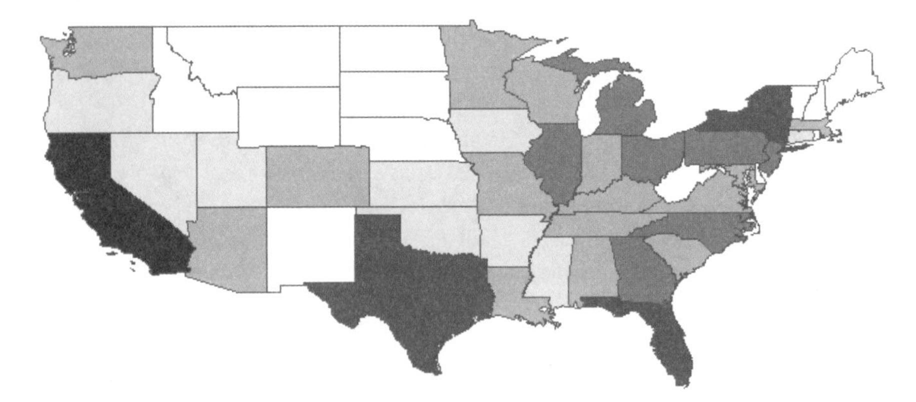

Note: You can also double-click each color symbol in the Symbology tab to change the classification colors individually.

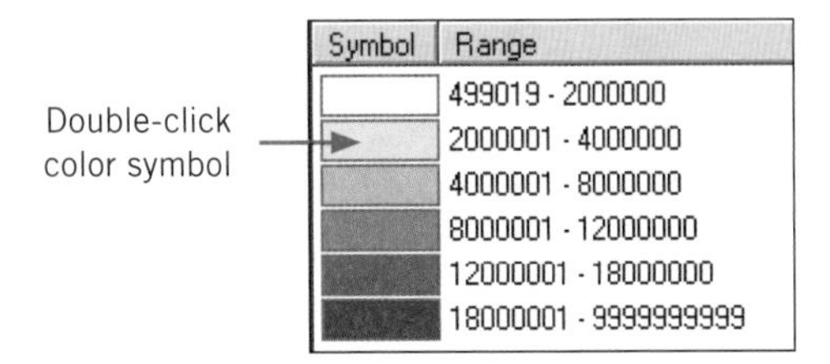

Pin (point) maps

Pin maps, otherwise known as point maps, show exact locations of data or events using individual point markers for each record. In this example, you will create a pin map showing the range of populations in U.S. cities using graduated symbols.

Create a pin map of U.S. cities by population

1 Turn off all the layers in your map and create a new group layer called Population By City.

2 Add the data layers **\Gistutorial\UnitedStates\States.shp** and **\Gistutorial\UnitedStates\Cities.shp**.

3 Double-click States in the Population By City group layer to open its Layer Properties window.

4 Click the Symbology tab, change the symbol to a hollow fill with a black outline of 1.5, and click OK, then click OK again.

5 Double-click Cities in the Population By City group layer to open its Layer Properties window.

6 Click the Symbology tab and change the layer's symbology from Single Symbol to Quantities, Graduated Symbols.

7 In the Fields panel, change the Value to POP2000, the template symbol to a red circle, symbol size to 2–18, and assign the break points and legend labels as shown below.

8 Click OK.

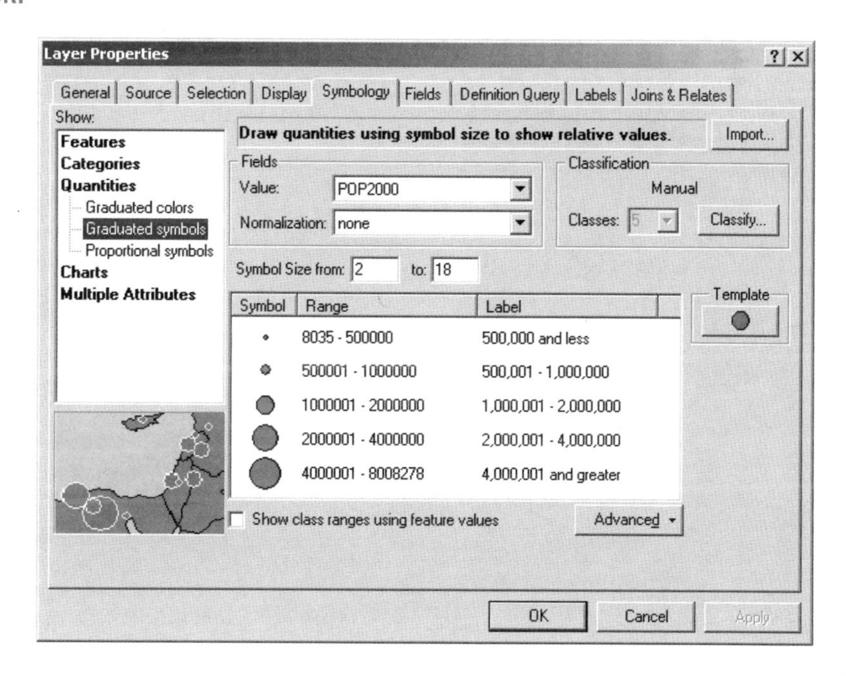

The resultant pin map shows U.S. cities classified by population.

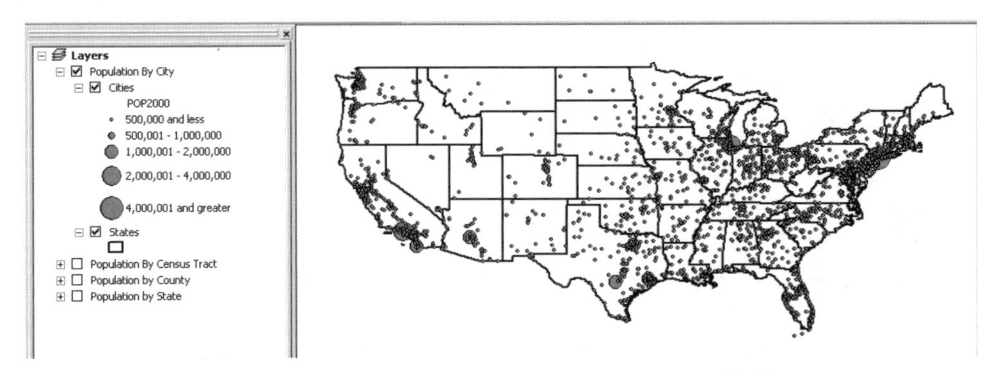

9 Click File, Save.

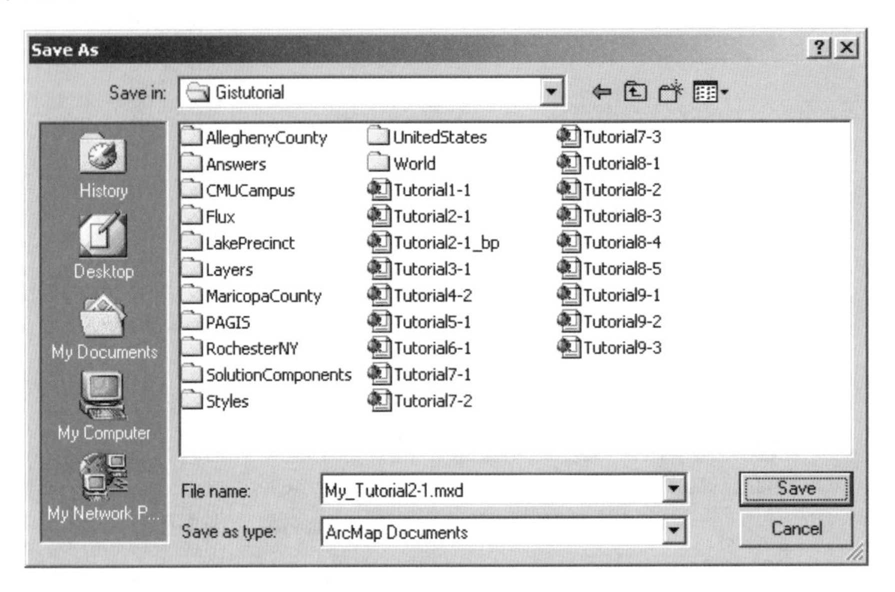

Create a pin map based on feature query

Pin maps can be created by selecting a subset of features from an existing layer. For example, suppose you have a layer containing all the cities in Pennsylvania, but you only want to display the cities with populations between 10,000 and 49,000. To display the correct cities, you can create a definition query to filter out all the cities with population values outside the desired range.

Create a new map

1 **Click File, New.**

2 **Click the My Templates tab and Blank Document from the New dialog box.**

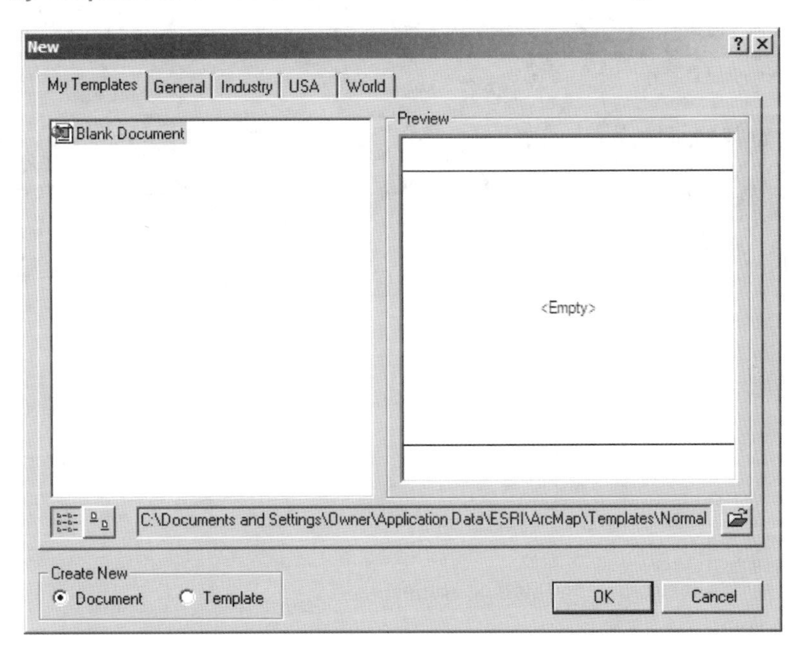

3 **Click OK.**

Add data to the map

1 **Click the Add Data button.** 

2 **Navigate to folder where you have the Gistutorial data installed, click \Gistutorial\UnitedStates\Pennsylvania, and add the following layers: PACounties.shp and PACities.shp.**

This displays a map showing county polygon features for Pennsylvania and detailed cities. ArcMap picks an arbitrary color fill and point marker for the polygons and points.

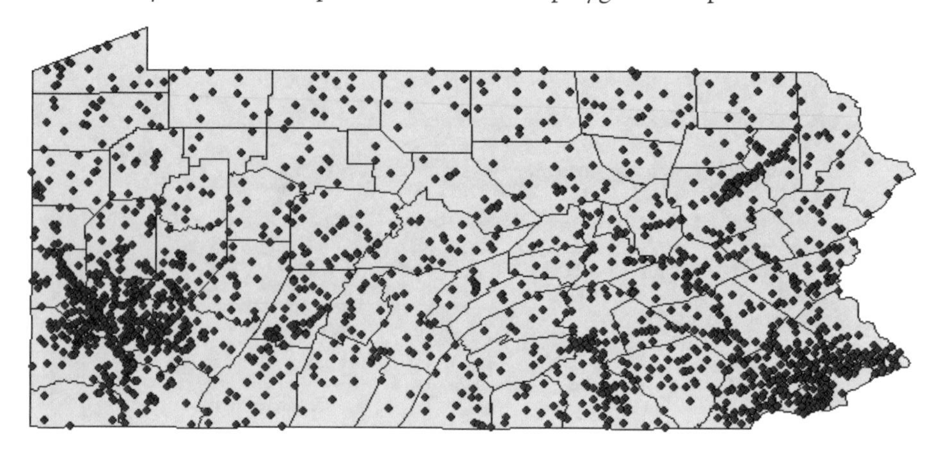

Create ground polygons

To draw attention away from a feature, colors should be very light or, in the case of these polygons, have no color at all.

1 **Right-click the PACounties layer and click Properties.**

2 **Click the General tab and change the name of the layer to Pennsylvania Counties.**

3 **Click the Symbology tab, click the symbol, and click the Hollow fill style from the Symbol Selector.**

4 **Click OK and OK again.**

Display a queried subset of Pennsylvania cities

1 **Right-click the PACities layer and click Properties.**

2 **Click the Definition Query tab and Query Builder button.**

3 **In the Query Builder window double-click "FEATURE".**

4 **Click "=" as the logical operator.**

5 **Click the Get Unique Values button.**

The resulting list has all unique values in the FEATURES attribute. Note that the attribute stores classification values for a numerical scale.

6 **In the Unique Values List, double-click '10,000 to 49,999'.**

The completed query ("FEATURE" = '10,000 to 49,999') will yield a layer with only the cities in Pennsylvania with populations between 10,000 and 49,999. If the query has an error, just edit it in the lower panel of the Query Builder or delete it by clicking Clear and then repeat steps 3 through 6.

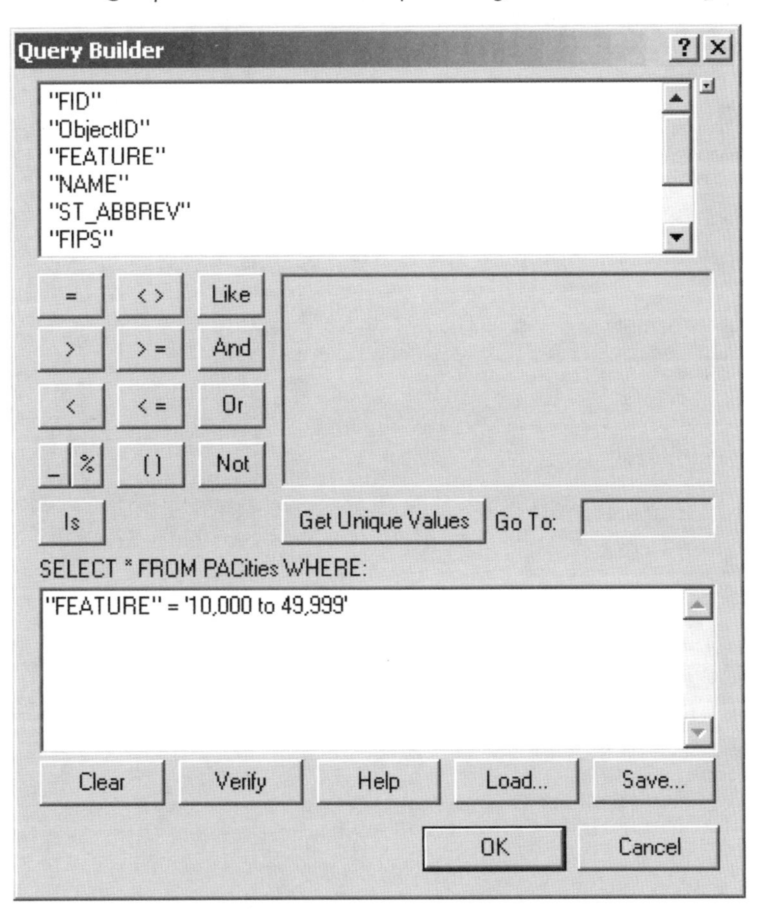

8 **Click OK twice to execute your query and close the Layer Properties dialog box.**

Change the layer's name and symbol

1 Right-click the PACities layer and click Properties.

2 Click the General tab and change the name of the layer to **Population 10,000 to 49,999**.

3 Click the Symbology tab.

4 Click the Symbol button.

5 Click the Circle 2 symbol icon.

6 Change the color to Ultra Blue and the size to 8.

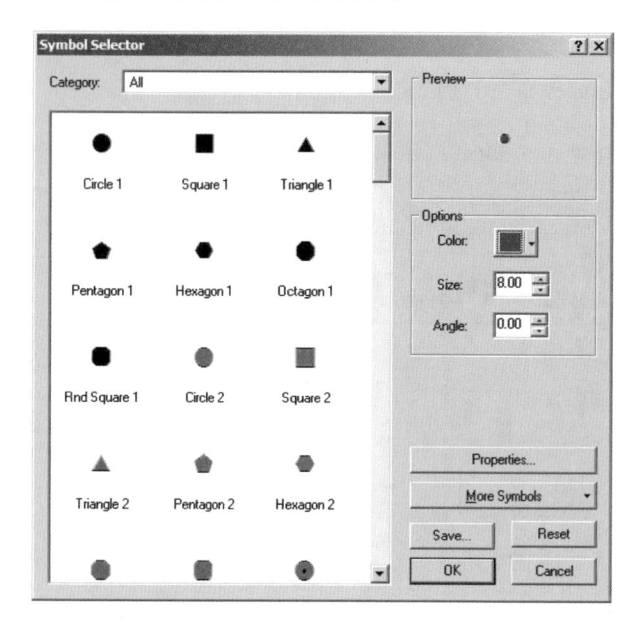

7 Click OK, then click OK again.

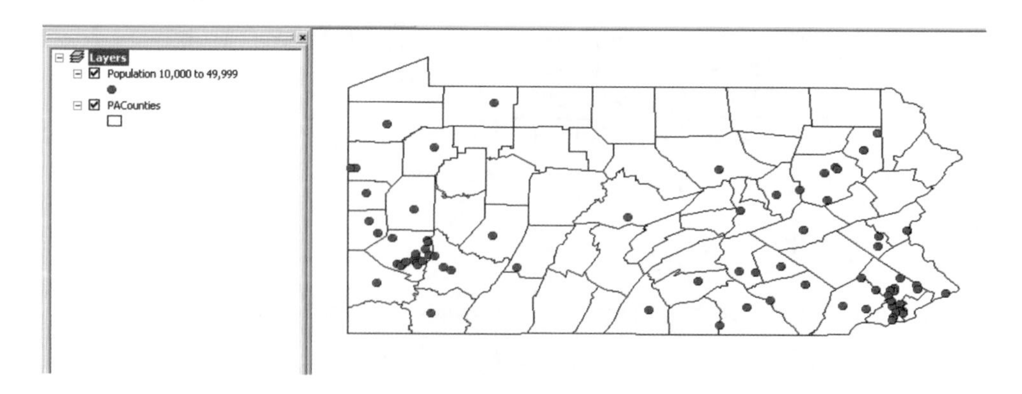

Add Pennsylvania's state capital city

1 Click the Add Data button. 

2 Navigate to the folder where you have the Gistutorial data installed, click **\Gistutorial\UnitedStates\Pennsylvania**, and add the **PACities.shp** again.

3 Right-click the PACities layer and click Properties.

4 Click the Definition Query tab.

5 Click the Query Builder button.

6 Scroll down in the fields list of the Query Builder window and double-click "STATUS".

7 Click "=" as the logical operator.

8 Click the Get Unique Values button.

9 Double-click 'State Capital County Seat'.

The completed query ("STATUS" = 'State Capital County Seat') yields a layer with only one city, Pennsylvania's state capital of Harrisburg.

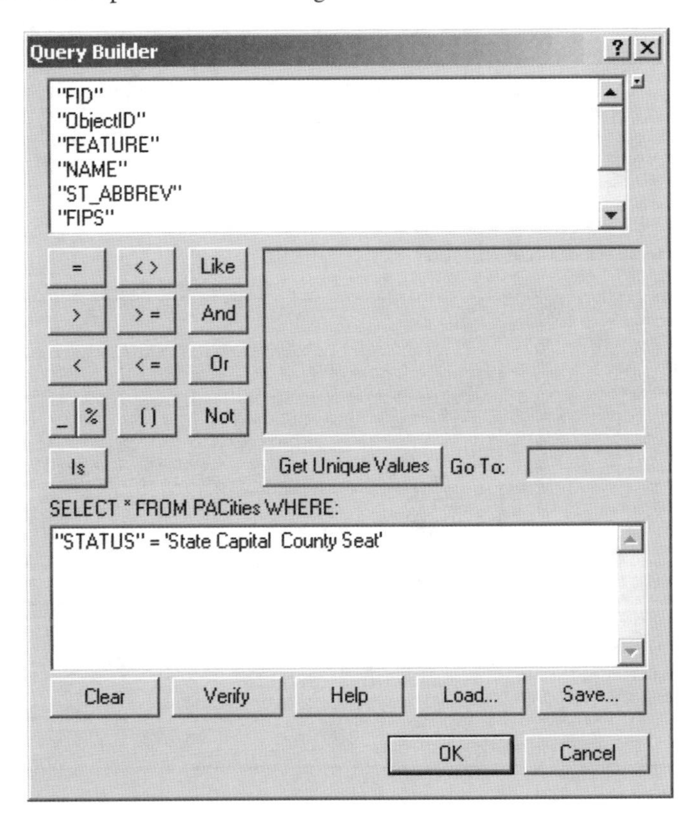

10 Click OK and OK again.

Change the layer's name and symbol

1. **Right-click the PACities layer and click Properties.**
2. **Click the General tab and change the name of the layer to State Capital.**
3. **Click the Symbology tab.**
4. **Click the Symbol button.**
5. **Scroll down and click the Star 3 symbol icon.**
6. **Change the color to Solar Yellow and the size to 25.**
7. **Click OK and OK again.**

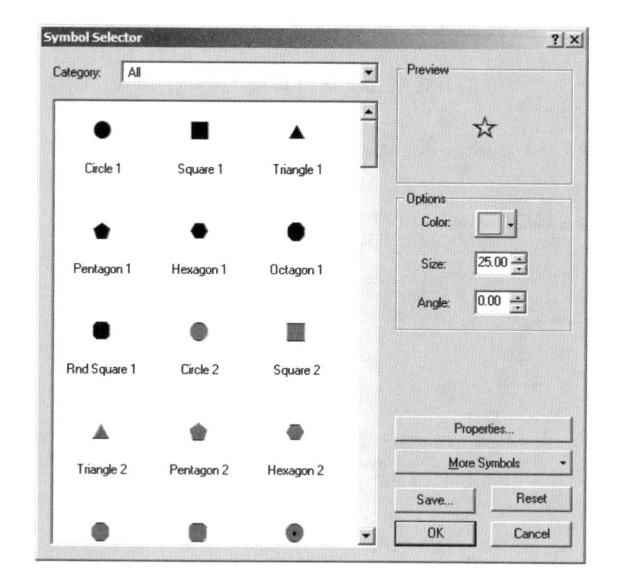

The resultant map shows the state capital of Harrisburg.

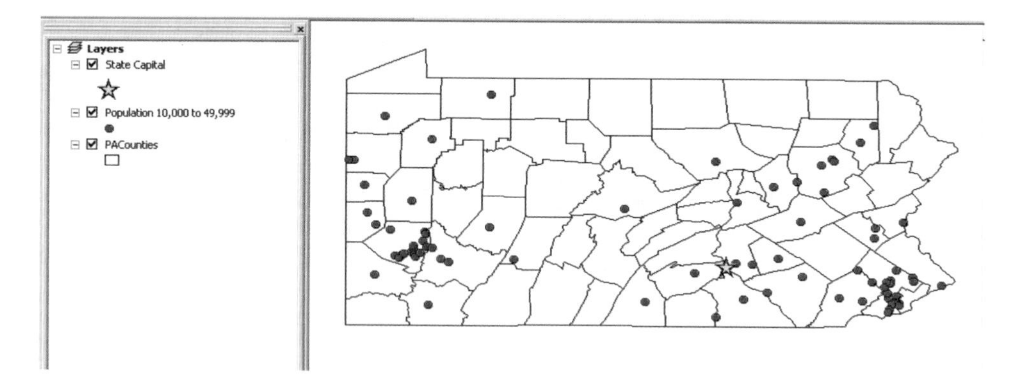

YOUR TURN

Add the PACities layer once more and create a definition query that displays Pennsylvania's two largest cities, Philadelphia and Pittsburgh, and show the two cities with a symbol that makes them stand out on the map. *Hint: One solution is to use the query criterion "NAME" = 'Pittsburgh' OR "NAME" = 'Philadelphia'.* Turn Label Features on for this layer and State Capital.

Create hyperlinks

The Hyperlink tool allows access to documents or Web pages by clicking features. There are three types of hyperlinks: documents, URLs, and macros.

Create a dynamic hyperlink

1 **From the Tools toolbar, click the Identify button.** 

2 **Click the point symbol for Harrisburg.**

3 **In the Identify Results window, right-click Harrisburg in the left panel and click Add Hyperlink from the context menu.**

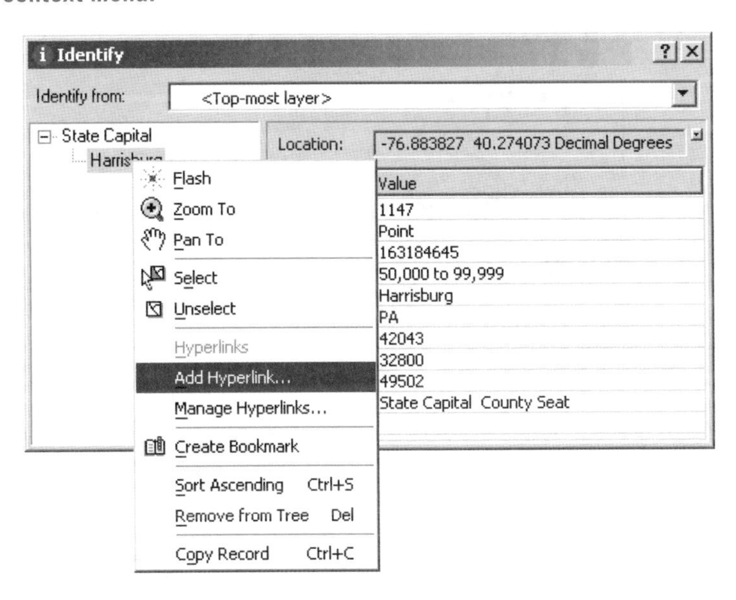

4 **Click the Link to a URL radio button and type http://www.harrisburgpa.gov.**

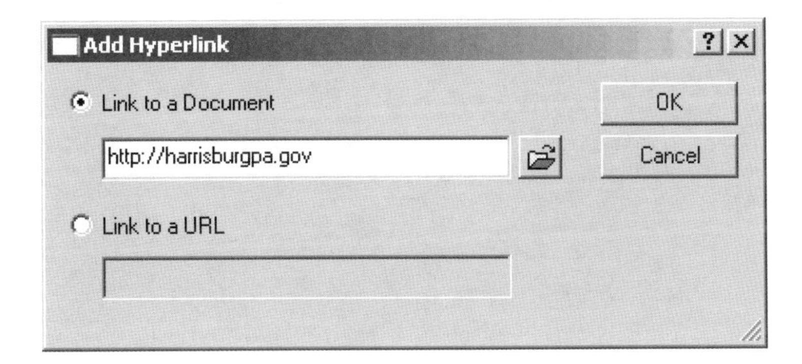

5 **Click OK, then close the Identify window.**

Launch the hyperlink

1 From the Tools toolbar, click the Hyperlink button. 

Features that have hyperlinks get a small blue circle drawn on them. In this case, the Harrisburg star point marker is the only such feature.

2 Move the cursor to the city of Harrisburg feature.

When you are over a feature for which a hyperlink exists, the cursor turns from a yellow to a black lightning bolt and you see a pop-up tip with the name of the target. Place the tip of the lightning bolt on the hyperlink's small blue circle.

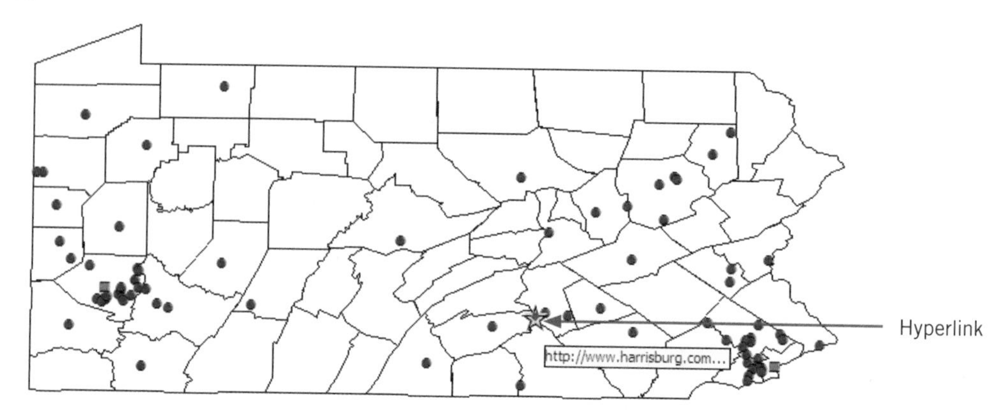

3 Click the feature to go to the Harrisburg Web site.

YOUR TURN

Add hyperlinks to the Pittsburgh and Philadelphia point markers in the layer that just has those two cities. You can surf the Web for sites to use as hyperlinks or use the following: *http://www.phipps.conservatory.org* for Pittsburgh and *http://www.longwoodgardens.org* for Philadelphia.

Create MapTips

When you hover your pointer over a feature on a map, it is possible to have an attribute of that feature automatically displayed as a MapTip.

1 Right-click the Population 10,000 to 49,999 layer in the table of contents.

2 Right-click Properties and click the Fields tab.

3 If necessary, select Name to be the Primary Display Field.

This is the field that will be used for the MapTip.

4 Click the Display tab and click Show MapTips.

5 Click OK.

6 From the Tools toolbar, click the Select Elements button.

7 Hover over any city in the Population 10,000 to 49,999 layer to see its name.

8 Click File, Save as, browse to \Gistutorial, type Tutorial2-2.mxd as the name, and click Save.

YOUR TURN

Add MapTips to PACounties. *Note: if the Show MapTips check box is ever disabled, you will need to add a spatial index for the shapefile.* Browse to the shapefile in ArcCatalog, right-click the shapefile, click Properties, the Index tab, the check box for the desired attribute, and Add.

Assignment 2-1

Map showing schools in the city of Pittsburgh by enrollment

Suppose that the City of Pittsburgh school board wants to do an extensive evaluation of local schools. They have collected data about all schools, public and private. The initial project identifies schools as either public or private and shows their enrollment. Your task is to make a map for the school board comparing the enrollment of public and private schools. You will use point features of different sizes to show this.

Start with the following:

- **C:\Gistutorial\PAGIS\Neighborhoods.shp**—polygon layer of Pittsburgh neighborhoods.
- **C:\Gistutorial\PAGIS\Schools.shp**—point layer of all schools.
- **Attributes of Schools table** for Pittsburgh Schools using the following fields:
  - **DISTRICT =** school type ("City of Pittsburgh" is a public school, "Pittsburgh Diocese" and "Private School" are private schools)
  - **ENROLL95 =** number of students enrolled
  - **STATUS =** open or closed

Create pin map with hyperlink

Create a map document called C:\Gistutorial\Answers\Assignment2\Assignment2-1.mxd showing separately the enrollment of students in public and in private schools that are open. Include Pittsburgh neighborhood polygons for reference. Hyperlink the Web site *http://www.pps.k12.pa.us/schenleyhighschool.asp* to the point for Schenley High School (high school of Andy Warhol).

Hints

- Add two copies of Schools.shp to your map document. Use one copy for public schools and the other for private schools.
- Use the same increasing-width interval scale for both public and private schools.
- Use MapTips for schools and label neighborhoods. *Tip: Use a small, dark gray font for the labels.* Under the Labels tab of the Layer Properties window, click Placement Properties and type 1 in the Buffer field to improve appearance.

Assignment 2-2

Map showing K–12 population versus school enrollment

In this assignment, you will create a choropleth map showing the population by census tract for the entire state of Pennsylvania, and also a map zoomed into the city of Pittsburgh for the K–12 school-age population. Layers will turn on or off depending on the zoom level. You will also show schools by enrollment.

Start with the following:

Note: First, add PATractStatePlane.shp. Then add the remaining layers. The tracts shapefile is the only one used in this assignment that has projected coordinates (in State Plane). ArcMap uses the projection of the first-added map layer for the coordinate system of the data frame, by default. You will learn about projections in chapter 5.

- **C:\Gistutorial\UnitedStates\Pennsylvania\PATractStatePlane.shp**—polygon layer of Pennsylvania census tracts, 2000.
- **C:\Gistutorial\UnitedStates\Pennsylvania\PACounties.shp**—polygon layer of Pennsylvania counties.
- **C:\Gistutorial\PAGIS\BlockGroups.shp**—polygon layer of Pittsburgh census block groups, 2000 that will be shown when zoomed into the Pittsburgh area.
- **C:\Gistutorial\PAGIS\Neighborhoods**—polygon layer of neighborhoods.
- **C:\Gistutorial\PAGIS\Schools.shp**—point layer of Pittsburgh schools. The value "City of Pittsburgh" for DISTRICT identifies public schools.

Create choropleth maps with scale thresholds

Create a new map document called C:\Gistutorial\Answers\Assignment2\Assignment2-2.mxd that shows the Pennsylvania census tracts for K–12 school-age population (ages 5–17) and county outlines for the entire state. For Pittsburgh, show the K–12 population using census block groups and neighborhood outlines. Include the point layer for Pittsburgh public schools that are open but with low enrollment (over 0 and under 200 students). Use MapTips for schools. Label counties and neighborhoods.

When zoomed to the entire state, do not show the city of Pittsburgh details, but turn on these layers when zoomed into that area. Have the Pennsylvania details turned off when zoomed into the Pittsburgh details. Create a bookmark to help you easily zoom into the Pittsburgh details.

Hints

- Create two layer groups: one for the state of Pennsylvania and one for Pittsburgh details so you can turn them on or off as necessary.
- Add a halo to labels to make them easier to read. In the Labels tab of the Layer Properties window, click Symbol, Properties, the Mask tab, the Halo radio button. Then type 1.5 for the size and click Symbol to use a light gray halo. Use a size 7 or 8 text symbol.

Questions

Create a Microsoft Word file called C:\Gistutorial\Answers\Assignment2\Assignment2.doc with answers to the following questions:

1. **The seven public schools meeting the criteria (over 0 and under 200 enrollment) are in what neighborhoods?**
2. **Name a school that may close. Explain why you picked this school.**

What to turn in

If you are working in a classroom setting with an instructor, you may be required to submit the assignments. Below are the files to turn in. Be sure to use a compression program such as PKZIP or WinZip to include all three files as one .zip document for review and grading. Include your name and assignment number in the .zip document (YourNameAssn2.zip). Use relative paths for map documents.

ArcMap documents

C:\Gistutorial\Answers\Assignment2\Assignment2-1.mxd
C:\Gistutorial\Answers\Assignment2\Assignment2-2.mxd

Word document

C:\Gistutorial\Answers\Assignment2\Assignment2.doc with answers to the above questions.

OBJECTIVES

Use interactive GIS
Produce print layouts
Create a custom map template and map series
Create a custom map template for multiple maps
Add reports to layouts
Export layouts as files
Generate other outputs

GIS Tutorial 3

GIS Outputs

GIS can produce many forms of output, from interactive desktop projects similar to maps in tutorials 1 and 2, to printed maps for distribution, to image files for placement in presentations or on Web sites. Final map compositions created in ArcMap are constructed in Layout mode. While in this mode, users see their current map on a virtual page (the layout) and can add map elements to it such as a title, map legend, north arrow, or scale bar. In addition, it is possible to add graphs or tabular reports to layouts. Layouts can be designed in various sizes, from regular letter-sized paper to very large sheets used in plotters.

Launch ArcMap

1 From the Windows taskbar, click Start, All Programs, ArcGIS, ArcMap.

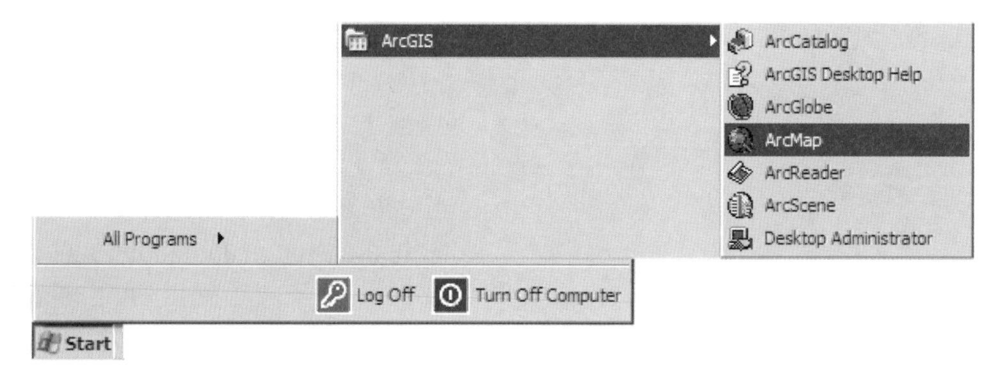

2 Click the An existing map radio button in the ArcMap dialog box.

3 Click OK.

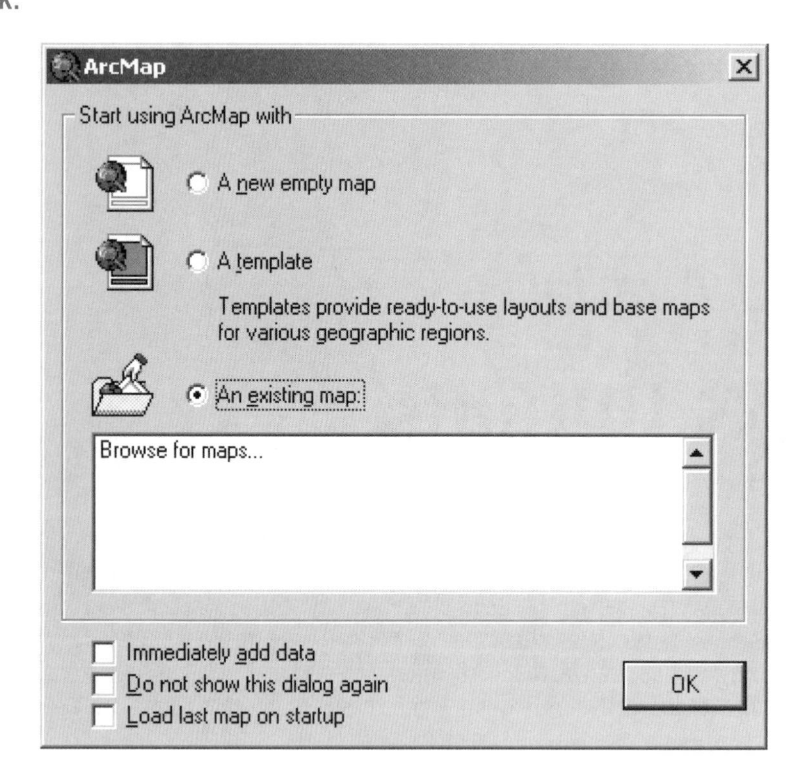

Open an existing map

1 **Browse to the drive on which the Gistutorial folder has been installed (e.g., C:\Gistutorial), select Tutorial3-1.mxd, and click the Open button.**

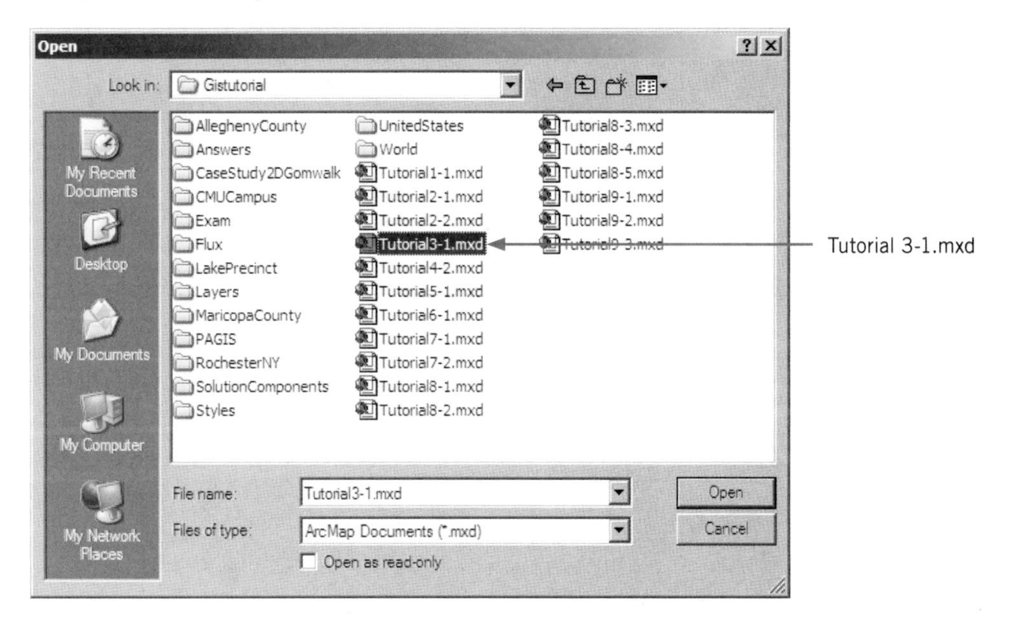

Tutorial3-1.mxd opens in ArcMap showing a map of the United States with state capitals displayed.

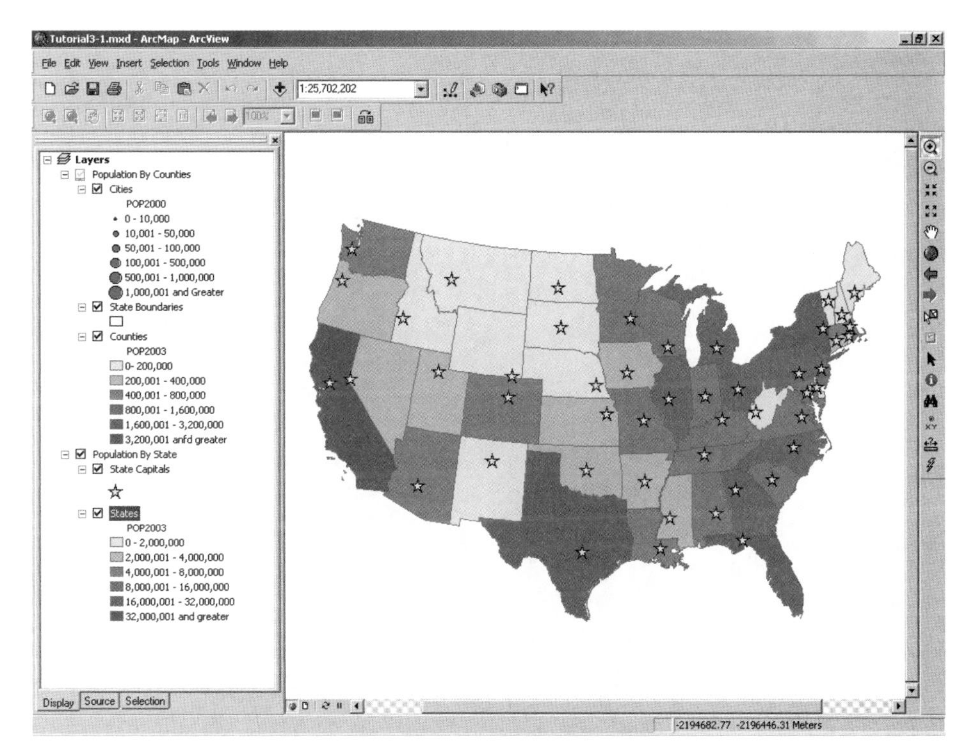

Use interactive GIS

The map document you just opened builds on work that you did in tutorials 1 and 2. It provides the ability to scan the entire area of interest, the continental United States; to get information on state e-government Web sites; and to pick an area to zoom into for more detailed information. Feature labels and MapTips provide additional information. Applying GIS in this way is quite powerful and popular. A good example is crime mapping by municipal police departments. The area-wide scan uses choropleth maps of crime counts by area to provide general information on neighborhoods where crimes are increasing or decreasing. The analyst can then drill down to small areas to see detailed information on individual crime points.

YOUR TURN

Start by using the Hyperlink tool (the lightning bolt on the Tools toolbar) to access some state e-government Web sites by clicking state capitals. Try any state along the west or south coast of the United States.

Zoom into an area about the size of a state to see the threshold scales at work. Try the MapTips over any city or county. Bookmarks are available to help you move around and get back to the continental United States.

Produce print layouts

Often, it is desirable to produce a paper copy or file copy of a map for distribution. ArcMap has a Layout View for this purpose and several built-in templates for producing layouts.

Choose a built-in layout template

1 **Click View, Bookmarks, Continental U.S.**

2 **Click View, Layout View.**

3 **Click View, Toolbars and make sure that Layout is selected.**

4 **Click the Change Layout button.** 

5 **Click the General tab, LandscapeClassic.mxt.**

The .mxt extension is the file type for map layout templates.

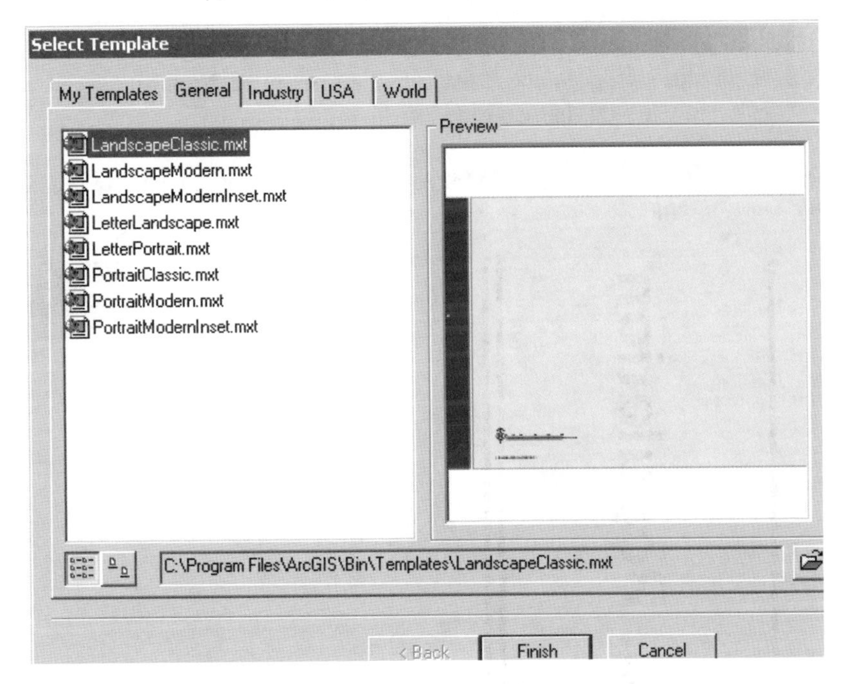

6 **Click Finish.**

Set up and customize the template page

1 Right-click the top border of the layout, outside any rectangles containing map elements, and click Page and Print Setup.

2 In the Map Page Size panel, click the Standard Sizes drop-down list and click Letter.

3 Click the Landscape Orientation option in the Map Page size frame and again in the Paper frame, click the Scale Map Elements proportionally to changes in the Page size check box, and click OK.

4 On the left side of the map in the brown area, double-click the title place holder (< Double-click to enter map title >) and type **U.S. Population and Cities**.

5 Click the Change Symbol button, change the color from white to black, click OK, then click OK again.

6 Double-click inside the brown left panel of the layout, outside the rectangles containing the legend and map title.

7 In the Properties window, change the Fill Color from brown to white and click OK.

8 Right-click the light blue background surrounding the map, click Properties, click the Frame tab, change the Background Color from light blue to white, and click OK.

9 Double-click the small text placeholder in the lower left corner of the map, type **Map Designed by "Your Name"**, and click OK.

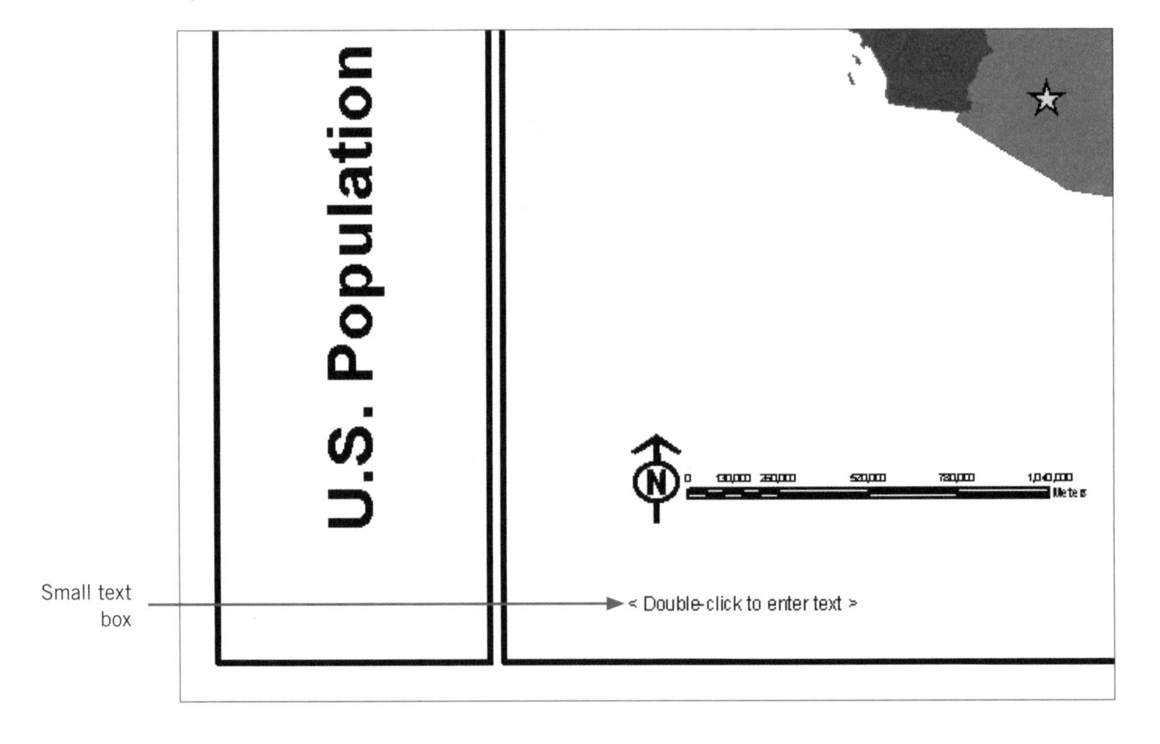

10 Click File, Save As, browse to the **\Gistutorial** folder, name the map **Tutorial3-2.mxd**, and click Save.

The resulting layout has a new title, colors, and your name.

YOUR TURN

Return to Data View and turn off the Population By State group (but leave the Population By Counties group on). Your map will be blank.

Use a bookmark to zoom to one of the available cities (Los Angeles, for example).

Change back to Layout View and change the layout template to LetterLandscape.mxt.

Make modifications and additions as necessary. *Hint: If you double-click the map scale bar, you can change its units to miles and its style.*

Save the results as Tutorial3-3.mxd.

Create a custom map template and map series

Sometimes you'll want to produce a number of maps, each with the same design and layers, but with different attributes displayed. In this case, each map displays population of a different racial or ethnic group. To facilitate comparisons, you'll need to use the same numerical scale for each map.

Start a new map

1 **Click File, New, My Templates tab and double-click the Blank Document template.**

Starting with a blank document is the only way to create a custom template.

2 **Click View, Data View.**

3 **Click the Add Data button,**  **browse to the \Gistutorial\UnitedStates\ folder, and double-click States.shp.**

4 **Right-click the States layer in the table of contents and click Properties.**

5 **Click the General tab, and change the Layer Name to Population.**

6 **Click the Symbology tab, Quantities, Graduated Colors, a blue Color Ramp, and in the Fields panel click AMERI_ES as the Value.**

7 **Click Classify, change the number of Classes to 7, change the Method to Manual, and type the following Break Values starting at the bottom of the list: 9999999999, 10000000, 5000000, 1000000, 500000, 100000, 50000.**

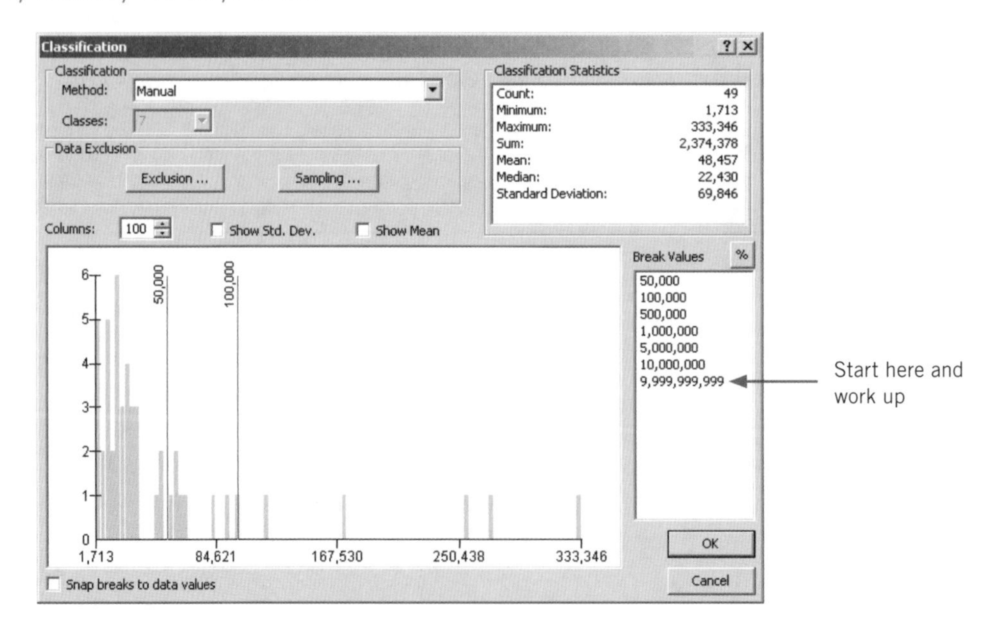

Complete the new map project

1 Click OK to return to the Symbology tab of the Layer Properties window.

2 Click the gray Label heading to the right of the Range heading, click Format Labels, Show thousands separators, and OK.

3 In the Label column, change the first label to 50,000 and less and change the last label to 10,000,001 and greater.

4 Click OK.

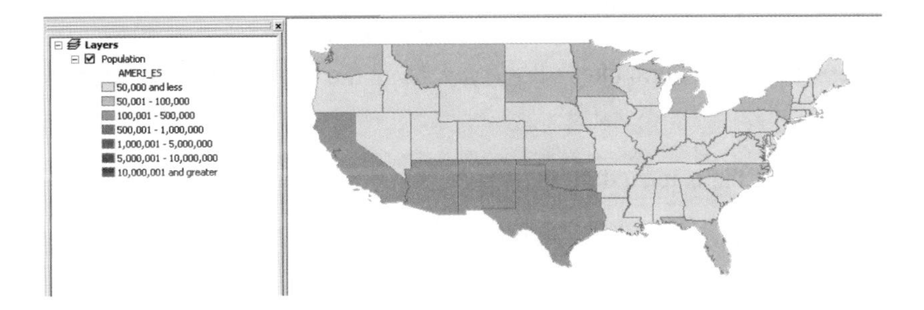

5 Right-click the Population layer in the table of contents, click Save As Layer File, browse to the **\Gistutorial\Layers** folder, type **PopulationStatesNativeAmericans.lyr**, and click Save.

Set up the layout view

1 Click View, Layout View.

2 Click Tools, Options, and the Layout View tab.

3 Make selections as shown below.

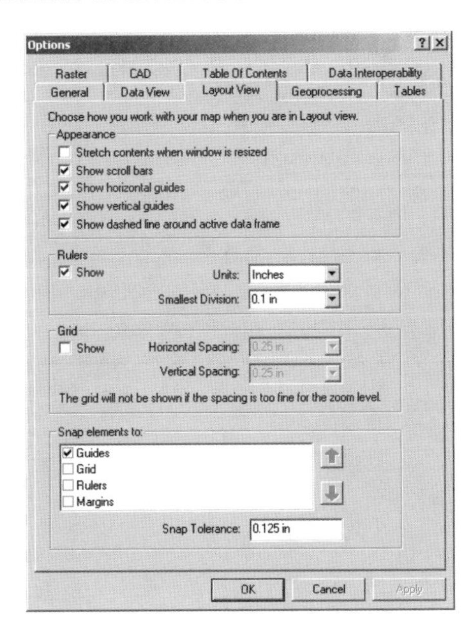

4 Click OK.

Set up guidelines in the layout view

In the next steps, you will use the vertical and horizontal rulers to set guides for positioning map elements on the layout page.

1 **Click at 8.5 inches on the top horizontal ruler to create a vertical blue guide line at that location.**

If you place your guide at the wrong place, right-click its arrow, click Clear Guide, and start again.

2 **Do the same at 7 inches on the left vertical ruler.**

3 **Right-click the map and click Properties, the Size and Position tab.**

4 **Click the Preserve Aspect Ratio check box, type 7.5 in the Size Width field, press the Tab key, and click OK.**

5 **Click, hold, and drag the map so that its upper right corner is at the intersection of the two guides, and release.**

The grab handle will snap precisely to the intersection of the guides when you release.

Add elements

1 Click Insert, Title.

2 Click outside the resulting title, then double-click the title, type **Native American Population (2001)** in the text box, click OK, and center the title above your map.

3 Click the horizontal ruler at 10.5 inches to create a new vertical guide.

4 Click Insert, Legend, click Next four times, and click Finish.

5 Click, hold, and drag the legend so that its right side snaps to the 10.5-inch guide.

6 Click the horizontal ruler at 9 inches to create a new vertical ruler.

7 Click the top left grab handle of the legend and drag to the right and down to make the legend smaller and so that it snaps to the 9-inch guide.

8 Click Insert, Text, then click outside the resulting small text box, click it again, and drag it away from the map to the lower-right corner of the layout.

9 Double-click the text box and type **Map designed by "Your Name"**.

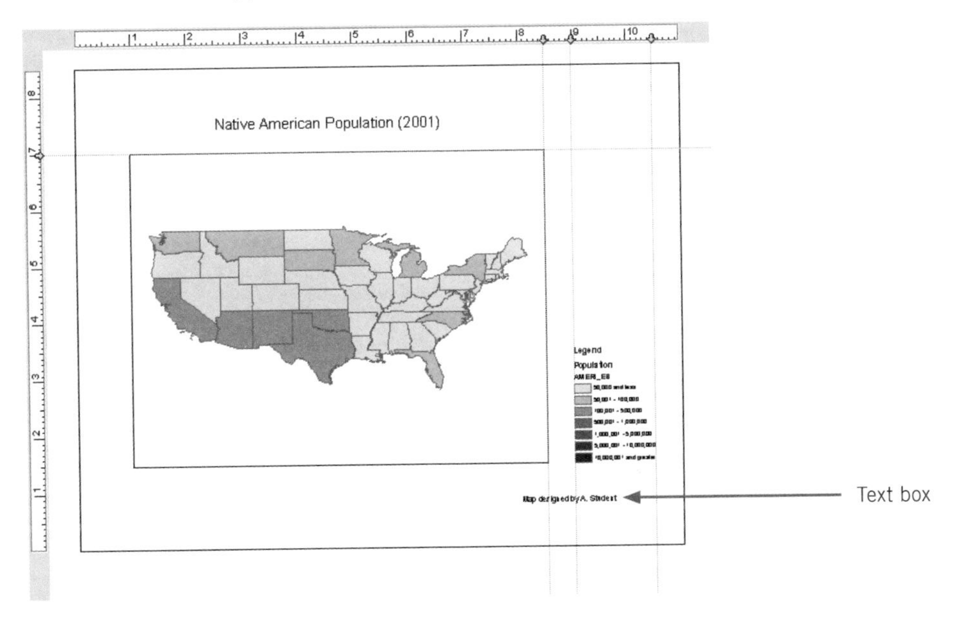

10 Click File, Save As, make sure the file type is ArcMap Document, name the map **Tutorial3-NativeAmericans.mxd**, and click Save.

11 Click File, Save As, browse to the **\Gistutorial** folder, change the Save As Type to ArcMap Template, name the template **Tutorial3-Template**, and click Save.

Note: Once you save the layout as a template, your selection for ArcMap Document will no longer be available in the Save As dialog box.

Use the custom template for a new map

1. Click File, New.

2. In the New window, click the Browse button,  browse to the **\Gistutorial** folder, and double-click **Tutorial3-Template.mxt**.

3. Click View, Data View.

4. Right-click the Population layer in the table of contents, click Properties, the Symbology tab, and the Import button.

5. In the Import Symbology dialog box, click the Browse button, browse to the **\Gistutorial\Layers** folder, double-click **PopulationStatesNativeAmericans.lyr**, and click OK.

6. In the Import Symbology Matching Dialog, click the Value Field drop-down arrow, click ASIAN, and click OK.

7. In the Layer Properties window, change the color ramp to a green color ramp and click OK.

8. Right-click the Population layer in the table of contents, click Save As Layer File, browse to the **\Gistutorial\Layer** folder, key in **PopulationStatesAsians.lyr**, and click Save.

9. Click View, Layout View.

10. Double-click the map title, change Native American to Asian and click OK.

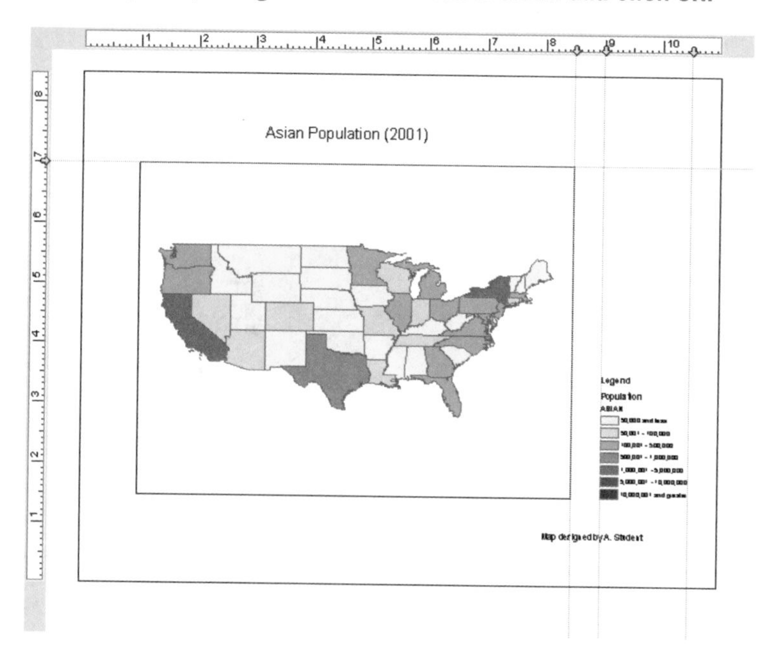

11. Click File, Save As, browse to the **\Gistutorial** folder, name the file **Tutorial3-Asians.mxd**, and click Save.

YOUR TURN

Complete the map series by making maps for at least one of the following: Blacks, Whites, or Hispanics.

Name the maps Tutorial3-Blacks.mxd, Tutorial3-Whites.mxd, and Tutorial3-Hispanics.mxd.

Save layer files for each Population layer.

Create a custom map template for multiple maps

To facilitate comparisons, it is a good idea to place two or more maps on the same layout. Our population maps by racial/ethnic groups are ideal for this because they all share the same break points, making comparisons easy.

Create a new map

1. **Click File, New, then, in the My Templates tab, click Blank Document, and click OK.**
2. **If necessary, click View, Layout View.**
3. **Right-click the border of the layout, click Page and Print Setup, change the standard page size to Letter, click both Portrait radio buttons in the Page and Paper frames, and click OK.**
4. **Right-click the vertical ruler, click Clear All Guides, and do the same to the horizontal ruler.**
5. **Click the horizontal ruler at the 0.5-, 6.5-, and 8.0-inch marks.**
6. **Click the vertical ruler at the 0.5-, 5.4-, 5.6-, and 10.5-inch marks.**
7. **Click and drag the layers frame so that its upper left corner snaps to the intersection of 10.5-inch horizontal guide and 0.5-inch vertical guide.**
8. **Click and drag the lower right grab handle of the data frame to snap it at the 5.6-inch horizontal guide and 6.5-inch vertical guide.**
9. **Click Insert, Data Frame and drag/modify the new data frame to fit in the guides below the original frame.**

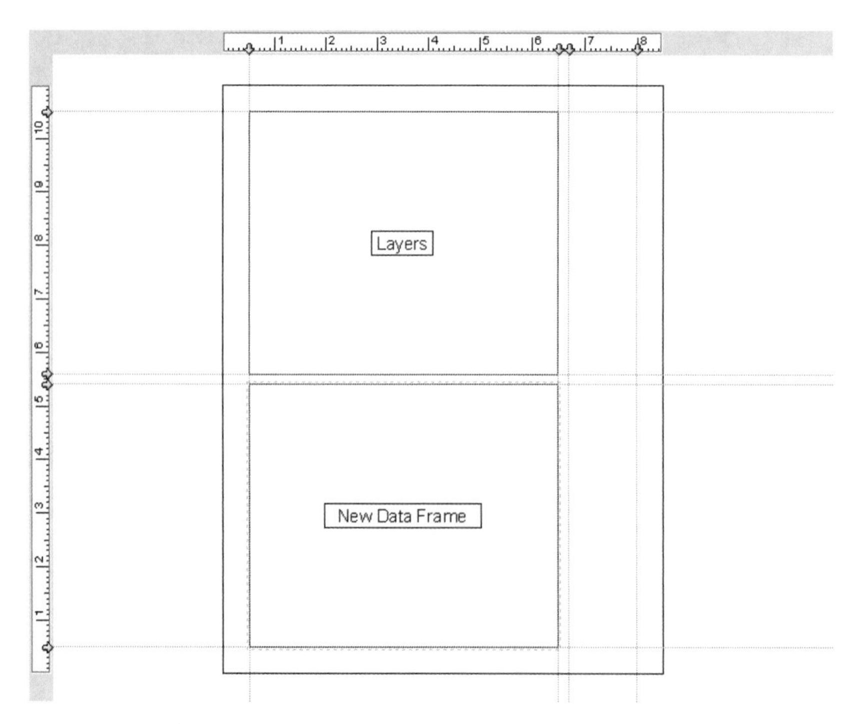

Add elements

1 In the table of contents, right-click the Layers data frame, click Add Data, browse to the **\Gistutorial\Layers** folder, and double-click **PopulationStatesNativeAmericans.lyr**.

2 In the table of contents, right-click the New Data Frame, click Add Data, and double-click **PopulationStatesAsians.lyr**.

3 Click Insert, Legend.

4 Click Next four times, and then Finish.

5 Drag the legend so that it snaps on the lower right to the 8-inch vertical guide and 0.5-inch horizontal guide intersection, then resize it to fit the available space.

6 Right-click the Layers layer, click activate, and repeat steps 3–5 so that the second legend's lower right is at the intersection of the 8-inch vertical and 5.6-inch horizontal guides.

7 Click Insert, Title, click outside the Title frame, and then double-click the Title frame.

8 In the Properties window, type **Population**, press the Enter key, type **of Native Americans**, press the Enter key, type **And Asians**, change the angle to 90, click the Change Symbol button, change the Size to 20, and click OK twice.

9 Position the Title frame on the top right of the layout.

10 Click File, Save As, name the map **Tutorial3-NativeAmericansAndAsians** and click Save.

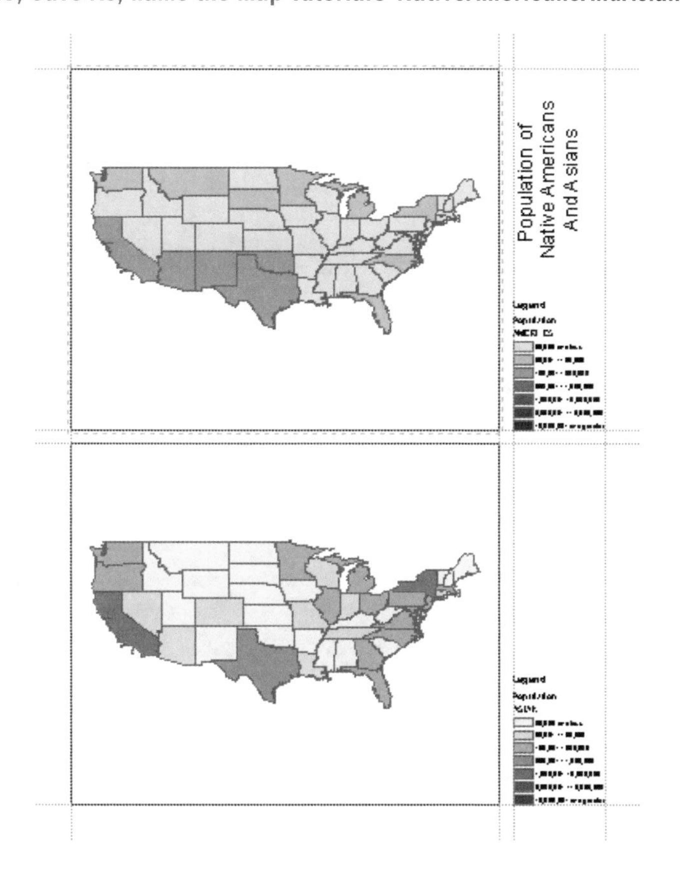

Add reports to a layout

ArcMap has a built-in capability to make tabular reports. These can be added to layouts to provide detailed information.

Open an existing map

1 **Click File, Open, browse to the \Gistutorial folder, and double-click Tutorial3-NativeAmericans.mxd.**

If you do not have this file in the \Gistutorial folder, browse to \Gistutorial\SolutionComponents\Tutorial3\ and you will find a copy there.

2 **Click File, Save As, and type Tutorial3-NativeAmericansReport.mxd for the File Name, then click Save.**

3 **Click View, Data View.**

Make a selection of records

1 **Right-click the Population layer in the table of contents and click Open Attribute Table.**

2 **Scroll to the right in the Attributes of Population to find the AMERI_ES column, right-click the AMERI_ES column heading, and click Sort Descending.**

3 **Scroll left in the table until you see the STATE_NAME column.**

4 **If necessary, make the table large enough so you can see the first twelve state records. Click the row selector for the top row, then hold and drag down to select the top twelve rows in the table (California through Minnesota).**

You will generate a report for the selected records only.

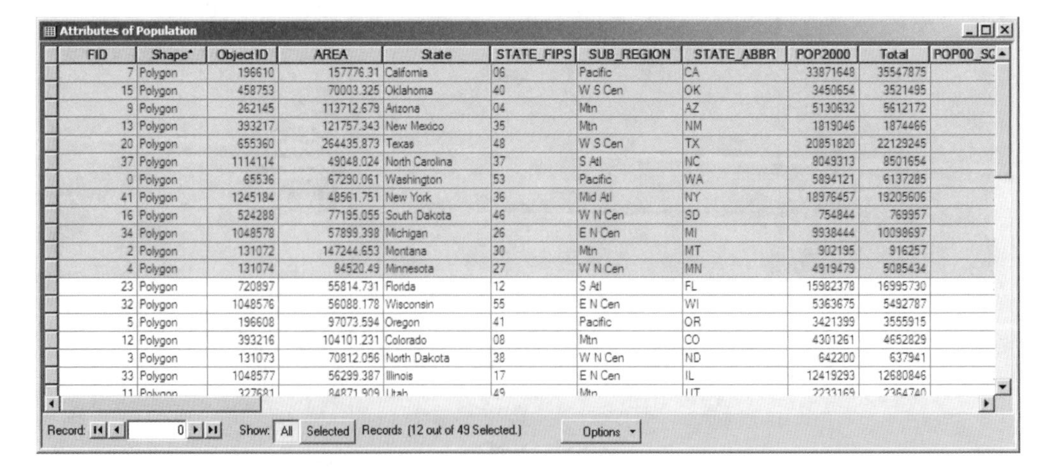

Attributes of Population

FID	Shape*	ObjectID	AREA	State	STATE_FIPS	SUB_REGION	STATE_ABBR	POP2000	Total	POP00_SC
7	Polygon	196610	157776.31	California	06	Pacific	CA	33871648	35547875	
15	Polygon	458753	70003.325	Oklahoma	40	W S Cen	OK	3450654	3521495	
9	Polygon	262145	113712.679	Arizona	04	Mtn	AZ	5130632	5612172	
13	Polygon	393217	121757.343	New Mexico	35	Mtn	NM	1819046	1874466	
20	Polygon	655360	264435.873	Texas	48	W S Cen	TX	20851820	22129245	
37	Polygon	1114114	49048.024	North Carolina	37	S Atl	NC	8049313	8501654	
0	Polygon	65536	67290.061	Washington	53	Pacific	WA	5894121	6137285	
41	Polygon	1245184	48561.751	New York	36	Mid Atl	NY	18976457	19205606	
16	Polygon	524288	77195.055	South Dakota	46	W N Cen	SD	754844	769957	
34	Polygon	1048578	57899.398	Michigan	26	E N Cen	MI	9938444	10098697	
2	Polygon	131072	147244.653	Montana	30	Mtn	MT	902195	916257	
4	Polygon	131074	84520.49	Minnesota	27	W N Cen	MN	4919479	5085434	
23	Polygon	720897	55814.731	Florida	12	S Atl	FL	15982378	16995730	
32	Polygon	1048576	56088.178	Wisconsin	55	E N Cen	WI	5363675	5492787	
5	Polygon	196608	97073.594	Oregon	41	Pacific	OR	3421399	3555915	
12	Polygon	393216	104101.231	Colorado	08	Mtn	CO	4301261	4652829	
3	Polygon	131073	70812.056	North Dakota	38	W N Cen	ND	642200	637941	
33	Polygon	1048577	56299.387	Illinois	17	E N Cen	IL	12419293	12680846	
11	Polygon	327681	84871.909	Utah	49	Mtn	UT	2233169	2364740	

Record: 0 Show: All Selected Records (12 out of 49 Selected.) Options

5 **Close the Attributes of Population window.**

Start the report

1 Click View, Layout View.

2 Click Tools, Reports, Create Report.

3 In the Report Properties dialog box, in the Available Fields box, double-click STATE_NAME, AMERI_ES, and POP2003.

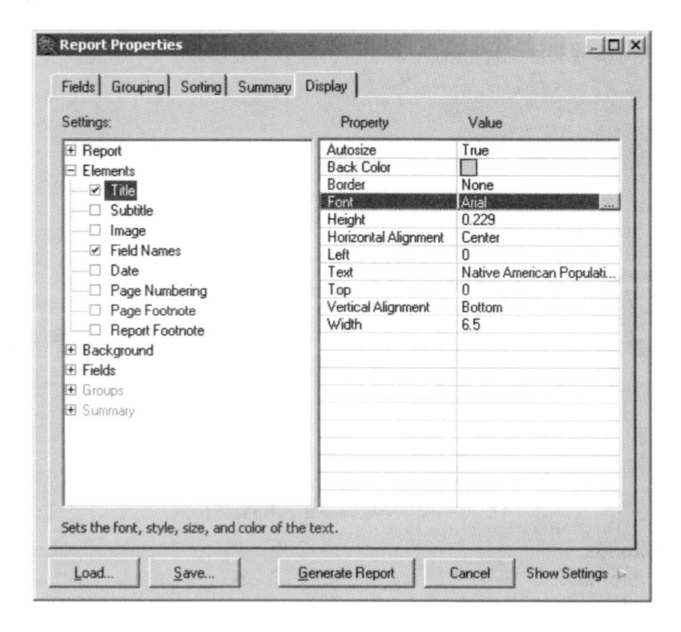

4 Click the Sorting tab and change None in the AMERI_ES row to Descending.

5 Click the Display tab, check the box to the left of the Title item, change the Text property from Report Title to Native American Population in the Top 12 States, change the Font to Arial by clicking the ... box and selecting the font from the resulting pop-up window, and then click OK.

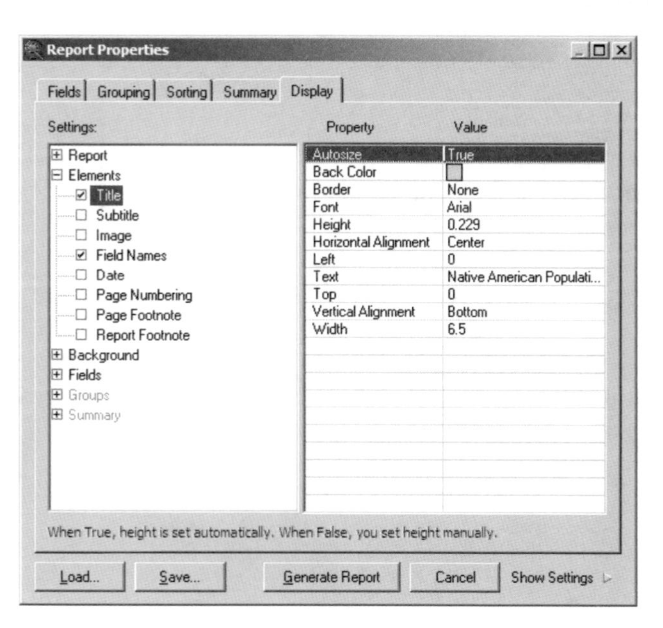

Finish the report

1 Click the plus sign (+) to the left of the Fields item to expand it.

2 Click the STATE_NAME field and change its text to **State**, click the AMERI_ES field and change its text to **Native Americans**, and click the POP2003 field and change its text to **Total**.

3 Change the fonts of these fields to Arial.

4 Click the AMERI_ES field, Number Format, the builder button, the Show Thousands Separators check box, and OK.

5 Repeat step 4 for POP2003.

6 Click the Generate Report button.

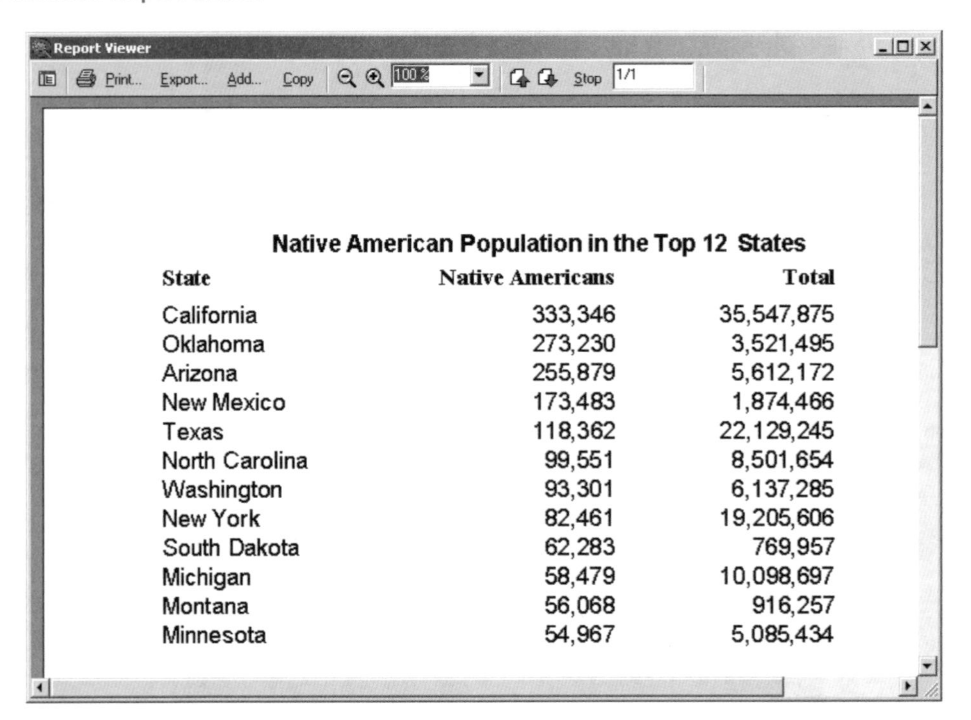

Native American Population in the Top 12 States

State	Native Americans	Total
California	333,346	35,547,875
Oklahoma	273,230	3,521,495
Arizona	255,879	5,612,172
New Mexico	173,483	1,874,466
Texas	118,362	22,129,245
North Carolina	99,551	8,501,654
Washington	93,301	6,137,285
New York	82,461	19,205,606
South Dakota	62,283	769,957
Michigan	58,479	10,098,697
Montana	56,068	916,257
Minnesota	54,967	5,085,434

7 In the Report Viewer window, click Export, browse to the **\Gistutorial** folder, select Rich Text Format (*.rtf) from the Save as type drop-down list, name the report **NativeAmericans**, click Save, and close the Report Viewer.

8 In the Report Properties window, click Save, browse to the Gistutorial folder, type **NativeAmericans** as the File Name, and click Save, Close.

This saves the report specification as an RDF (report), which allows you to modify it and generate a new report output. It can be loaded in the Report Properties window.

Add the report to a layout

1. **In the Layout View, click Insert then Object.**

2. **In the Insert Object dialog box, click the Create from File radio button, browse to the \Gistutorial folder, and double-click NativeAmericans.rtf (the rich text format report output).**

3. **Click OK in the Insert Object window.**

The report is now on the layout, but needs to be resized and moved.

4. **Click anywhere on the map to activate its frame, right-click the frame, click Properties, click the Frame tab, change the border color to no color, and click OK.**

5. **Click the vertical ruler on the left at 7.5 inches to create a horizontal guide, and click and drag the map frame up to this guide.**

6. **Click the vertical ruler at 1-inch and 3.2-inch locations to create horizontal guides and click and drag the report frame down so that its lower right corner is at the 1-inch horizontal, 8.5-inch vertical guide intersection.**

7. **Move and resize the report using the guidelines.**

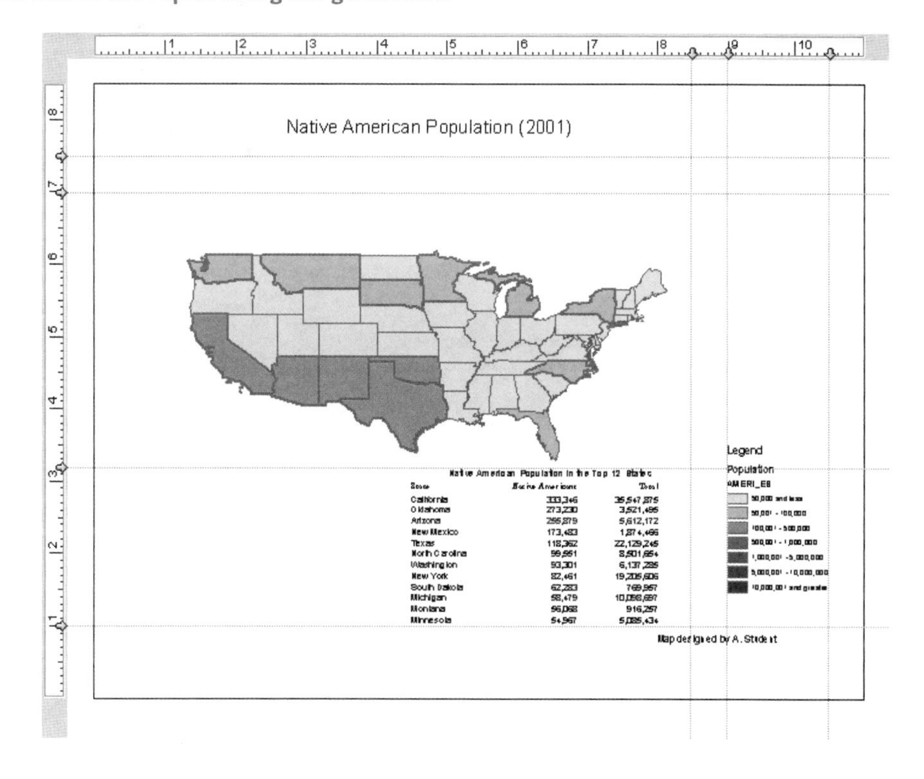

Add final touches to the layout

1 **Right-click the Population layer in the table of contents and click Properties.**

2 **Click the Labels tab, check the box next to Label features in this layer, and click OK.**

3 **Click File, Save.**

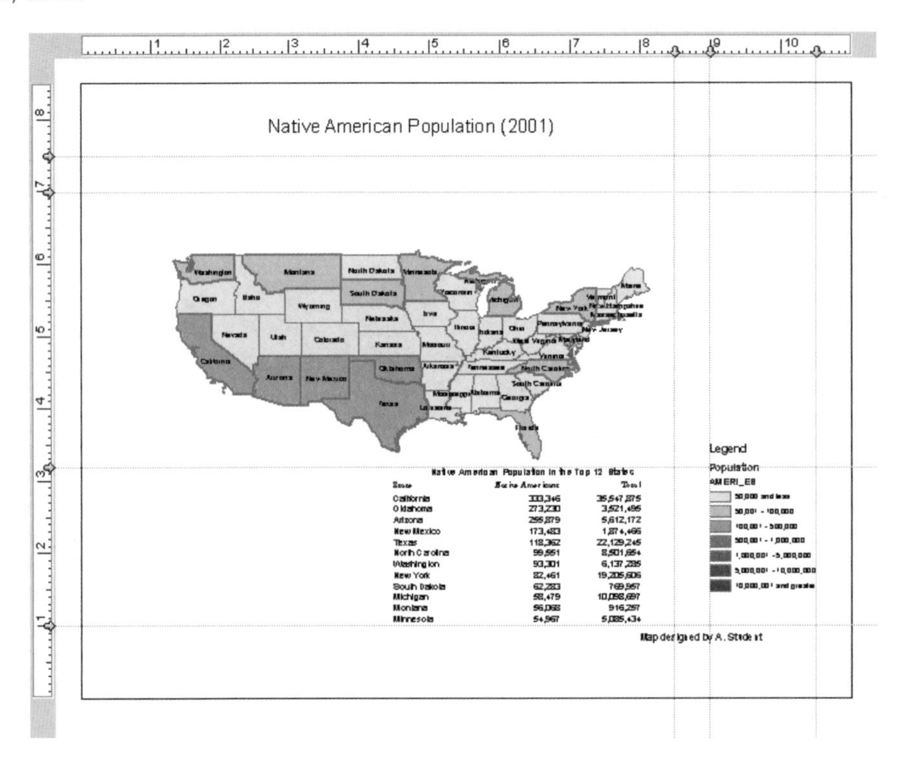

YOUR TURN

Create a report for another racial or ethnic group corresponding to a layout that you created earlier, and add it to the layout.

Save your layout using Save As and adding "Report" to the end of the .mxd file name.

Add graphs to a layout

The graphs available from ArcMap are easy to use, but have limitations. In this section you will build an ArcMap graph and then build the same graph in Microsoft Excel, both for use on an ArcMap layout. The Excel graph gives you more control over the graph design, but it takes several more steps than the ArcMap graph.

Open an existing map

1 **Click File, Open, browse to the \Gistutorial folder, and double-click Tutorial3-NativeAmericans.mxd.**

2 **Click View, Data View.**

Select records for graphing

1 **Right-click the Population layer in the table of contents and click Open Attribute Table.**

2 **In the Attributes of Population window, scroll to the right until you can see the AMERI_ES column.**

3 **Right-click the AMERI_ES column heading and click Sort Descending.**

4 **Scroll back to the left, to the beginning of the table, click the first row's record selector, hold and drag down to include Minnesota, then release to select the first twelve rows.**

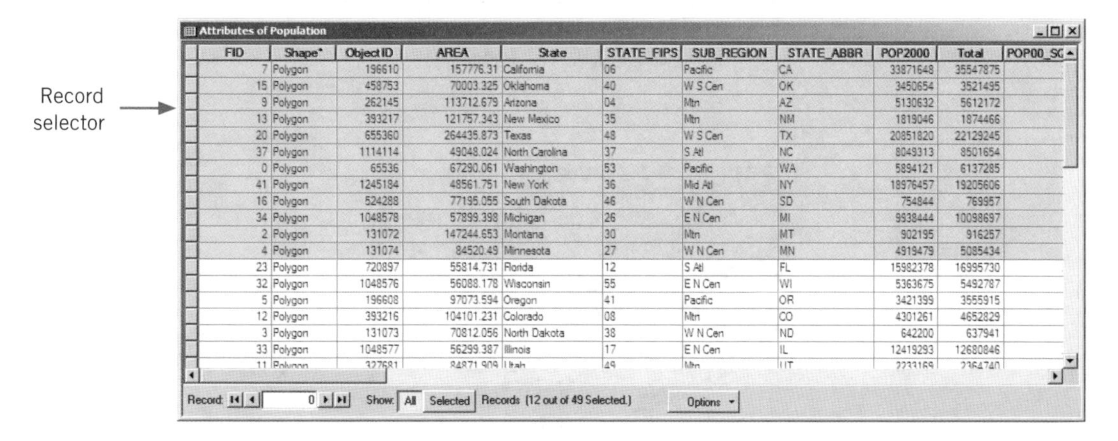

FID	Shape*	Object ID	AREA	State	STATE_FIPS	SUB_REGION	STATE_ABBR	POP2000	Total	POP00_SC
7	Polygon	196610	157776.31	California	06	Pacific	CA	33871648	35547875	
15	Polygon	458753	70003.325	Oklahoma	40	W S Cen	OK	3450654	3521495	
9	Polygon	262145	113712.679	Arizona	04	Mtn	AZ	5130632	5612172	
13	Polygon	393217	121757.343	New Mexico	35	Mtn	NM	1819046	1874466	
20	Polygon	655360	264435.873	Texas	48	W S Cen	TX	20851820	22129245	
37	Polygon	1114114	49048.024	North Carolina	37	S Atl	NC	8049313	8501654	
0	Polygon	65536	67290.061	Washington	53	Pacific	WA	5894121	6137285	
41	Polygon	1245184	48561.751	New York	36	Mid Atl	NY	18976457	19205606	
16	Polygon	524288	77195.055	South Dakota	46	W N Cen	SD	754844	769957	
34	Polygon	1048578	57899.398	Michigan	26	E N Cen	MI	9938444	10098697	
2	Polygon	131072	147244.653	Montana	30	Mtn	MT	902195	916257	
4	Polygon	131074	84520.49	Minnesota	27	W N Cen	MN	4919479	5085434	
23	Polygon	720897	55814.731	Florida	12	S Atl	FL	15982378	16995730	
32	Polygon	1048576	56088.178	Wisconsin	55	E N Cen	WI	5363675	5492787	
5	Polygon	196608	97073.594	Oregon	41	Pacific	OR	3421399	3555915	
12	Polygon	393216	104101.231	Colorado	08	Mtn	CO	4301261	4652829	
3	Polygon	131073	70812.056	North Dakota	38	W N Cen	ND	642200	637941	
33	Polygon	1048577	56299.387	Illinois	17	E N Cen	IL	12419293	12680846	
11	Polygon	327681	84871.909	Utah	49	Mtn	UT	2233169	2364740	

5 **Close the table and click View, Layout View.**

The graph that you will create next will use just the records that you selected.

Create a graph in ArcMap

1. Click Tools, Graphs, Create.

2. In the Create Graph wizard, select AMERI_ES for the Value field and STATE_ABBR for the XLabel field. Click Next.

3. Click on the Show only selected features/records on the graph radio button and click off the Graph legend check box.

4. Type **Native American Population** in the Title field, click the Graph in 3D View check box, and click Finish.

5. Click anywhere on the map to activate its frame, right-click the frame, click Properties, change the Border Color to No Color, and click OK.

6. Click the vertical ruler at 7.5 inches to create a new guide, and drag the map frame up so that its top snaps to the new guide.

7. Right-click the graph, click Add to Layout, and close the graph window.

8. Resize and position the graph object.

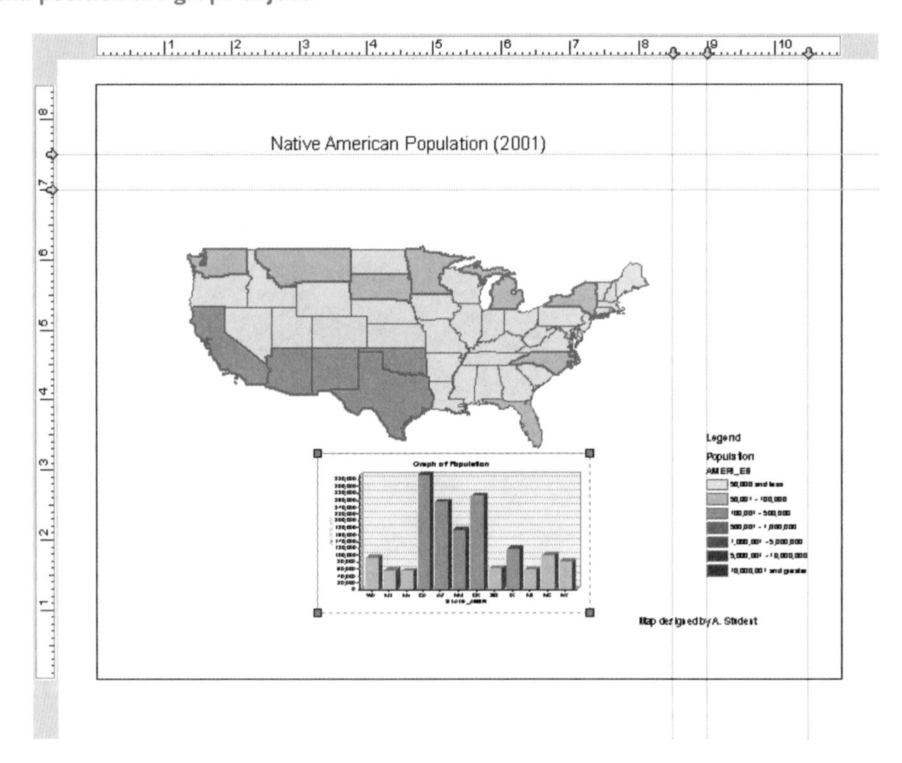

The result is an attractive graph, but it would be better if you could control the sorting of the bars, for example in descending order.

Export data

1 Click View, Data View.

2 Right-click the Population layer in the table of contents and click Open Attribute Table.

The top twelve states should still be selected, although they will not appear as a contiguous group of records.

3 In the lower right corner of the Attributes of Population window, click the Options button, then click Export, browse to the **\Gistutorial** folder, change the name of the output file from Export_Output.dbf to **NativeAmericanPopulation.dbf**, click Save, click OK, and click No when asked to add it to your map.

4 Close the Attributes of Population table.

Import data into Microsoft Excel

1 Start Microsoft Excel (you can minimize ArcMap, but do not close your project), click File, Open, browse to your **\Gistutorial** folder, change the Files of Type to dBase Files (*.dbf), and double-click NativeAmericanPopulation.dbf.

2 Click column headings and use Edit, Delete to delete all but STATE_NAME and AMERI_ES columns.

3 Change the column headings from STATE_NAME to State and AMERI_ES to Population.

4 Select all cells in the table and click Data, Sort, click the Descending radio button for Population, and click OK.

	A	B
1	State	Population
2	California	333346
3	Oklahoma	273230
4	Arizona	255879
5	New Mexico	173483
6	Texas	118362
7	North Carolina	99551
8	Washington	93301
9	New York	82461
10	South Dakota	62283
11	Michigan	58479
12	Montana	56068
13	Minnesota	54967

Create a graph in Microsoft Excel

1 With the table still selected, click the Chart Wizard button in Excel.

2 In the Chart Wizard, click Next until finished.

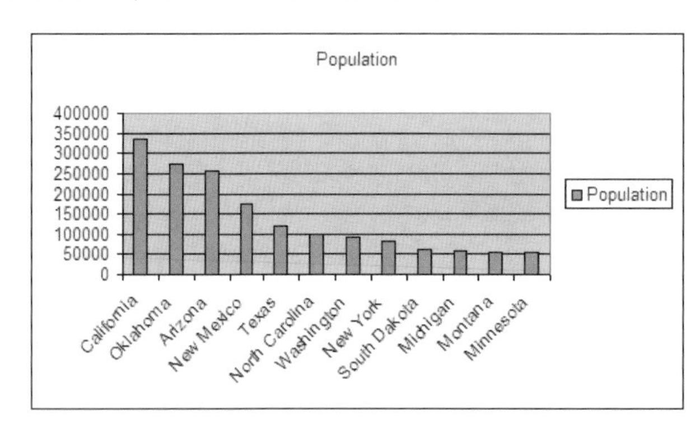

3 Click the frame of the Population legend to the right of the graph and press Delete on your keyboard.

4 Resize the graph so that it is about 3 inches wide and 2 inches tall.

5 Double-click the horizontal axis.

6 In the resulting Format Axis window, click the Alignment tab. Then click and hold the red dot at the end of the Text line and rotate up to vertical.

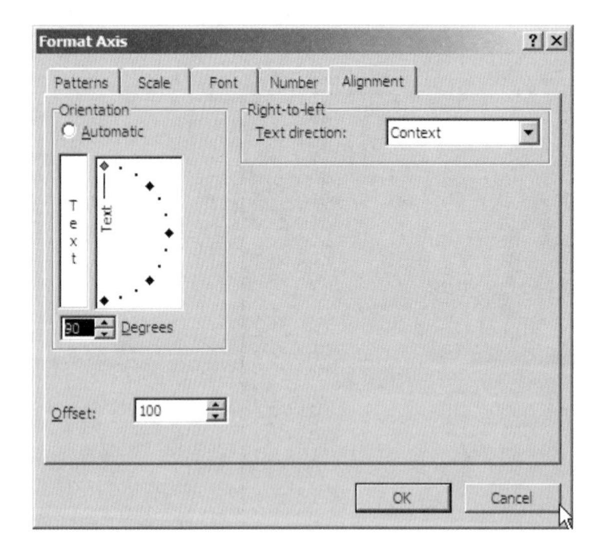

7 Click the Scale tab, enter values of 1 for the first three text boxes, and click OK.

8 Double-click the vertical axis of the graph, click the Scale tab, change the Major Unit to 100000, and click OK.

Add the Microsoft Excel graph to the layout

1 In Excel, click the boundary of the graph to activate its frame.

2 Click Edit, Copy.

3 Close Excel (save as an Excel workbook if you like).

4 Switch to ArcMap, which should still be open, and click View, Layout View.

5 Click the Population graph to activate its frame and press Delete on your keyboard.

6 Click Edit, Paste and relocate/resize your Excel graph.

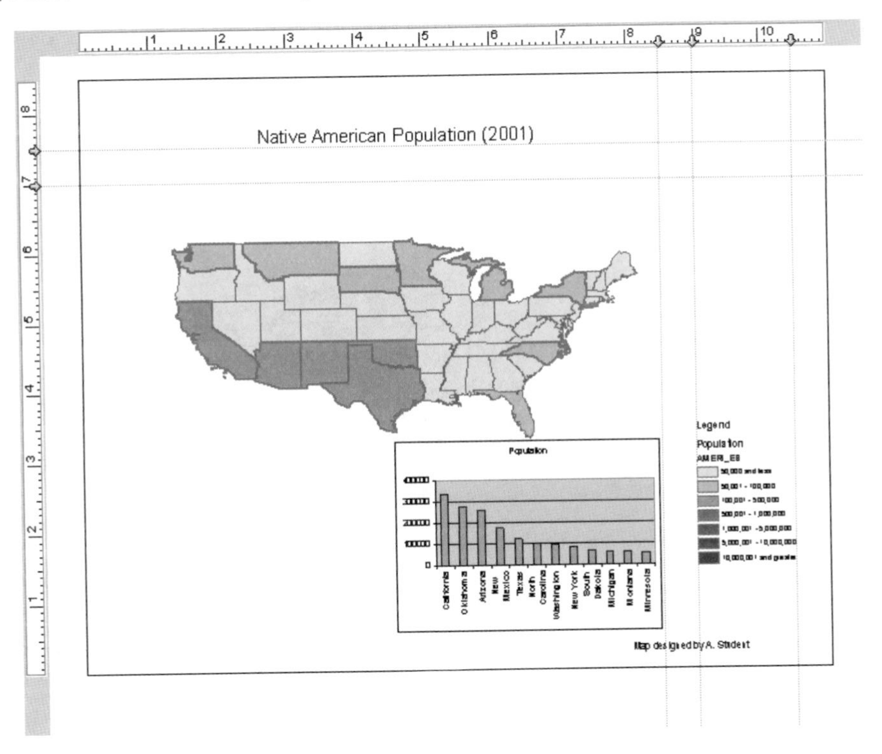

7 Click File, Save As, key in **Tutorial3-Graph.mxd** as the File Name, and click Save.

Export layouts as files

It is often desirable to include maps in Microsoft Word documents or Microsoft PowerPoint presentations, or on Web sites. ArcMap allows you to export layouts in a variety of file formats.

1 In Layout View of **Tutorial3-Graph.mxd**, click File, Export Map.

2 In the Export Map window, click the Save as type drop-down arrow and click JPEG (*.jpg).

3 Browse to the **\Gistutorial** folder and click Save.

You now have a very nice JPEG file, displayed below, that you can import as a picture into a variety of software packages or upload to a Web site.

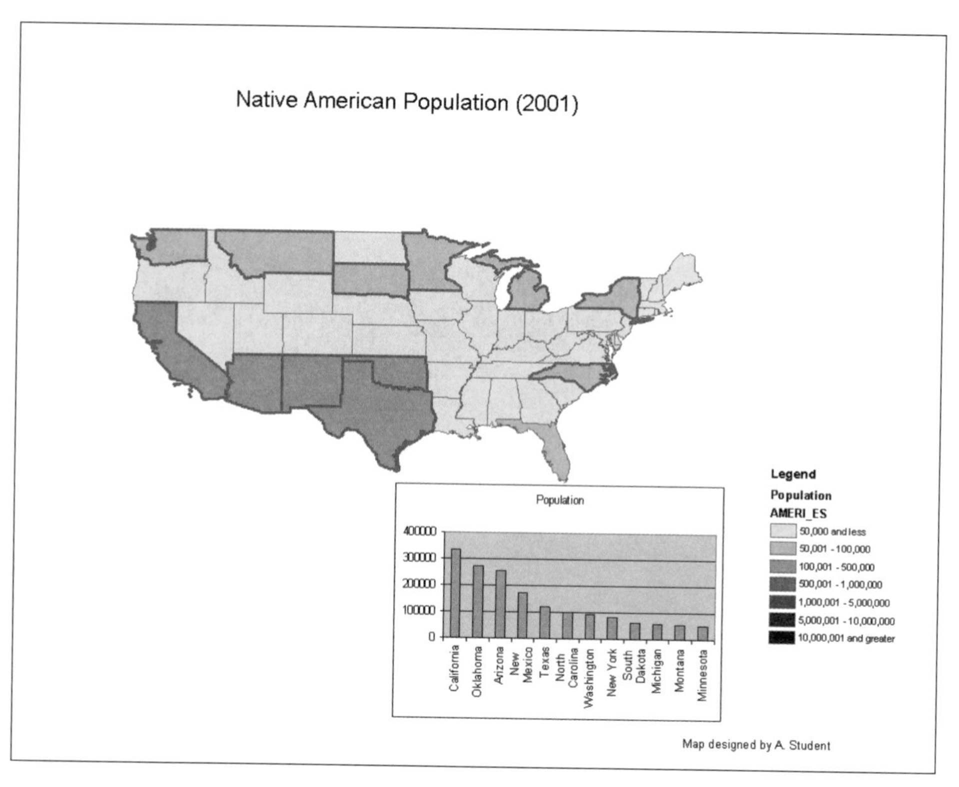

4 Save and close ArcMap.

Other outputs

We have covered several outputs of GIS in this tutorial, including interactive GIS, print layouts, and file exports of layouts. There are many more GIS outputs, and while not covered in this chapter, we will list a few more here for your information.

1. Spatial data processing—GIS provides unique outputs that no other kind of software can produce because it incorporates world coordinate systems and related algorithms. A major GIS output is a point on a map that represents a street address. Points are generated from street addresses (like 4800 Forbes Ave., Pittsburgh, PA 15213) through a unique GIS process called address matching. It is then possible to use another spatial data processing step, spatial overlay, to assign area identifiers (like county, census tract, and so on) to such points. We will cover geocoding, address matching, and spatial overlay in later chapters.

2. Web-based, interactive GIS—the ArcIMS package from ESRI—is an easy way to provide most of the functionality of tutorials 1 and 2 on the Web. You can try out sample Web sites at *www.esri.com/software/internetmaps/index.html.* This is not a difficult package to learn, given your current GIS knowledge, but ArcIMS is limited to organizations that can afford the package and have staff who can implement and maintain it.

3. Free GIS package for map display and querying—ArcExplorer—is a simple, free GIS package that includes functionality similar to tutorials 1 and 2. See *www.esri.com/software/arcexplorer/index.html.*

Assignment 3-1

Layout comparing males, females, and young population in Orange County, California

Sometimes you will want to compare two or more maps in layout view to visualize information about multiple attributes and analyze those attributes. In this problem, you will create a map layout with two maps with population percentages for youths and the elderly in Orange County, California.

Start with the following:

- **C:\Gistutorial\UnitedStates\California\OrangeCountyTracts.shp**—census tract polygon boundaries for Orange County, California, Census 2000.

Create a comparison map of census data

Create a new map document called C:\Gistutorial\Answers\Assignment3\Assignment3-1.mxd that includes an 11×8.5-inch portrait layout with two data frames: one with the percentage of 2000 population who are 5 to 17 years old and the second with percentage 65 or older. Use the same numerical scale for both maps. Include a graphic scale bar in miles.

Hints

- Use the Layer Properties, Symbology tab to show a population as a percentage of the total population for the year 2000 (use POP2000 field to normalize the data).
- Use your judgment as to the color, sizes, titles, and other map elements to add or modify.

Export the map as a JPEG file called C:\Gistutorial\Answers\Assignment3\Assignment3-1.jpg.

Assignment 3-2

Walking map of historic districts in downtown Pittsburgh

Many city planning departments are using GIS as a tool to create maps for their cities. These maps can be used in planning documents, tourist attraction documents, or Web sites for visitors in a city. Visit Pittsburgh's Department of City Planning Web site to see examples of maps *(www.city.pittsburgh.pa.us/cp)*. Click "Walking Tours" to see maps and photos of Pittsburgh's historic sites. In this exercise, you will create maps that the planning department can use to promote historic areas. In the layout, you will create an overall view of the historic sites in the Central Business District as well as a zoomed map for one area.

Start with the following:

- **C:\Gistutorial\PAGIS\CentralBusinessDistrict\CBDOutline.shp**—polygon feature of Pittsburgh's Central Business District neighborhood outline.
- **C:\Gistutorial\PAGIS\CentralBusinessDistrict\CBDBLDG.shp**—polygon features of Pittsburgh's Central Business District buildings.
- **C:\Gistutorial\PAGIS\CentralBusinessDistrict\CBDStreets.shp**—line features of Central Business District streets.
- **C:\Gistutorial\PAGIS\Histsite.shp**—polygon features of historic areas in Pittsburgh's Central Business District. This layer shows historic district polygons for the entire city of Pittsburgh. You will focus only on those historic areas within the Central Business District neighborhood.
- **C:\Gistutorial\PAGIS\Histpnts.shp**—point features of historic sites in Pittsburgh's Central Business District. This layer shows points for the entire city of Pittsburgh. You will focus only on those historic points within the Central Business District neighborhood.

Create a large-scale map

Create a new map called C:\Gistutorial\Answers\Assignment3\Assignment3-2.mxd with an 8.5×11-inch layout containing two data frame maps—one scaled at 1:14,000 showing all of the historic districts in the Central Business District, and one scaled at 1:2,400 showing one of the historic districts (you choose the focus) in detail. See hints for how to set a fixed scale in a layout.

Keep in mind basic mapping principles such as colors, ground features, and so forth, covered in previous chapters. Choose labels and other map elements that you think are appropriate for each map as well as the overall layout. Include a photograph of a building that you download from the City of Pittsburgh Web page and save to the C:\Gistutorial\Answers\Assignment3 folder.

Draw the Historic Sites polygons as a transparent layer (see hints) so you can see the buildings under the sites and the Central Business District as a thick outline.

Export your map as a PDF file called C:\Gistutorial\Answers\Assignment3\Assignment3-2.pdf.

Hints

Drawing a layer transparently

- Click the View menu, point to Toolbars, and click Effects.
- Click the Layer drop-down arrow and click the layer you want to adjust.
- Click the Adjust Transparency button.
- Drag the slider bar to adjust the transparency.

Setting a Fixed Scale in a Layout Data frame

- Click the black pointer tool and select a data frame in a layout.
- Right-click the data frame and click Properties.
- Click the Data Frame tab.
- Click the Fixed Scale radio tab and enter your new scale here.

What to turn in

If you are working in a classroom setting with an instructor, you may be required to submit the exercises you created in tutorial 3. Below are the files you are required to turn in. Be sure to use a compression program such as PKZIP or WinZip to include all three files as one .zip document for review and grading. Include your name and assignment number in the .zip document (YourNameAssn3.zip).

ArcMap documents

C:\Gistutorial\Answers\Assignment3\Assignment3-1.mxd
C:\Gistutorial\Answers\Assignment3\Assignment3-2.mxd

Exported maps

C:\Gistutorial\Answers\Assignment3\Assignment3-1.jpg
C:\Gistutorial\Answers\Assignment3\Assignment3-2.pdf

Downloaded image of a building

C:\Gistutorial\Answers\Assignment3\XX.XX

OBJECTIVES

Create a new personal geodatabase
Modify a personal geodatabase
Join tables
Aggregate data
Export data from a personal geodatabase
Use ArcCatalog utilities

GIS Tutorial 4

Geodatabases

Spatial data, such as you have seen in shapefiles, can have many formats. Tutorial 5 will examine several such formats, but in this tutorial we first provide an introduction to the powerful and modern relational database format. A geodatabase is a collection of maps and database tables stored in a relational database management system. You will learn how to build personal geodatabases, so named because they use the widely available Microsoft Access database. In the future, more GIS systems will use geodatabases because most organizations work with database packages and prefer them to file-based systems.

Launch ArcMap

1 From the Windows taskbar, click Start, All Programs, ArcGIS, ArcMap.

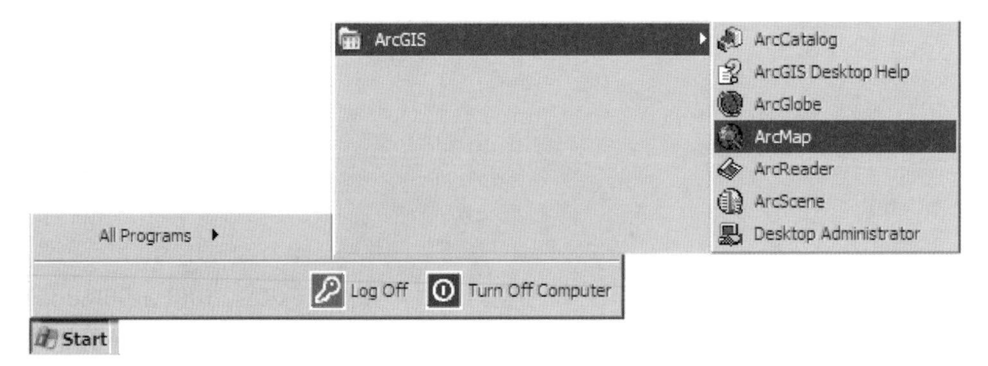

2 Click the A new empty map radio button in the ArcMap dialog box and click OK.

3 Click File, Save, browse to the Gistutorial folder, type **Tutorial4-1** in the File name field, and click Save.

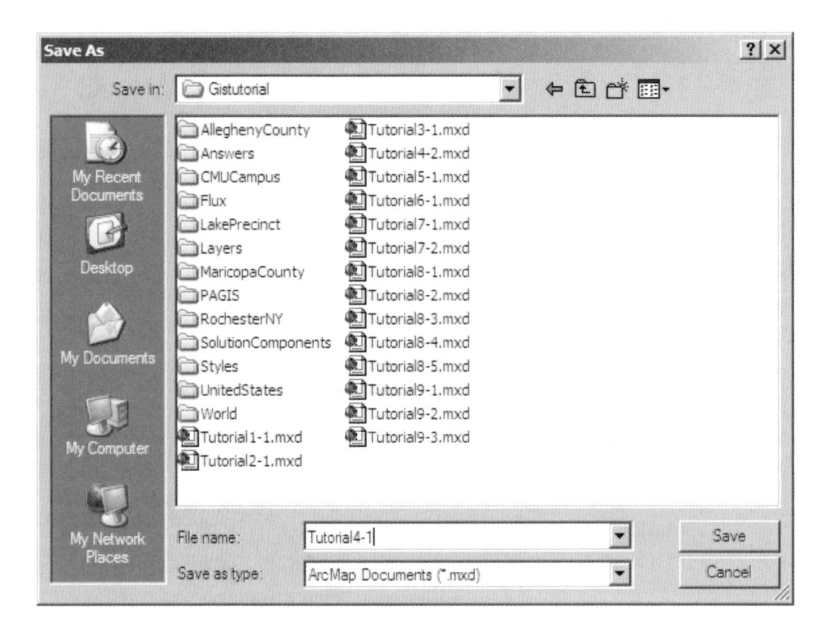

Launch ArcCatalog

1 **Click the ArcCatalog button.** 

ArcCatalog opens.

2 **In the catalog tree, navigate to your Gistutorial folder, then expand it by clicking the small box with the plus sign (+) to its left. Do the same for the MaricopaCounty folder, and then click the MaricopaCounty folder to expose that folder's contents in the Contents tab of the preview pane to the right.**

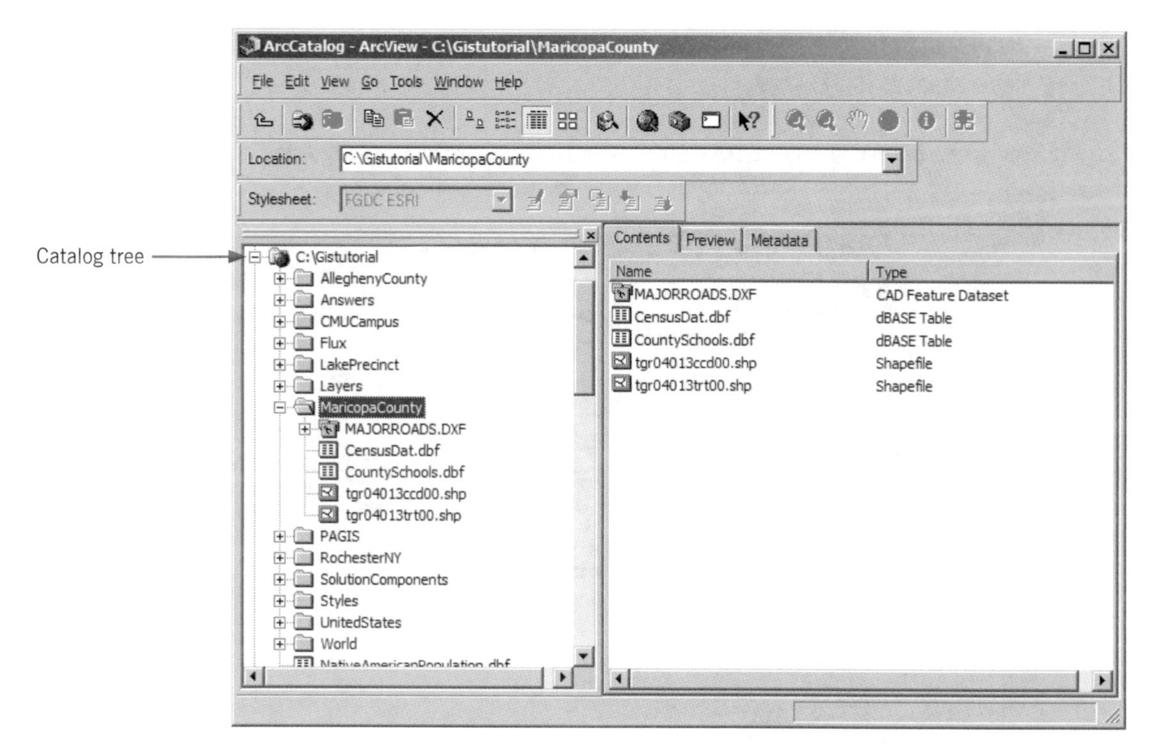

Create a new personal geodatabase

ArcCatalog shows that there are two shapefiles in the Maricopa County folder (Maricopa County, Arizona, includes the city of Phoenix and other cities). Both shapefiles were downloaded for free from the ESRI Geography Network (see tutorial 5). One is year 2000 census tract boundaries (tgr04013trt00.shp) and the other is county civil divisions (tgr04013ccd00.shp), which include city boundaries. The "tgr" stands for TIGER file, 04 is the FIPS code for Arizona, and 013 is the FIPS code for Maricopa County.

Create a new geodatabase

1 **In the catalog tree of the ArcCatalog window, right-click the MaricopaCounty folder, click New, Personal Geodatabase.**

ArcCatalog creates an empty geodatabase that you can now populate with feature classes and stand-alone tables. (Feature classes are map layers stored in a geodatabase.)

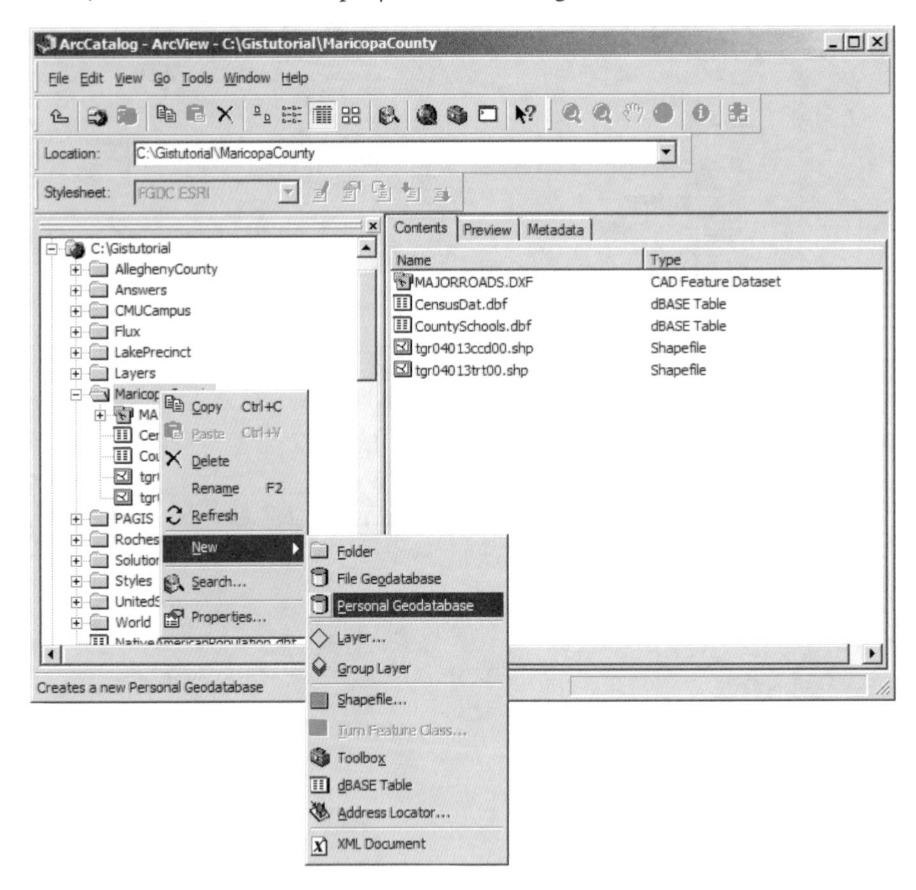

2 **In the catalog tree, right-click New Personal Geodatabase, click Rename, type MaricopaCountyIncome, and press Enter on your keyboard.**

Import shapefiles

1 **Right-click the MaricopaCountyIncome geodatabase, then click Import, Feature Class (single).**

2 **In the Feature Class to Feature Class dialog box, click the browse button next to the Input Features field, browse to Gistutorial\MaricopaCounty, double-click to open the folder, double-click tgr04013ccd00.shp.**

3 **Type tgr04013ccd00 as the Output Feature Class name, click OK, and Close after ArcCatalog finishes importing.**

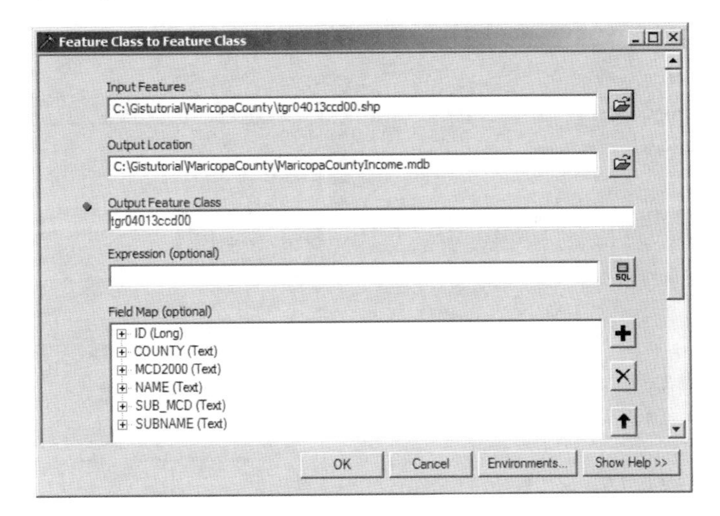

ArcCatalog imports the shapefile into the geodatabase.

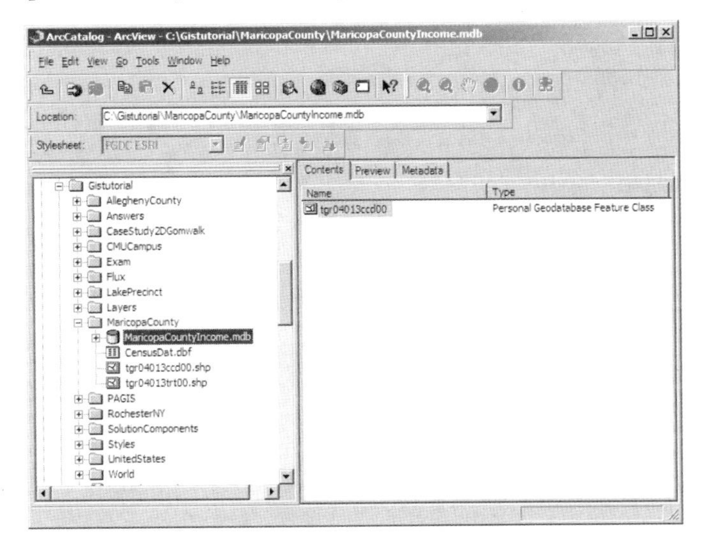

YOUR TURN

Import tgr04013trt00.shp into the MaricopaCountyIncome personal geodatabase.
Important note: You must complete this task to perform later steps of this tutorial.

Import a data table

Next you will import a 2000 census data table at the tract level that was downloaded from the *www.census.gov* Web site. See tutorial 5 for details on how to download data from this Web site. Eventually you will join this table to the tract map that you just imported, tgr04013trt00, and use it to map income disparities between Native Americans and Hispanics and Whites at both the city and tract levels.

1 In the ArcCatalog window, click the ArcToolbox button.

2 In the ArcToolbox window, expand the Conversion Tools toolbox by clicking the plus sign (+) to its left.

3 Expand the To Geodatabase toolset by clicking the box with a plus sign (+) to its left.

4 Double-click the Table to Geodatabase (multiple) tool.

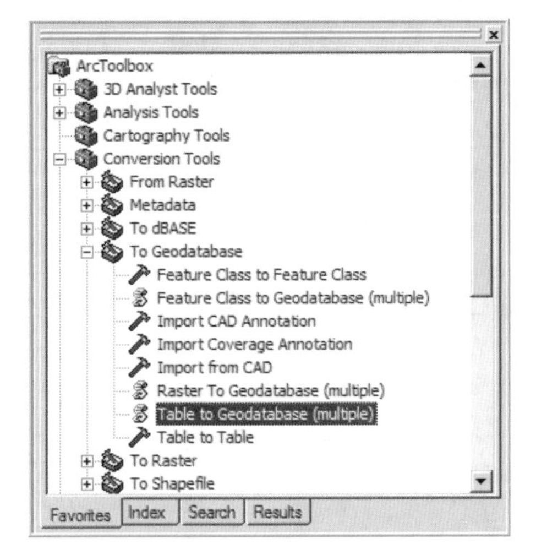

5 In the Table to Geodatabase (multiple) dialog box, browse to **Gistutorial\MaricopaCounty** for the Input table, click **CensusDat.dbf**, and click Add.

6 Browse to the output geodatabase in **\Gistutorial\MaricopaCounty**, then click **MaricopaCountyIncome.mdb**, Add, OK, and Close.

7 Close ArcToolbox and ArcCatalog.

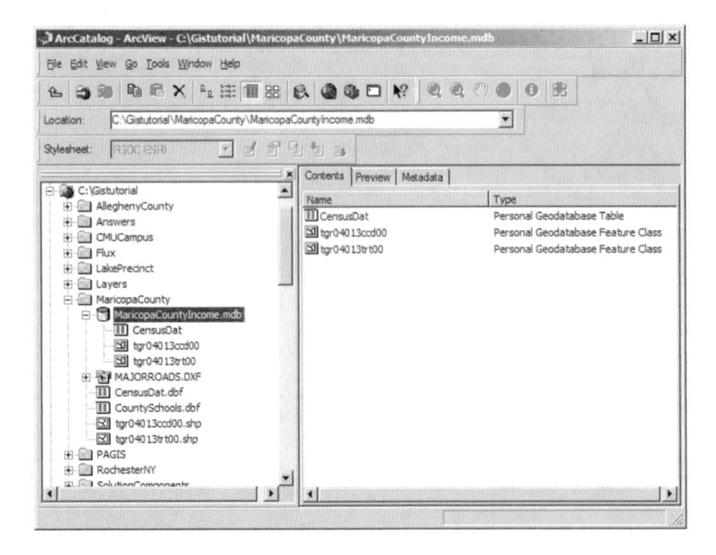

Add layers to ArcMap

1 In ArcMap, click the Add Data button. 

2 Browse to the **\Gistutorial\MaricopaCounty** folder, double-click **MaricopaCountyIncome.mdb**, hold down the Shift key, click tgr04013ccd00 and tgr04013trt00, and click Add.

Change layer properties

1 If necessary, click and hold the Cities (tgr04013ccd00) layer in the table of contents, drag it above Tracts (tgr04013trt00), and release.

2 Right-click tgr04013ccd00, click Properties, and change its Layer Name to Cities.

3 Click the Symbology tab, the Symbol button, Hollow color patch, make the Outline Width 2, and click OK.

4 Click the Labels tab and Label Features in this layer check box.

5 In the Text Symbol frame, click the Symbol button, change the font Size to 12, set the Style to bold, and click OK twice.

6 Right-click tgr04013trt00 in the table of contents, click Properties, click the General tab, change the Layer Name to Tracts, and click OK.

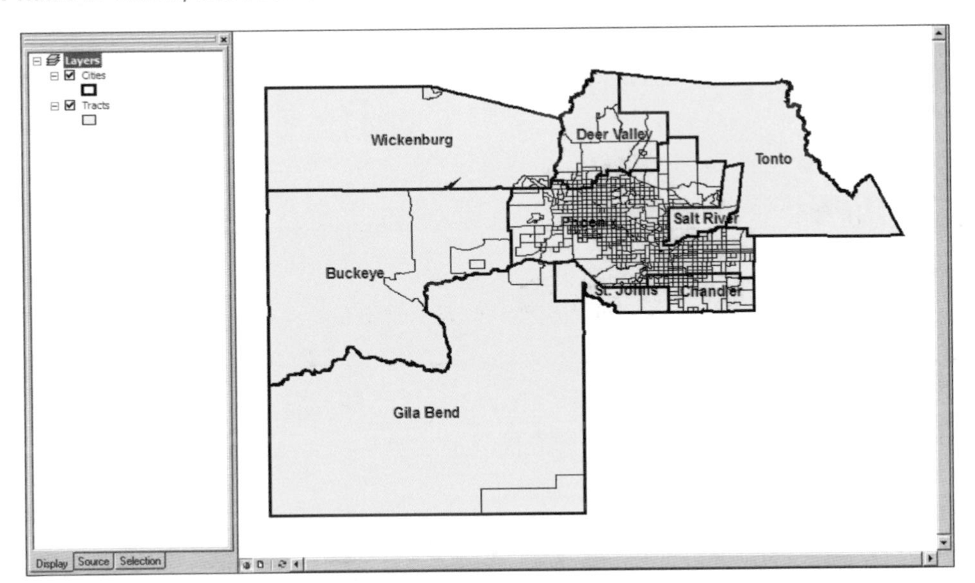

Modify a geodatabase

Examine and modify layer attribute tables

1 In ArcMap, right-click Tracts in the table of contents and click Open Attribute Table.

You will delete some unneeded columns. Note that most of the modifications you will make to geodatabases can also be made to shapefiles.

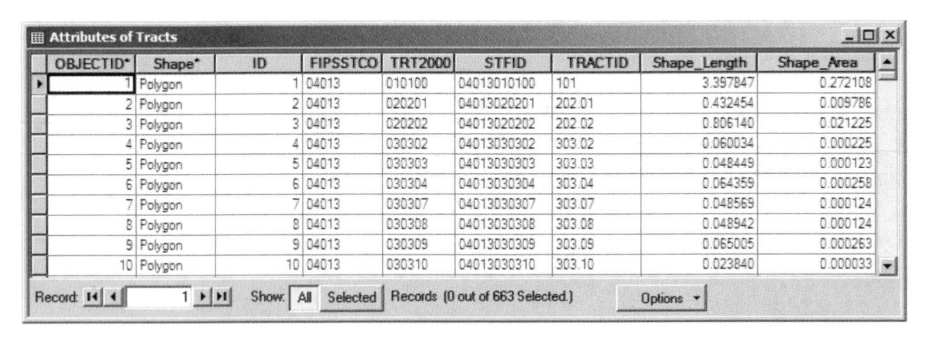

OBJECTID*	Shape*	ID	FIPSSTCO	TRT2000	STFID	TRACTID	Shape_Length	Shape_Area
1	Polygon	1	04013	010100	04013010100	101	3.397847	0.272108
2	Polygon	2	04013	020201	04013020201	202.01	0.432454	0.009786
3	Polygon	3	04013	020202	04013020202	202.02	0.806140	0.021225
4	Polygon	4	04013	030302	04013030302	303.02	0.060034	0.000225
5	Polygon	5	04013	030303	04013030303	303.03	0.048449	0.000123
6	Polygon	6	04013	030304	04013030304	303.04	0.064359	0.000258
7	Polygon	7	04013	030307	04013030307	303.07	0.048569	0.000124
8	Polygon	8	04013	030308	04013030308	303.08	0.048942	0.000124
9	Polygon	9	04013	030309	04013030309	303.09	0.065005	0.000263
10	Polygon	10	04013	030310	04013030310	303.10	0.023840	0.000033

2 In ArcMap, right-click the column heading for the ID field, click Delete Field, Yes.

ID is an extra key identifier that you do not need. The key identifier, or primary key, that you will retain is STFID, which is unique for every census tract in the United States.

3 Delete the FIPSSTCO, TRT2000, and TractID fields from the Attributes of Tracts table.

4 Close the Attributes of Tracts table.

YOUR TURN

Delete the following fields from the Cities layer: ID, County, SubMCD, and SubName. Keep MCD2000, which is a unique city identifier in the United States.

Modify a primary key

1 **Click the Add Data button.** 

2 **Browse to \Gistutorial\MaricopaCounty and double-click CensusDat.dbf.**

The table of contents tab automatically switches to the Source tab (bottom left of screen) to show that the table has been added to Tutorial4-1.mxd and the paths of all data sources.

3 **Right-click the CensusDat icon in the table of contents and click Open.**

4 **Right-click Tracts in the table of contents, click Open Attribute Table, and reposition the two tables so you can compare them.**

The STFID column of the Attributes of Tracts and the Geo_ID column of the Attributes of CensusDat are the unique identifiers of each table (primary keys with no duplicates or null values). The values of each primary key would match, but Geo_ID has the extra characters "14000US" at the beginning of each value. Next, you will use a string function, Mid([Geo_ID], 8,11), that extracts an 11-character string from Geo_ID starting at position 8 and creates a new column in Attributes of CensusDat to match STFID of Attributes of Tracts.

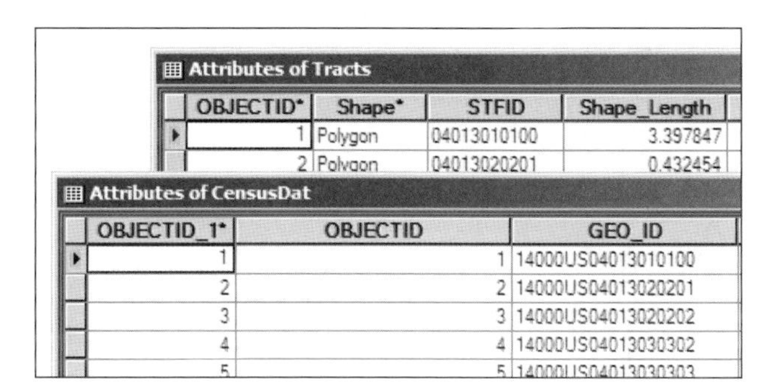

5 **In the Attributes of CensusDat table, click Options and Add Field.**

6 **In the Add Field dialog box, type STFID in the Name field, change the Type to Text and Length to 11, and click OK.**

7 **Scroll to the right in Attributes of CensusDat, right-click the STFID column heading, click Field Calculator, Yes.**

8 **In the Field Calculator dialog box, click the String radio button for Type, double-click the Mid() function, and in the STFID= box, edit the Mid() function to read Mid([Geo_ID], 8,11), and click OK.**

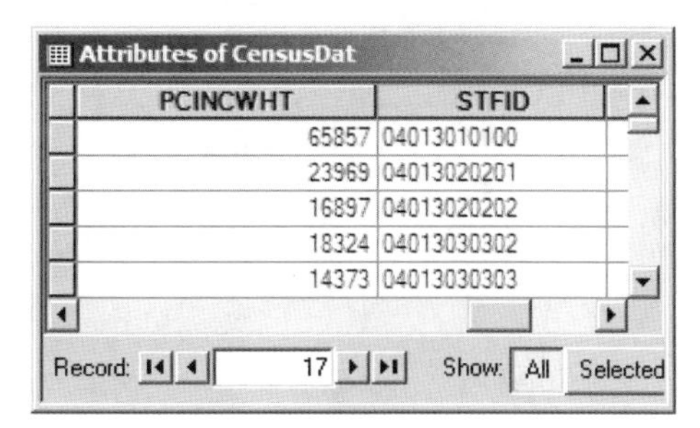

Calculate a new column

1 **In the Attributes of CensusDat table, click Options, Add Field.**

2 **In the Add Field window, type RNatWht in the Name field, change the Type to Float, and click OK.**

The new column will contain the Ratio of Native American per capita income to White per capita income. Next you must select only records where PCIncWht is greater than zero, because PCIncWht is the divisor for this ratio and will be used to calculate values for RNatWht.

3 **In the Attributes of CensusDat table, click Options, Select by Attributes.**

4 **In the Select by Attributes dialog box, scroll down in the list of fields, double-click PCINCWHT to add it to the lower Select panel, click the > symbol button, Get Unique Values, and double-click 0 in the Unique Values list (the middle panel list of numbers).**

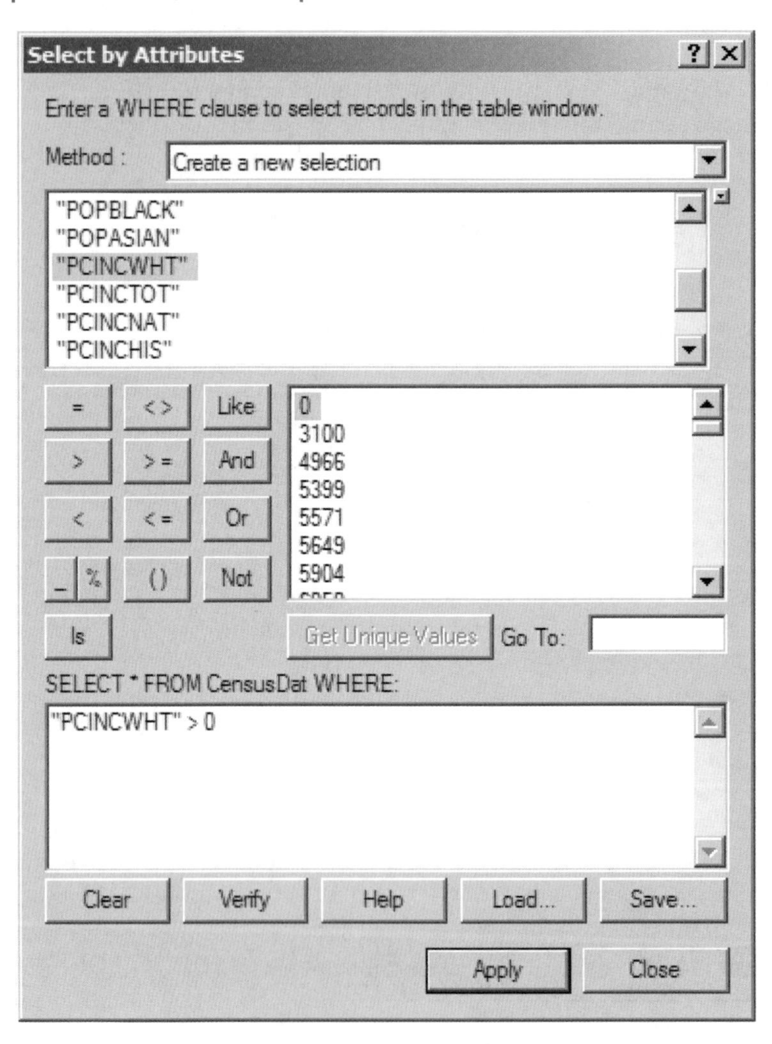

5 **Click Apply, Close.**

6 **Right-click the RNatWht column heading, click Field Calculator, Yes.**

7 **Delete the expression in the RNATWhite= panel.**

8 **In the Fields list, double-click PCINCNAT, click the / button, and double-click PCINCWHT in the Fields list.**

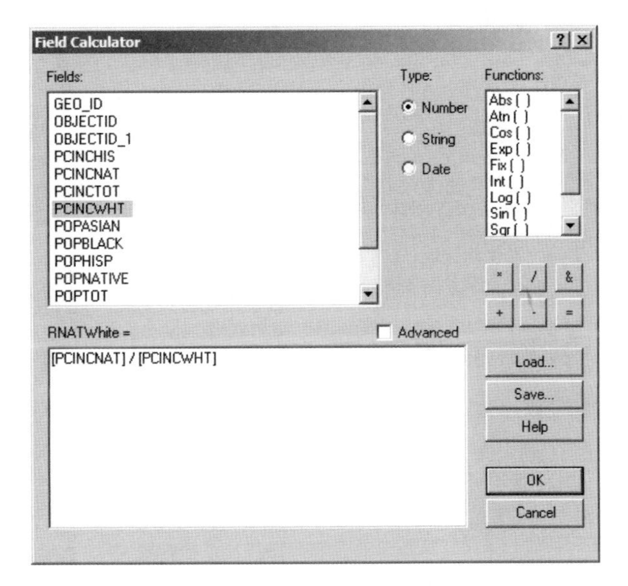

9 **Click OK.**

10 **In the Attributes of CensusDat table, click Options, Clear Selection.**

11 **Close the Attributes of Tracts table.**

YOUR TURN

Repeat the previous steps to calculate a new column in the Attributes of CensusDat table called RHisWht, which is the ratio of PCINCHIS divided by PCINCWHT. This is the ratio of per capita income of Hispanics divided by the per capita income of whites.

Join tables

Data comes from a variety of sources, and you may want to display information on your map that is not directly stored with your geographic data. For instance, you might obtain data from other departments in your organization, purchase commercially available data, or download data from the Internet. If this data is stored in a table such as a dBASE, Excel, INFO, or geodatabase table, you can join it to your geographic features and display the data on your map.

Join a table to a map

This task allows you to join data from the CensusDat table to the polygon shapefile for the Census Tracts.

1 **In the ArcMap table of contents, right-click the Tracts layer, then click Joins and Relates, Join.**

2 **In the Join Data dialog box, make sure that Join attributes from a table is chosen from the drop-down list at the top.**

3 **For item 1, choose the STFID field from the drop-down list.**

4 **For item 2, choose the CensusDat table from the drop-down list.**

5 **For item 3, choose the STFID field, click OK, Yes.**

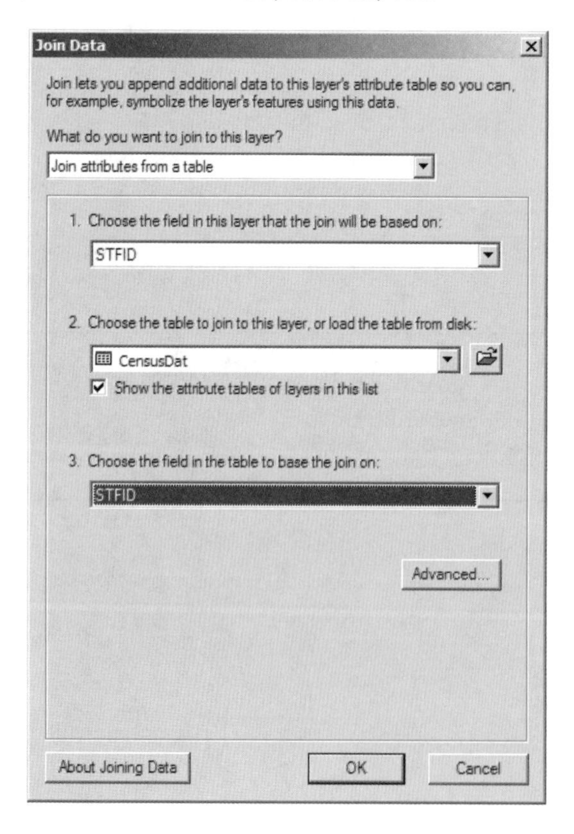

6 Right-click the Tracts layer in the table of contents, click Open Attribute Table, scroll to the right in the table, and verify that your CensusDat table has been joined to the Attributes of Tracts table.

This is an on-the-fly join that is not permanent, but will always be active when your Tutorial4-1.mxd document is open.

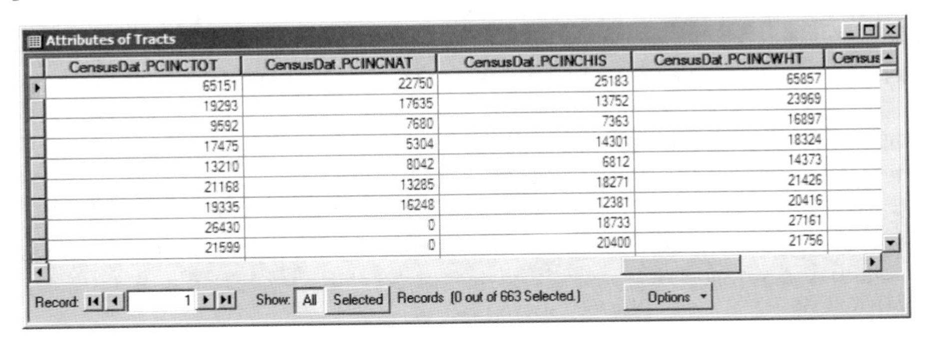

Attributes of Tracts

CensusDat.PCINCTOT	CensusDat.PCINCNAT	CensusDat.PCINCHIS	CensusDat.PCINCWHT	Census
65151	22750	25183	65857	
19293	17635	13752	23969	
9592	7680	7363	16897	
17475	5304	14301	18324	
13210	8042	6812	14373	
21168	13285	18271	21426	
19335	16248	12381	20416	
26430	0	18733	27161	
21599	0	20400	21756	

Record: 1 Show: All Selected Records (0 out of 663 Selected) Options

7 Close the Attributes of Tracts table.

Symbolize a map

Now that the CensusDat data is joined to the census tracts, the census data can be symbolized using the census polygons.

1 **Right-click the Tracts layer in the table of contents, click Properties, General tab, and in the Layer Name field type Ratio of Native Americans to White Per Capita Income.**

2 **Click the Symbology tab, then under Show, click Quantities, Graduated Colors, and under Fields change Value to CensusDat.RNatWht, then click Classify.**

3 **In the Classification dialog box, change Classes to 7, change the Method to Manual, and in the Break Values box change the existing values to the following: 0.2, 0.4, 0.6, 0.8, 1.0, 1.2, 99999 and click OK.**

4 **In the Symbol column, double-click each color chip and change colors to range from brown to pink for values of 1 or less and light to bright green for the two remaining categories, then click OK.**

5 **Right-click anywhere in the Label column and click Format Labels. In the Rounding panel of the Number Format dialog box, make sure that the Number of decimal places option is selected, then change the number of decimal places to 2, and click OK.**

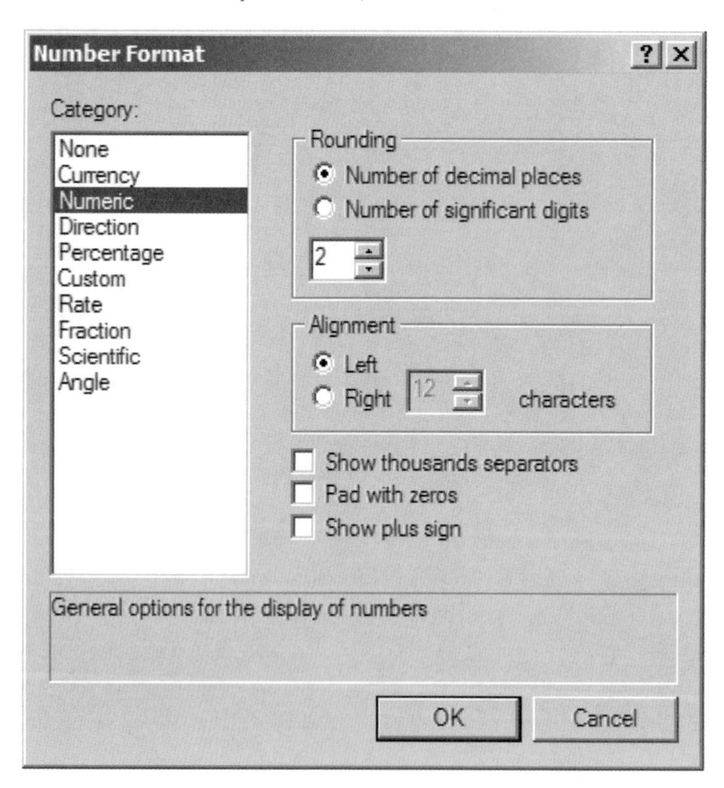

The resulting map shows that Native Americans generally have less per capita income than whites except in some urban areas.

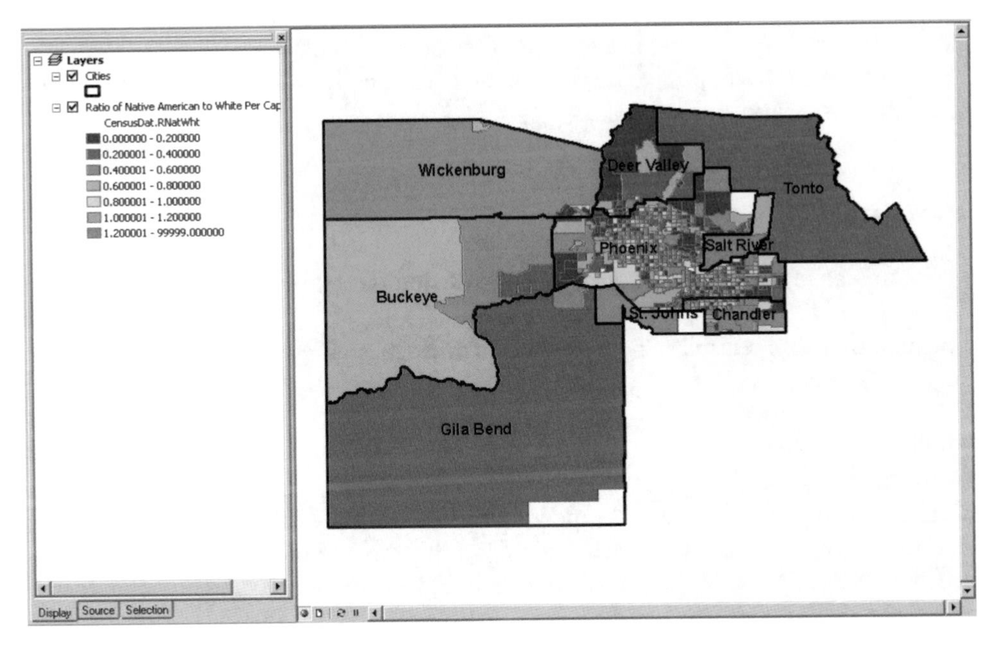

YOUR TURN

Save the Tracts layer as a layer file called Maricopatracts.lyr and add it to get a second copy in the table of contents.

Change the name and symbolize the new layer to show the Ratio of Hispanic to White Per Capita Income. Use the same numerical scale as for Native Americans. *Hint: In the symbology tab, click Import and import symbology from the Native American layer.* Use CensusDat.RHisWht as the value field.

When finished, save Tutorial4-1.mxd.

Aggregate data

The next part of this tutorial has you aggregate (count) points within police administrative areas (car beats) and then display the results on a map of the car beats. A car beat is the patrol area of a single police car. In this case, the end result will be a choropleth map of car beats displaying the number of crime-prone businesses of a certain kind—eating and drinking places.

Examine tables to join

1 In ArcMap, open Tutorial4-2.mxd from the \Gistutorial folder.

The map that opens displays police car beats in Rochester, New York, as polygons and all businesses as points.

2 Right-click the Businesses layer in the table of contents and click Open Attribute Table.

Notice that this table has latitude and longitude coordinates in it. As explained in the next tutorial, these coordinates allow the records in this table to be directly mapped as points. The SIC attribute is the Standard Industrial Code, a U.S. Census Bureau classification for all private-sector enterprises.

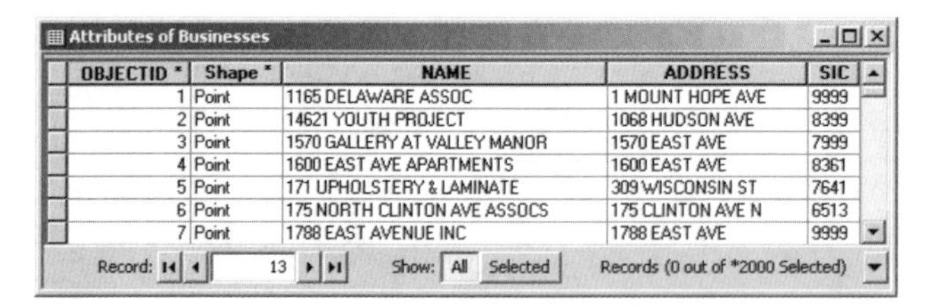

Attributes of Businesses

OBJECTID *	Shape *	NAME	ADDRESS	SIC
1	Point	1165 DELAWARE ASSOC	1 MOUNT HOPE AVE	9999
2	Point	14621 YOUTH PROJECT	1068 HUDSON AVE	8399
3	Point	1570 GALLERY AT VALLEY MANOR	1570 EAST AVE	7999
4	Point	1600 EAST AVE APARTMENTS	1600 EAST AVE	8361
5	Point	171 UPHOLSTERY & LAMINATE	309 WISCONSIN ST	7641
6	Point	175 NORTH CLINTON AVE ASSOCS	175 CLINTON AVE N	6513
7	Point	1788 EAST AVENUE INC	1788 EAST AVE	9999

Record: 13 Show: All Selected Records (0 out of *2000 Selected)

3 Click the Source tab at the bottom of the table of contents, right-click SIC in the table of contents, and click Open.

This is a code table with a definition for each SIC value. Next you will join this table to the Attributes of Businesses table, so that you can see the nature of each business.

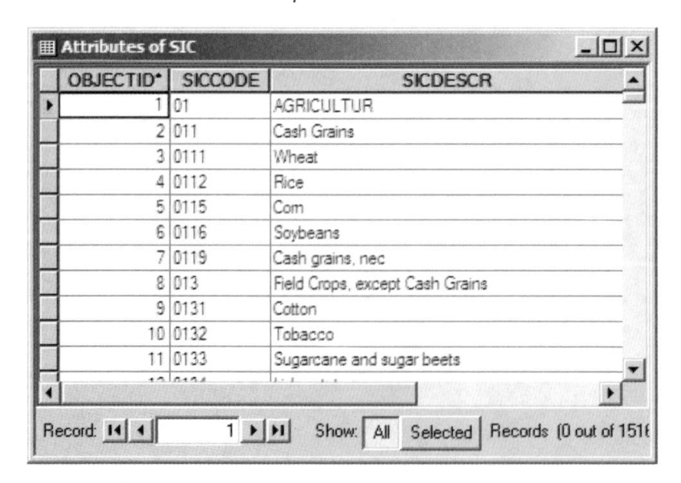

Attributes of SIC

OBJECTID*	SICCODE	SICDESCR
1	01	AGRICULTUR
2	011	Cash Grains
3	0111	Wheat
4	0112	Rice
5	0115	Corn
6	0116	Soybeans
7	0119	Cash grains, nec
8	013	Field Crops, except Cash Grains
9	0131	Cotton
10	0132	Tobacco
11	0133	Sugarcane and sugar beets

Record: 1 Show: All Selected Records (0 out of 151

4 Close the Attributes of SIC table.

Join tables

1 **Right-click the Businesses layer in the table of contents and click Joins and Relates, Join.**

2 **For item 1 in the Join Data window, choose SIC as the field in this layer on which the join will be based.**

3 **For item 2, choose SIC from the drop-down list.**

4 **For item 3, choose the SICCODE field, then click OK, Yes.**

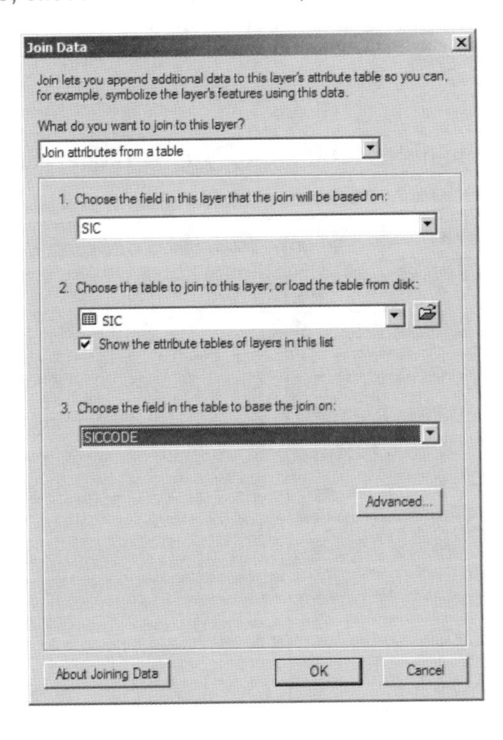

5 **Scroll to the right in the Attributes of Businesses and see that the SIC code table has been joined to it.**

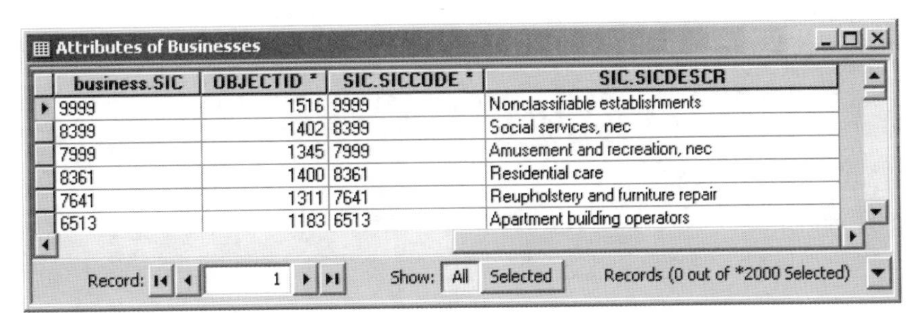
Attributes of Businesses

business.SIC	OBJECTID *	SIC.SICCODE *	SIC.SICDESCR
9999	1516	9999	Nonclassifiable establishments
8399	1402	8399	Social services, nec
7999	1345	7999	Amusement and recreation, nec
8361	1400	8361	Residential care
7641	1311	7641	Reupholstery and furniture repair
6513	1183	6513	Apartment building operators

Record: 1 Show: All Selected Records (0 out of *2000 Selected)

6 **Scroll down this table and see that business names and code table values for SICDESCR match up reasonably well.**

7 **Close the Attributes of Businesses table.**

Extract a subset of points

Suppose that police want to map retail establishments, regardless of what is sold. These establishments have SIC code values 5200 through 5990. You will extract these next.

1 From the ArcMap menu, click Selection, Select by Attributes.

2 In the Select By Attributes dialog box, select Businesses from the Layer drop-down list.

3 In the list of fields, double-click SIC.SICCODE, click the >= button, in the expression box type '5200', click the And button, double-click SIC.SICCODE, click the <= button, and in the expression box type '5990'.

Be sure to place single quotes (') around the numbers '5200' and '5990'. Without them, the query will not work.

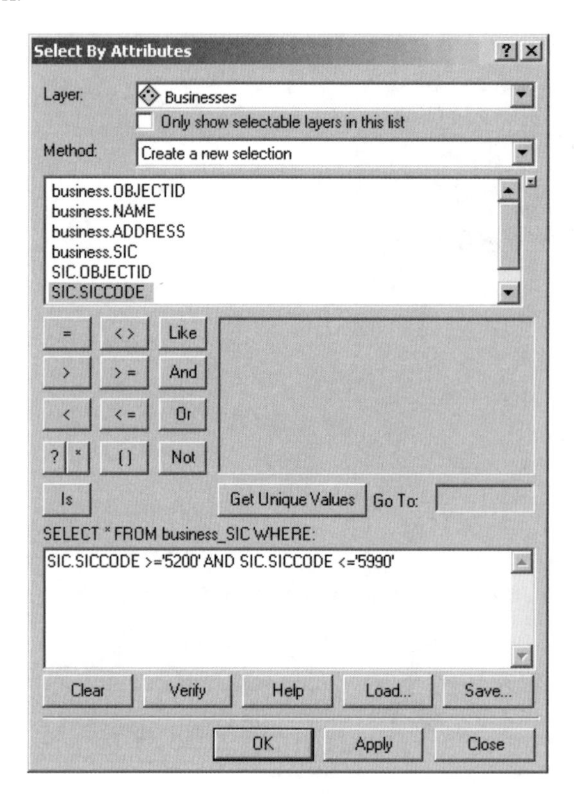

4 Click Apply and Close.

5 **Right-click Businesses in the table of contents, click Selection, Create layer from Selected Features.**

6 **In the table of contents, uncheck the Businesses layer to turn it off in the map display.**

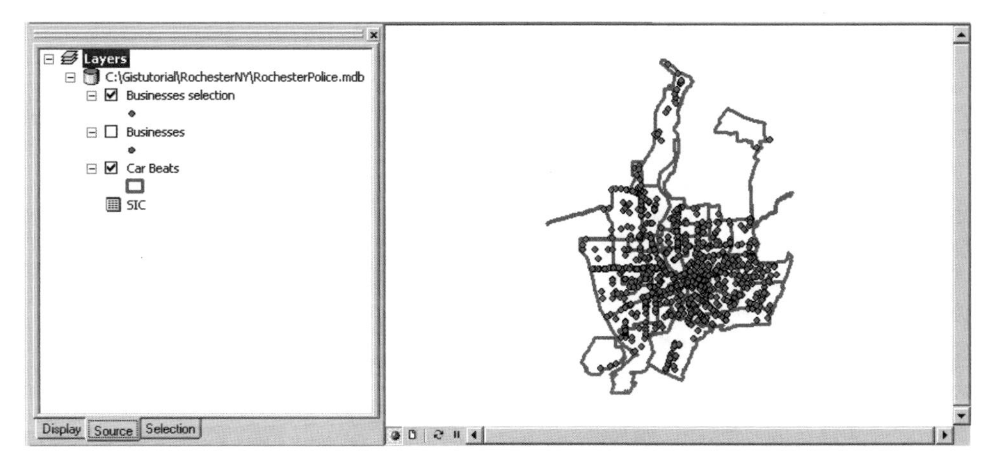

7 **Right-click the Businesses selection layer, click Properties, click the General tab, and change the Layer name to Retail Businesses.**

YOUR TURN

Extract a second set of businesses, those having to do with automobiles. You can use the following query expression:

SIC.SICDESCR LIKE '*auto*'

Call the new layer Automobile Businesses.

This expression includes any business record that has its SIC description containing the string 'auto'. The * is a wild card representing zero, one, or other additional characters. Look at selected Businesses records to see examples of SIC descriptions selected; for example, Auto and home repair shops, Automobile parking, and General automotive repair shops.

Spatially join point and polygon layers

Now that you have a layer of retail businesses, you can assign each of these points the Car Beat identifier within which it lies. To do this you will use a spatial join. Once you have joined the Car Beat identifier to each retail business, you will be able to summarize how many retail businesses are in each car beat.

1 **Right-click the Retail Businesses layer in the table of contents, click Joins and Relates, and click Join.**

2 **In the Join Data dialog box, from the What do you want to join to this layer? drop-down list, choose Join data from another layer based on spatial location.**

3 **For item 1, choose the Car Beats layer from the drop-down list.**

4 **For item 2, click the it falls inside radio button.**

5 **For item 3, click the Browse button. In the Saving Data dialog box choose File and Personal Geodatabase feature classes from the Save as type drop-down list.**

6 **Then browse to your Gistutorial\RochesterNY folder, double-click the RochesterPolice.mdb, name the output RetailBusinessesJoined, and click Save, OK.**

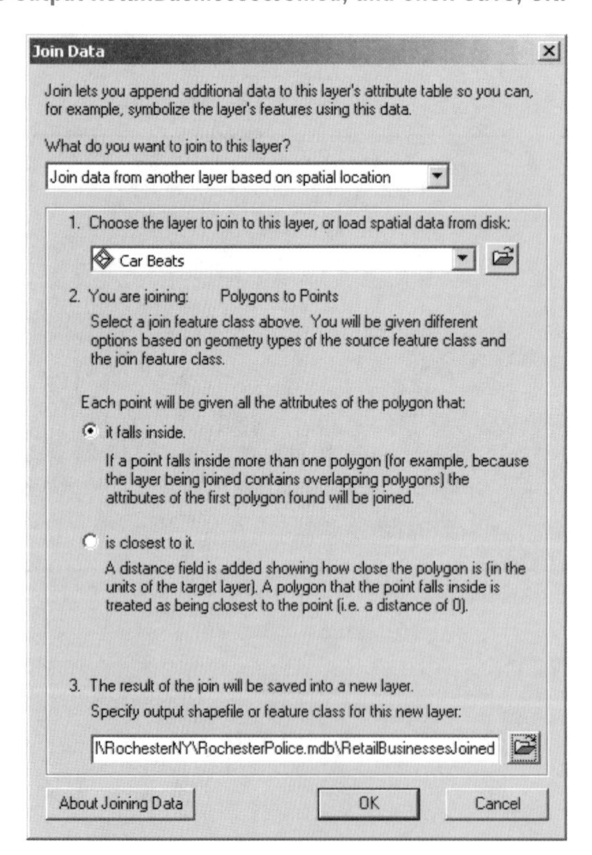

7 **Right-click RetailBusinessesJoined, click Open Attribute Table, and verify that each business has been assigned a car beat number (field BEAT). Leave this table open.**

Count points by polygon ID

Now you can summarize the number of retail businesses by car beat.

1 **In the Attributes of BusinessesJoined table, scroll to the right, right-click the column heading of the BEAT column, and click Summarize.**

2 **In the Summarize dialog box, for item 3, change the output table name from Sum_Output to BusCount, save it in your \Gistutorial\RochesterNY folder, click OK, and click Yes to add the table to the map.**

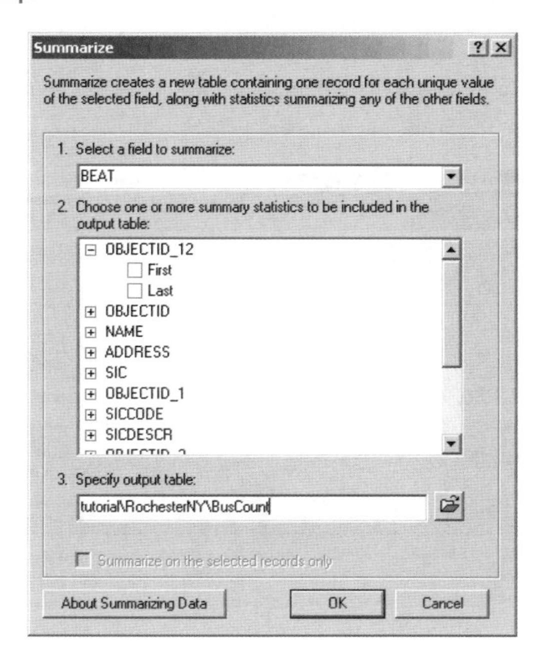

3 **Close the Attributes of RetailBusinessesJoined table.**

4 **Right-click the BusCount table in the table of contents and click Open.**

The Count_BEAT field contains the total number of retail business points in each police beat polygon.

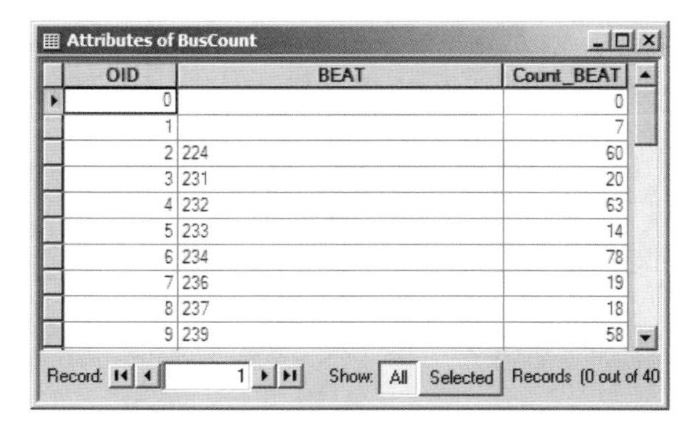
Attributes of BusCount

OID	BEAT	Count_BEAT
0		0
1		7
2	224	60
3	231	20
4	232	63
5	233	14
6	234	78
7	236	19
8	237	18
9	239	58

Record: 1 Show: All Selected Records (0 out of 40

5 **Close the table.**

Join a count table to a polygon map

1 **Right-click Car Beats in the table of contents, click Joins and Relates, Join.**

2 **In the Join Data dialog box, click the What do you want to join to this layer? drop-down list and choose Join attributes from a table.**

3 **For item 1, choose BEAT.**

4 **For item 2, choose BusCount.**

5 **For item 3, choose BEAT and click OK. Click Yes if prompted to create an index.**

6 **Right-click the Car Beats layer in the table of contents, click Open Attribute Table, scroll to the right in the table, and verify that each beat has a count of retail businesses.**

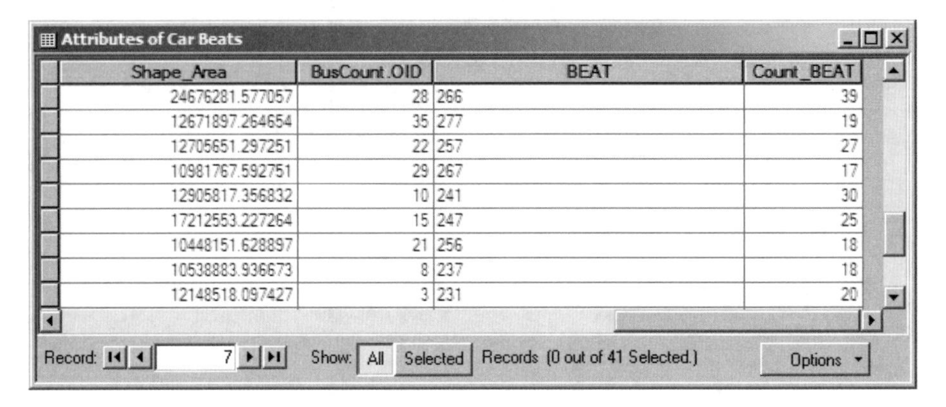

Shape_Area	BusCount.OID	BEAT	Count_BEAT
24676281.577057	28	266	39
12671897.264654	35	277	19
12705651.297251	22	257	27
10981767.592751	29	267	17
12905817.356832	10	241	30
17212553.227264	15	247	25
10448151.628897	21	256	18
10538883.936673	8	237	18
12148518.097427	3	231	20

7 **Close the Attributes of Car Beats table.**

YOUR TURN

Starting with the Automobile Businesses layer that you created in the previous Your Turn, carry out the following steps:

Add a second copy of the carbeats feature class to the table of contents.

Spatially join the new carbeat polygons to the Auto Businesses to create the AutoBusinessesJoined layer.

Summarize the count of Auto Businesses by BEAT to create the table AutoCount.

Join the new count table to the new carbeats layer.

When finished, click the new carbeats layer off.

Symbolize the choropleth map

With aggregate data on retail business points now available as counts per car beat, you are ready to create a car beat choropleth map. The resulting map will provide a good means for scanning the entire city for areas with high concentrations of retail businesses. Then the user can zoom to see details of the businesses.

1 **Right-click Car Beats in the table of contents, click Properties, the Labels tab, and the Label Features in this layer check box.**

2 **Click the General tab and change the Layer Name to Number of Retail Establishments.**

3 **Click the Symbology tab, Quantities, Graduated Colors.**

4 **Change the Value field to Count_BEAT, choose a monochromatic color ramp, and click Classify.**

5 **In the Classification dialog box, choose 7 classes, set the Method drop-down list to Quantile, and click OK twice.**

6 **Turn off all the layers except the Car Beats.**

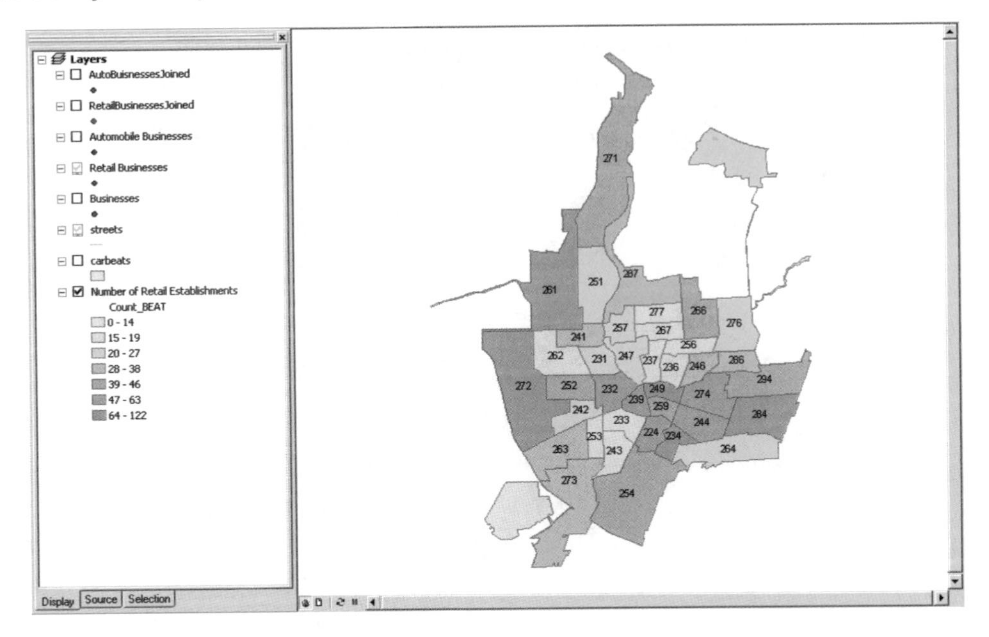

Symbolize the point map for drill down

1 Click the Add Data button, browse to **\Gistutorial\RochesterNY\RochesterPolice.mdb**, click streets, and click Add.

2 In the table of contents, right-click the line symbol just below the streets layer and choose a light gray color from the drop-down color palette.

3 Click the Zoom In button and zoom to the center of the city.

4 Turn on the Retail Businesses layer.

5 Right-click the Retail Businesses layer, click Properties, click the Labels tab, check the Label features in this layer box, choose SIC.SICDESCR from the Label field drop-down list, and click OK.

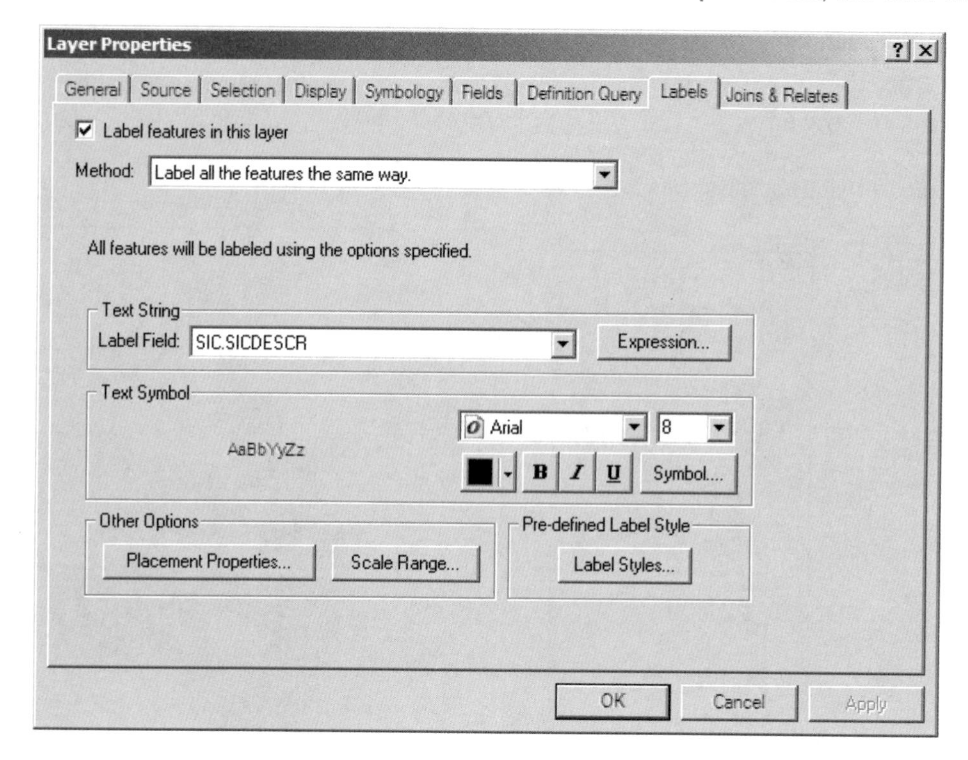

6 Right-click the Streets layer and click Visible Scale Range, Set Minimum Scale.

7 Right-click the Retail Businesses layer, click Visible Scale Range, and click Set Minimum Scale.

The map will show the businesses when zoomed to a scale larger than your current scale (zoom in a bit more).

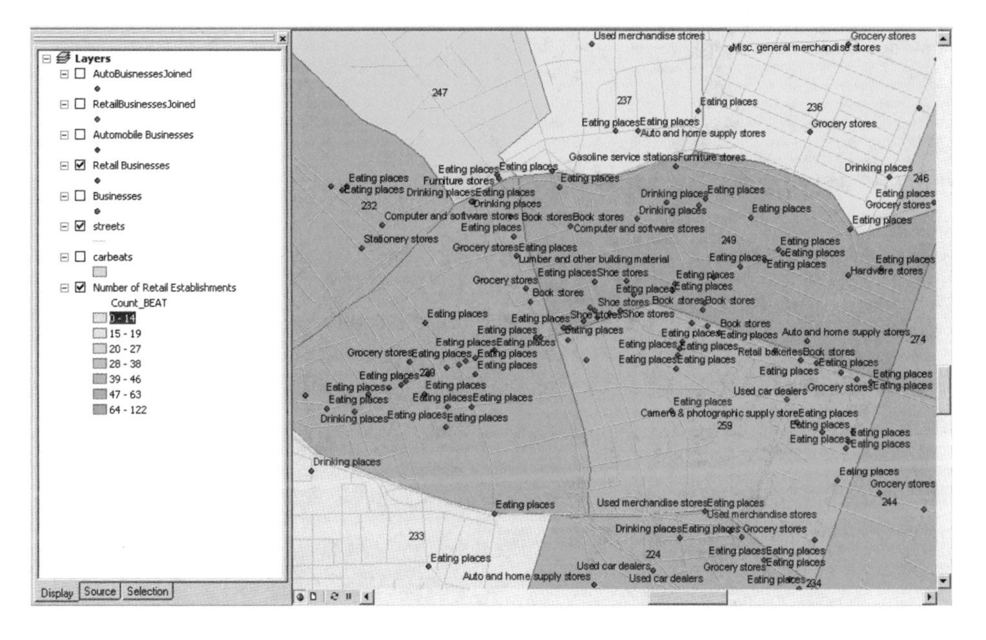

8 **Click the Full Extent button.** 

9 **Click File, Save, and close ArcMap.**

YOUR TURN

Symbolize the carbeats polygons for the count of Auto Businesses.

Symbolize the Auto Businesses point map for drill down.

Hint: You can copy and paste the value you set for minimum scale on the Retail Businesses layer by opening the Properties for that layer and selecting the General tab.

Export data from a geodatabase

Sometimes you will need to send another GIS user a map layer or data file that you have in a geodatabase. One way to do this is to use ArcCatalog to export the feature classes as shapefiles and the geodatabase tables as dBASE tables.

Export a table from a personal geodatabase

1 **Start ArcCatalog by clicking Start, Programs, ArcGIS, ArcCatalog.**

2 **In the catalog tree, navigate to and expand the Gistutorial\RochesterNY\RochesterPolice.mdb.**

3 **Right-click the SIC table in the RochesterPolice geodatabase, click Export, To dBase (single).**

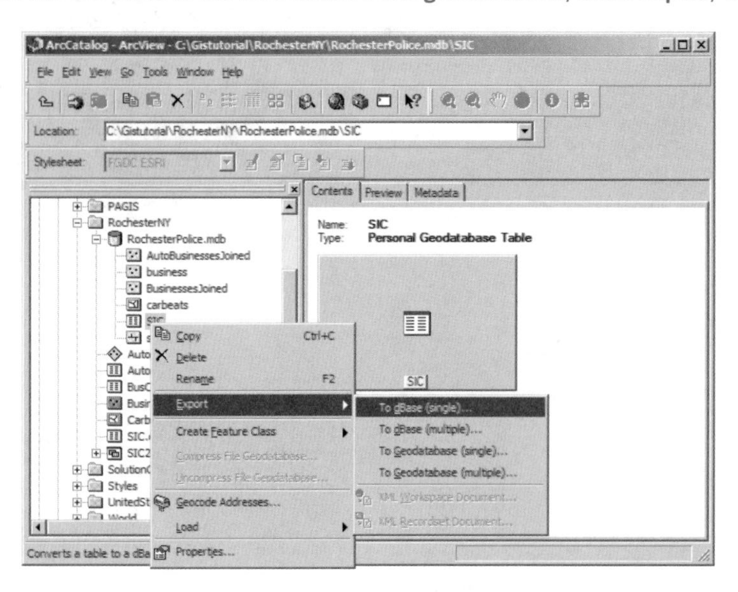

4 **In the Table to Table dialog box, browse to \Gistutorial\RochesterNY for the Output Location and change the output table from SIC to SIC2.**

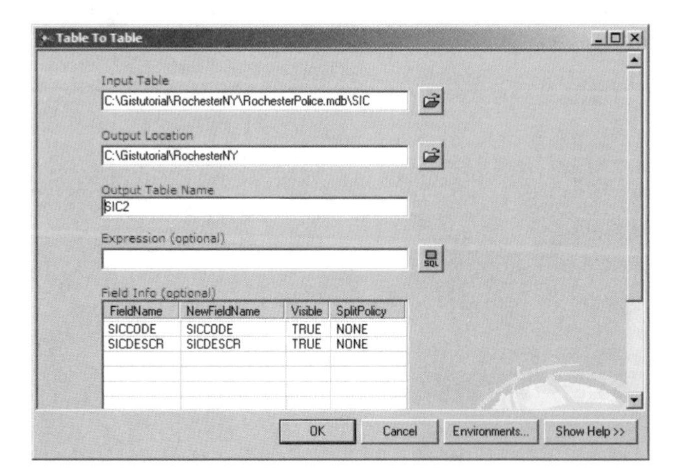

5 **Click OK, Close.**

Export a layer from a personal geodatabase

1 **In the catalog tree, expand the RochesterPolice geodatabase.**

2 **Right-click the RetailBusinessesJoined layer and click Export, To Shapefile (single).**

3 **In the Feature Class to Feature Class dialog box, click the Browse button to the right of the Output Location box, browse to your \Gistutorial\RochesterNY\ subfolder, then click Add.**

4 **Name the Output Feature Class RetailBusinessesJoined and click OK, Close.**

Your new shapefile is added to the RochesterNY folder.

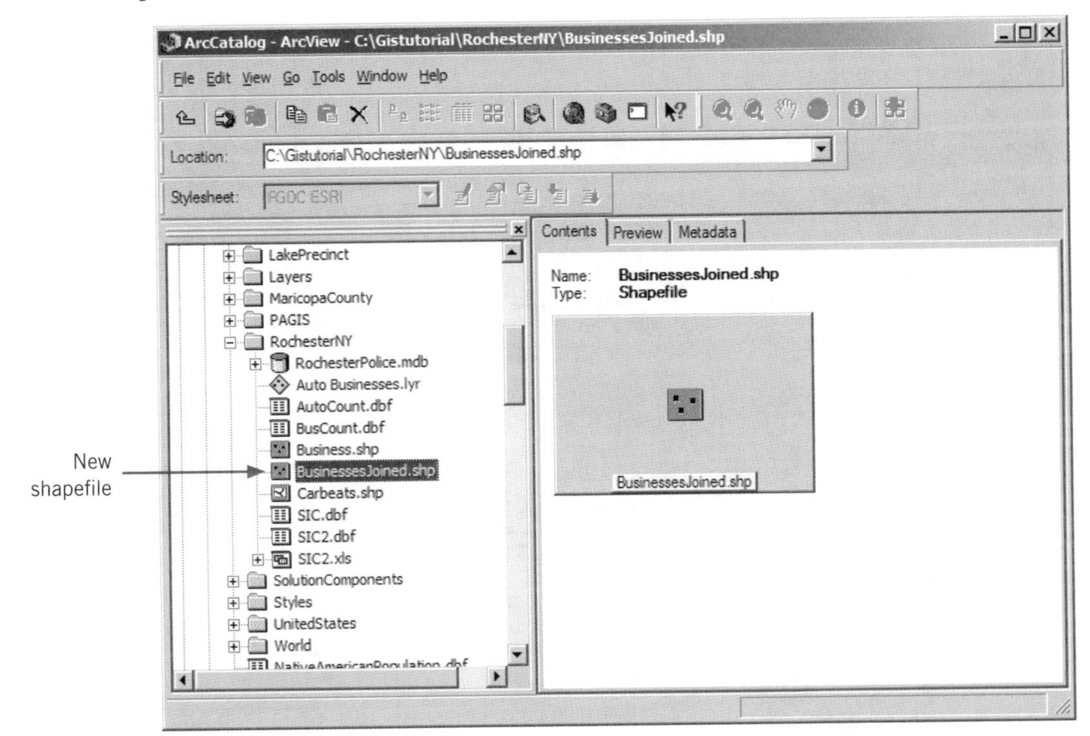

5 **In ArcCatalog, right-click the new RetailBusinessesJoined shapefile, click Properties, click the Fields tab, scroll down, see that the spatially joined attributes for car beats have been included in the shapefile, and click Cancel.**

The Car Beat attributes that were joined to the Retail Businesses layer are now a permanent part of the new shapefile. This is another reason for exporting: to make a spatial join permanent.

Use ArcCatalog utilities

Many of the file maintenance tasks for which you would normally use My Computer or Windows Explorer must be done using ArcCatalog. This is because a change in a single layer often affects several tables in the geodatabase, and ArcCatalog will maintain these relationships.

Copy and paste geodatabase layers

1 **If necessary, start ArcCatalog by clicking Start, Programs, ArcGIS, ArcCatalog.**

2 **In the catalog tree, navigate to \Gistutorial\RochesterNY, then expand the RochesterPolice geodatabase.**

3 **Right-click the carbeats layer that is inside the RochesterPolice geodatabase and click Copy.**

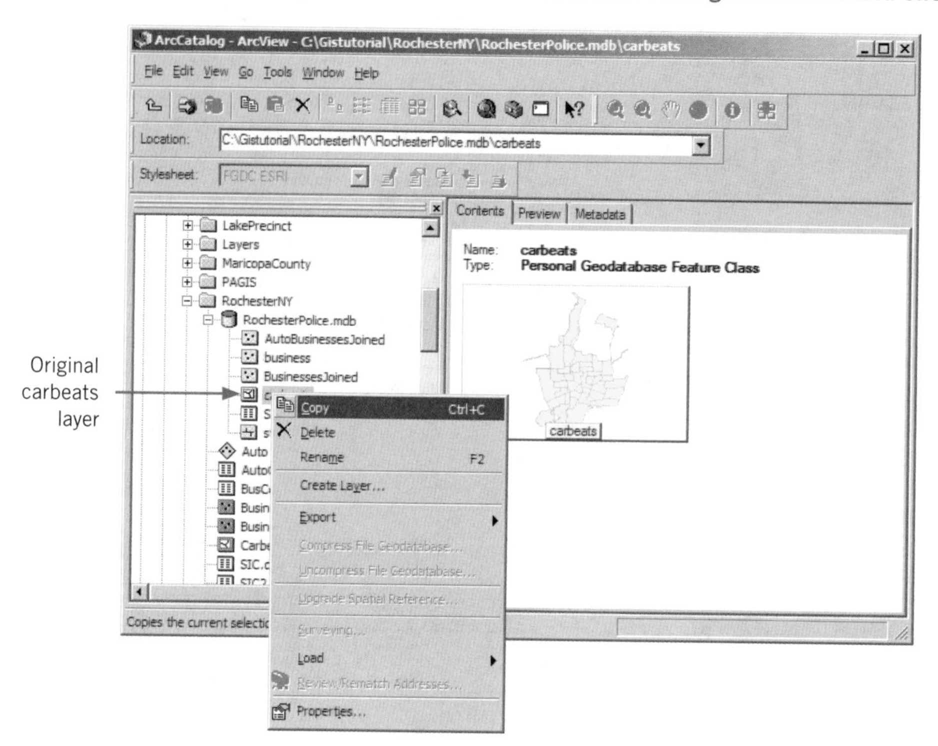

4 **Right-click the RochesterPolice geodatabase, click Paste, OK.**

Now you have a copy of the carbeats layer called carbeats_1.

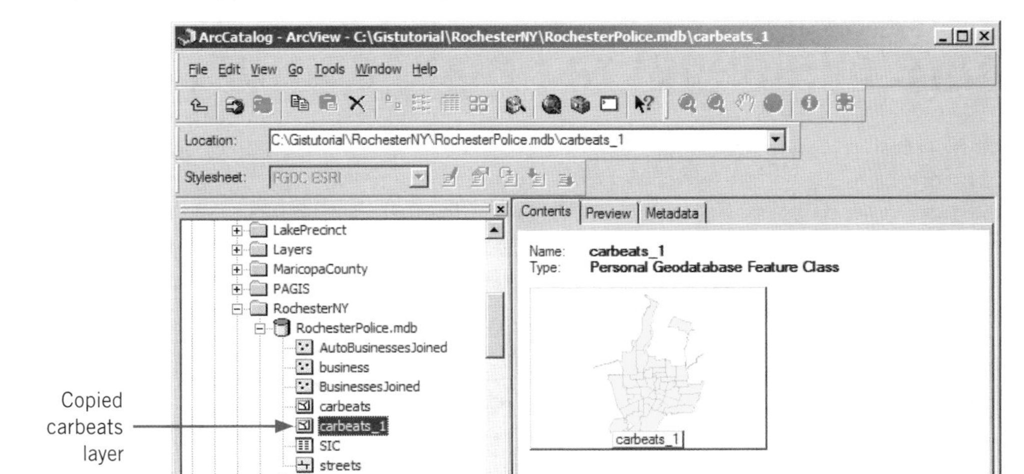

Rename and delete geodatabase layers

1 **In the catalog tree, right-click carbeats_1, click Rename, and change the layer's name to RochesterCarBeats.**

2 **Right-click RochesterCarBeats, Delete, and Yes.**

Compact a geodatabase

1 **Right-click the RochesterPolice geodatabase and click Compact Database.**

To maintain a fast and efficient database, you should compact it periodically.

2 **Close ArcCatalog.**

Assignment 4-1

Compare municipal tax compositions

Public finance experts prefer that municipalities obtain a larger share of taxes from earned income rather than real estate (or property) taxes. They reason that real estate taxes could be unfair in that they place a larger burden on low- and fixed-income families. For example, property values and real estate taxes increase over time, but retired people generally have fixed incomes. So over time, retired people use a larger proportion of their income for real estate taxes. It is fairer to tax earned income—whether it is fixed or increasing over time.

As a guide for municipalities to be fairer to low- and fixed-income inhabitants, create two maps of municipalities in Allegheny County, Pennsylvania—one with percentage of tax collections that are from real estate and the other with percentage of tax collections that are from earned income.

Start with the following:

Revenue table

- **C:\Gistutorial\AlleghenyCounty\Revenue.xls**—Excel table of 2004 municipal revenue data from the Pennsylvania Department of Economic Development.
  NAME = municipality name and primary key
  SHORTNAME = municipality name used for labeling
  TAX = total tax revenues
  REAL = tax revenues from real estate
  INCOMETAX = tax revenues from earned income

Allegheny County map layers

- **C:\Gistutorial\AlleghenyCounty\Munic.shp**—polygon layer of Allegheny County municipalities.
  NAME = municipality name and primary key

- **C:\Gistutorial\AlleghenyCounty\Rivers.shp**—polygon layer of three major rivers.

Create a personal geodatabase

Create a new personal geodatabase called CountyFinancials and stored as C:\Gistutorial\Answers\Assignment4\CountyFinancials.mdb. Import the above table and map layers into the geodatabase.

Create a map document

Create a map document called C:\Gistutorial\Answers\Assignment4\Assignment4-1.mxd that has two data frames. Both data frames need the two map layers and table from the geodatabase with the table joined to the municipality map. The data frames differ only by the attribute used to symbolize a choropleth map of municipalities:

- One data frame displays REAL/TAX
- The other data frame displays INCOMETAX/TAX

Use the normalization option in the Symbology tab of map layer properties to divide by TAX. Use quantiles with five categories for a numerical scale. Create a map layout with portrait page orientation and the two data frames, two legends, title, and other map elements that you choose. Label the municipalities. Use guidelines and design the layout carefully. Export your layout as JPEG image, C:\Gistutorial\Answers\Assignment4\Assignment4-1.jpg. Insert the image in a Word document, Assignment4-1.doc, saved in the same folder.

Hints
Start with a single data frame. Add the table and map layers to the data frame and join the table to the municipality map layer. Label the municipalities. Then copy and paste the data frame (right-click it) to create the second data frame.

You can edit map legends by converting them to graphics. For a legend in layout view, do the following:

- Activate the legend by clicking it.
- Right-click the legend and click Convert to Graphics.
- Right-click the legend and click Ungroup. You can further ungroup paint chips and labels in the legend, if needed to edit labels.
- For example, double-click a textbox, such as one that is too wide. Place the insertion point where you would like to split the line into two, press the Enter key, and click OK.
- For another example, click a textbox that you wish to delete and press the delete key.
- When finished editing, select all elements of a legend, right-click them and click Group to turn the legend back into a single graphic.

Note: You will find patterns in the two maps. To help you with interpretation, you should know that the poorer municipalities are along the rivers, in the old industrial parts of the county. The wealthier, suburban municipalities are in the northern and southern parts of the county. Does it look like taxation is fair?

Assignment 4-2

Compare youth population and total school enrollment

GIS has the unique ability to assign polygon identifiers to point features, otherwise known as a spatial join.

In an earlier exercise, you studied the schools in the city of Pittsburgh by enrollment using a pin map. Another way to study the same data is to spatially join the school points to a polygon layer (census tracts), and then sum the number of students in each polygon. After a few more steps, the end result is a choropleth map symbolizing census tracts with the newly summarized school data.

Start with the following:

- **C:\Gistutorial\PAGIS\PghTracts.shp**—polygon layer of Pittsburgh census tracts, 2000.
- **C:\Gistutorial\PAGIS\Schools.shp**—point layer of all schools with student enrollment data.

Create a personal geodatabase

In ArcCatalog, create a personal geodatabase called SchoolStudy and stored as C:\Gistutorial\Answers\Assignment4\SchoolStudy.mdb with the above layers imported.

Create a map document

In ArcMap, create a map document saved as C:\Gistutorial\Answers\Assignment4\Assignment4-2.mxd with a copy of each of the above layers added from the geodatabase. Carry out the spatial overlay process using the tip provided below, creating a new layer called TractSchoolJoin. Symbolize this layer as a choropleth map with SUM_ENROLL95, created in the spatial join process, using 5 classes and quantiles. Symbolize the original tract layer with size-graduated point markers for AGE_5_17, also with 5 classes and quantiles. Create an 8.5 × 11-inch landscape layout with map, legend, and heading. Export the layout to a .jpg image file, C:\Gistutorial\Answers\Assignment4\Assignment4-2.jpg, and insert it into a Word document, Assignment4-2.doc, in the same folder.

Spatial joining tip

Below is a shortcut for spatially joining points to polygons that automatically counts and summarizes data.

1. Right-click the PghTracts layer and click Joins and Relates, Join.
2. Spatially join the Schools point layer to the PghTracts layer.
3. Click Sum in the join dialog box. This will sum all of the numerical fields in the Schools point layer, including ENROLL95.
4. Save the new layer in your personal geodatabase as TractSchoolJoin.
5. Open the attribute table of the new TractSchoolJoin layer and examine the fields that were created by joining the points to the polygons. Of particular interest will be the fields "COUNT_", which is the number of schools (points) in each census polygon, and "SUM_ENROLL95", which is the sum of students enrolled in each school. Some fields will be <null> because those census tracts have no schools or student enrollment.

Question

Does it appear that schools are well-located relative to the youth population?

What to turn in

If you are working in a classroom setting with an instructor, you may be required to submit the exercises you created in tutorial 4. Below are the files you are required to turn in. Be sure to use a compression program such as PKZIP or WinZip to include all three files as one .zip document for review and grading. Include your name and assignment number in the .zip document (YourNameAssn4.zip).

ArcMap documents

C:\Gistutorial\Answers\Assignment4\Assignment4-1.mxd
C:\Gistutorial\Answers\Assignment4\Assignment4-2.mxd

Exported maps

C:\Gistutorial\Answers\Assignment4\Assignment4-1.doc
C:\Gistutorial\Answers\Assignment4\Assignment4-2.doc

Personal geodatabases and all files imported into them

Note: Before turning in the personal geodatabases, compact them in ArcCatalog by right-clicking and choosing “Compact Database.” This will reduce the file size.
C:\Gistutorial\Answers\Assignment4\CountyFinancials.mdb
C:\Gistutorial\Answers\Assignment4\SchoolStudy.mdb

OBJECTIVES

Familiarize yourself with sources of maps and data
Learn vector spatial data formats
Identify projections and learn how and when to change them
Store metadata on projections
Prepare attribute data

GIS Tutorial 5

Importing Spatial and Attribute Data

This tutorial serves as a guide to using GIS data sources, including some of the major free resources on the Internet. This chapter compares the major file formats used to create GIS basemaps that you will encounter, including shapefiles, ArcInfo coverages, CAD files, and raster images. You will learn how to export and import files to and from other applications. Because file formats and map projections vary, you will learn how to manipulate and transform them.

Sources of maps and data

ESRI Web site

ESRI, the world's leader in GIS and mapping software, maintains a Web site that is a useful resource for obtaining information and data used to create GIS maps.

Note: If you have difficulty downloading the tutorial files for this section, they are available in the \Gistutorial\SolutionsComponents\Tutorial5 folder.

1 **Open your Web browser.**

2 **Go to *http://www.esri.com*.**

The content of this home page varies from time to time, so your page may be different.

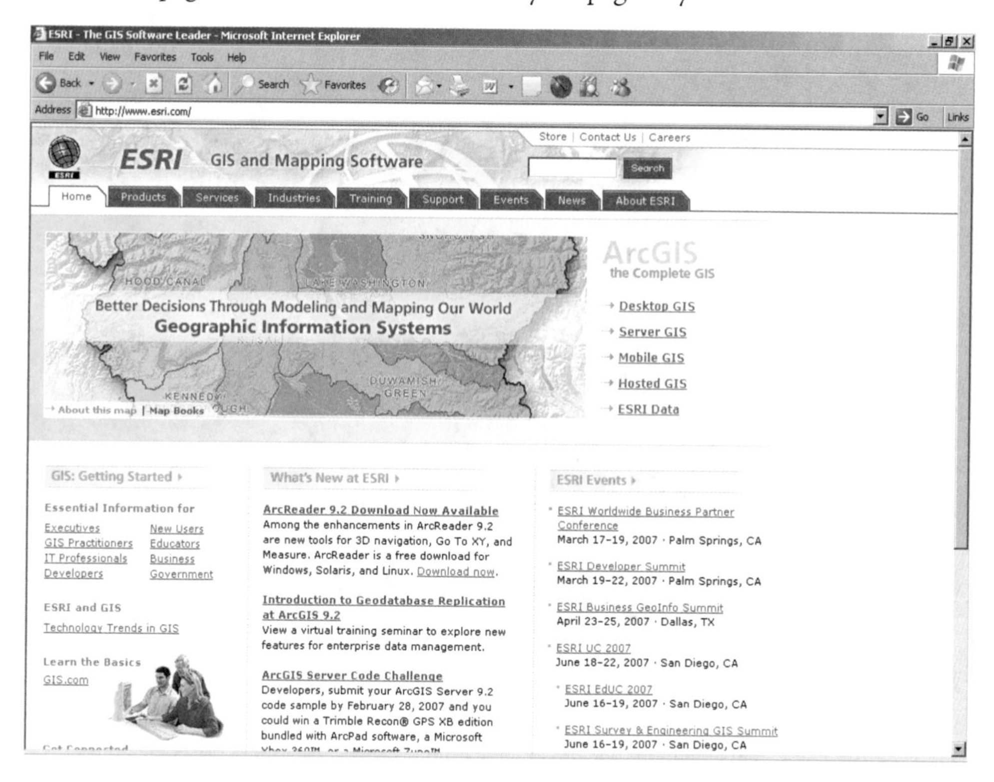

Navigate to free ESRI data

1 **Open your Web browser.**

2 **Go to *http://www.esri.com*.**

3 **From the Products tab, select ESRI Data.**

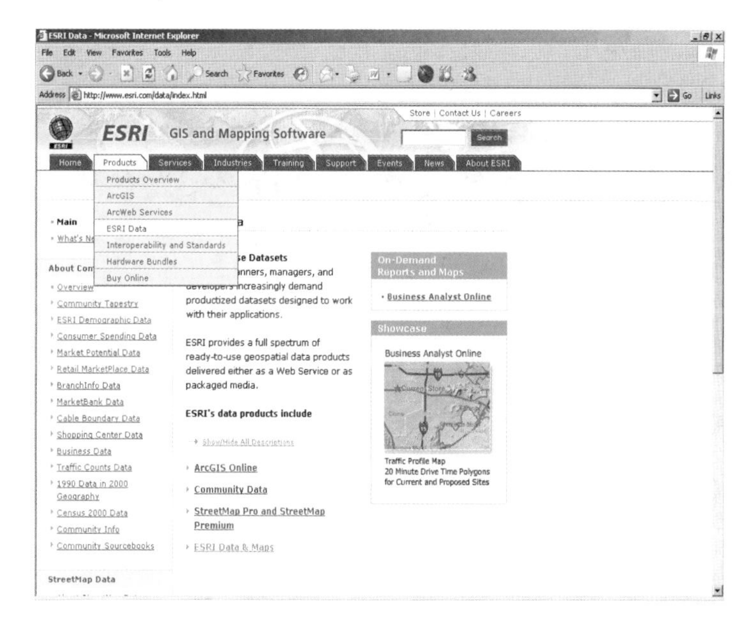

4 **From the left margin, under Resources, click the Geographic Data Portals link.**

5 **Click the ESRI World Basemap Data link under Featured Downloadable Data section.**

You should now be able to select the state, county, and urban area data layers needed for this tutorial. You may also access the Census 2000 TIGER/Line Data from this page that is required on page 147.

Download ESRI basemaps

1 Click the Preview and Download link on the left side of the page.

The Data Downloader window appears. Data can be downloaded by zooming to the area to download or by typing the area to download.

2 Use the Zoom In button  **to zoom to an area in Western Europe.**

3 Zoom in again to the area surrounding the United Kingdom and click Next.

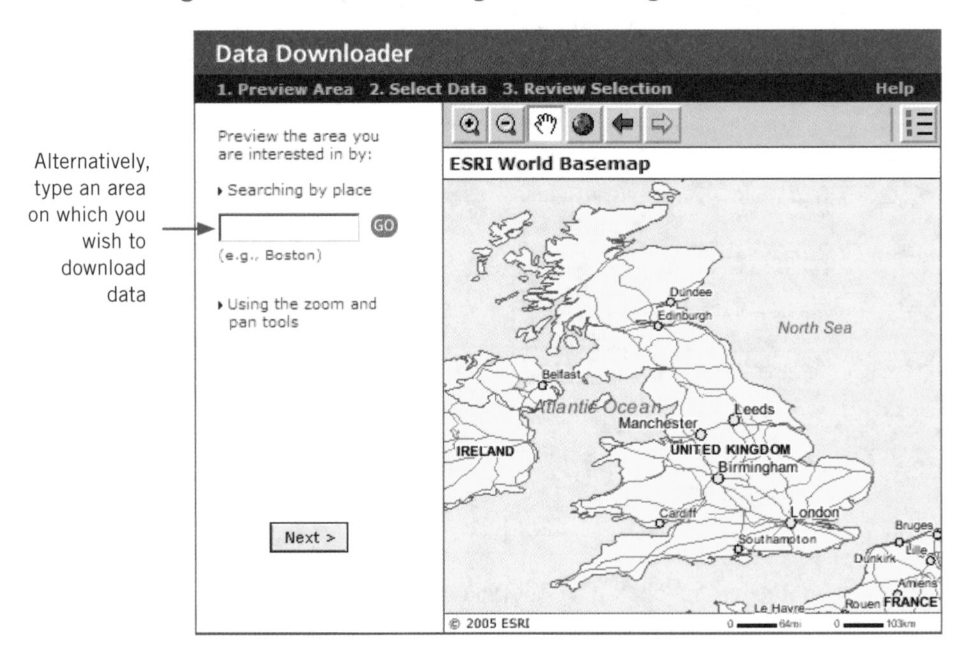

Three size selections will appear to help you choose the geographic range.

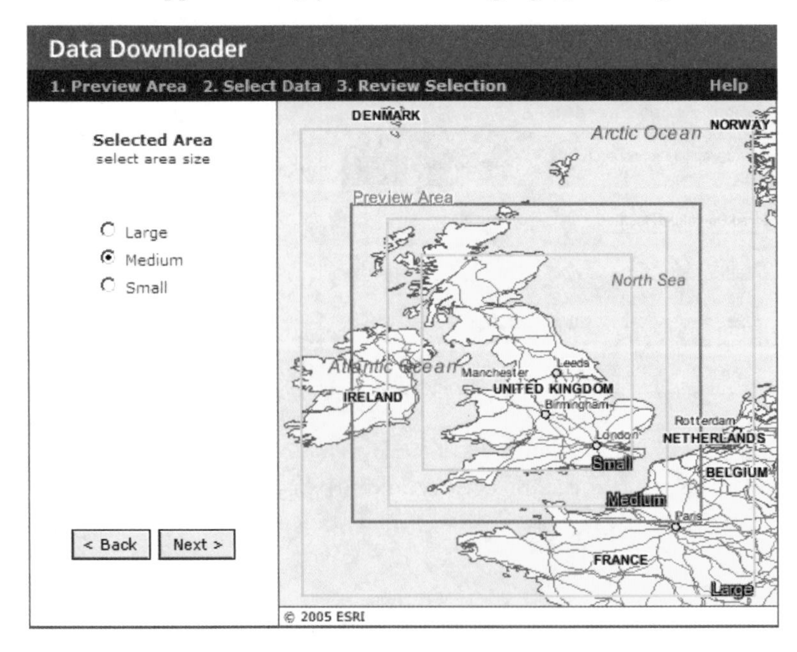

4 Click the radio button for Medium area and click Next.

5 From the Data Downloader window, if necessary select All Layers (all the other boxes will automatically be checked too).

6 Click Select.

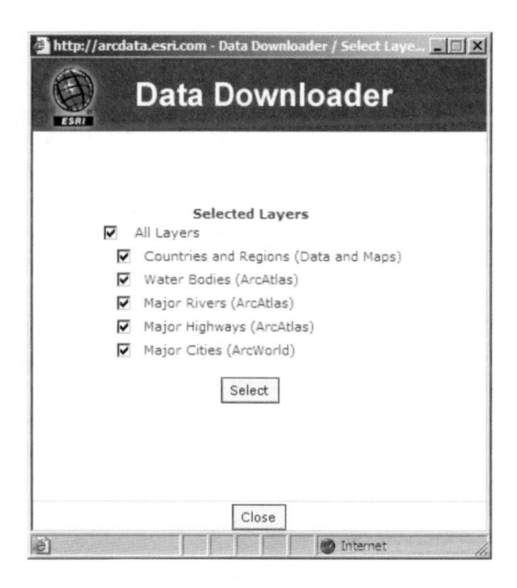

7 From the Data Downloader window, click Generate File.

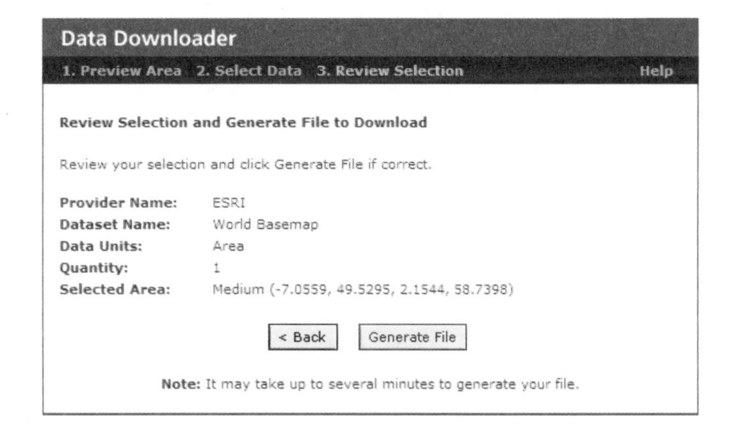

8 Click Download File when your data file is ready.

9 Click Save to save the data on your local machine.

Note that the generated file's name varies with each download.

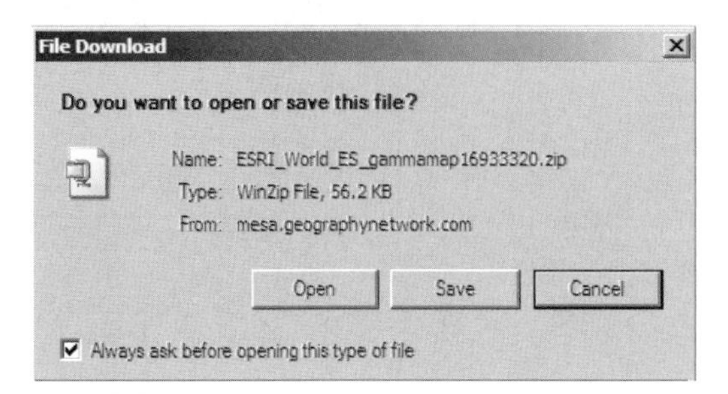

10 Create a folder called \Gistutorial\UnitedKingdom and save the downloaded file here.

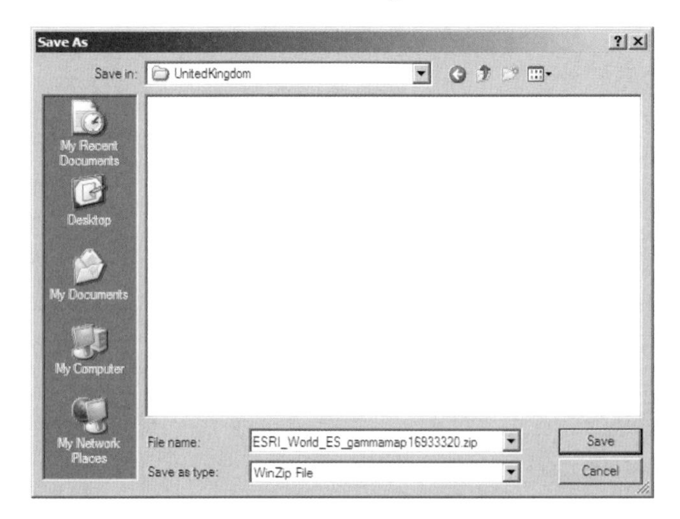

You will see a message indicating that the data has been downloaded. The files are stored in one .zip file that needs to be unzipped.

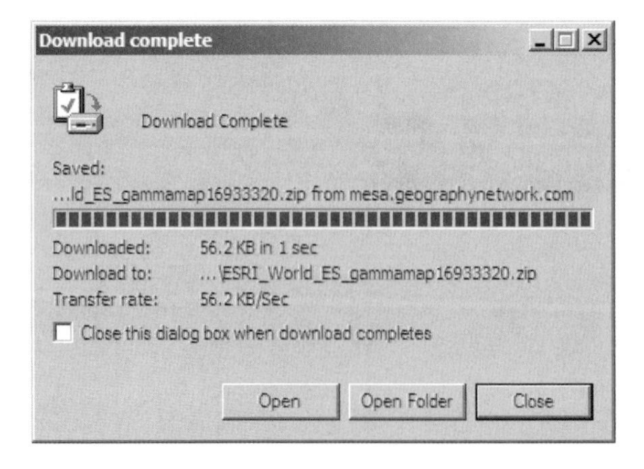

11 Close your Web browser, then unzip the files to the \Gistutorial\UnitedKingdom folder.

Add the data in ArcMap

1 Start ArcMap and open **\Gistutorial\Tutorial5-1.mxd**.

2 Click the Add Data button  to add all the shapefiles from the **\Gistutorial\UnitedKingdom** folder.

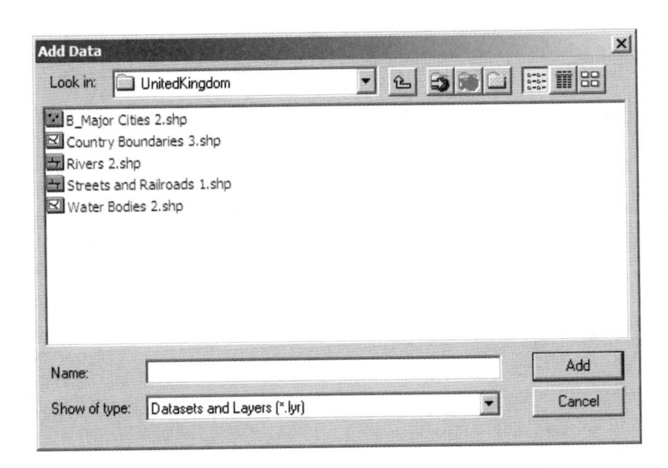

3 Change the symbology of the layers in the map to your liking.

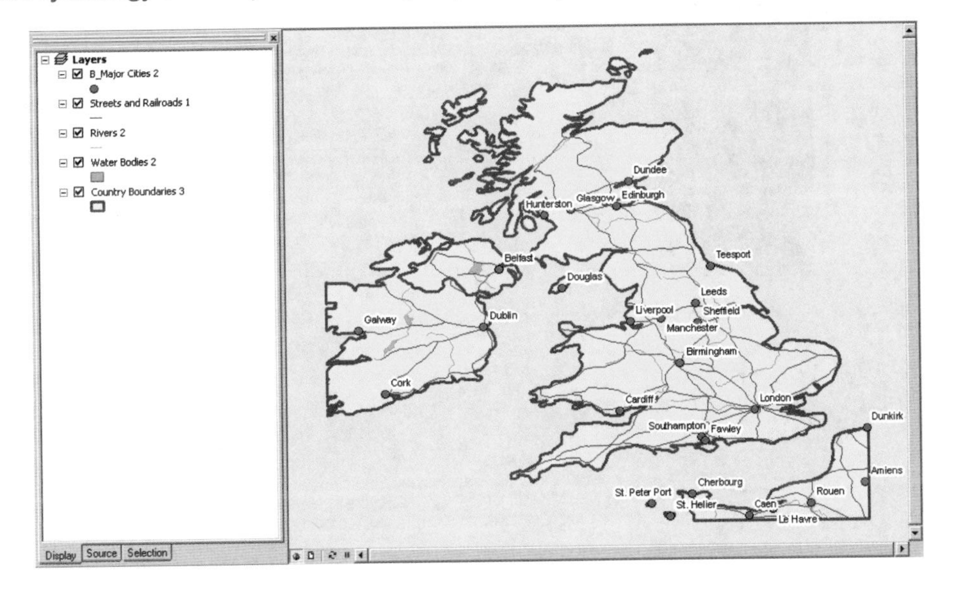

4 From the File menu, click Save, then close ArcMap.

YOUR TURN

Download another area of the world using other data from www.esri.com. Create a corresponding folder with the name of the region or country for storing the downloaded files. Also, create a map document, \Gistutorial\Tutorial5-1b.mxd.

Geography Network

The Geography Network is another valuable resource for free and inexpensive GIS data. The Geography Network is managed and maintained by ESRI.

1 **Open your Web browser.**

2 **Use the Web browser to go to *http://www.geographynetwork.com/freeresources.html*.**

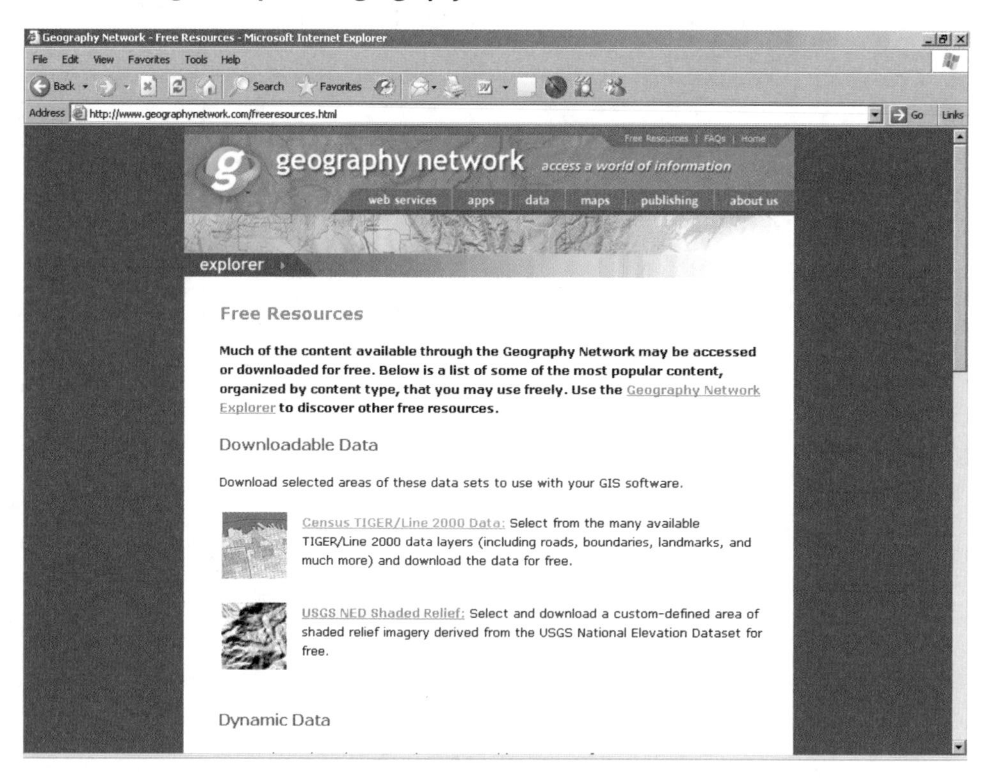

Download Census TIGER/Line Data

1 **From the Geography Network–Free Resources page, click the Census TIGER/Line 2000 Data link.**

2 **On the Census 2000 TIGER/Line Data page, click the Preview and Download link on the left.**

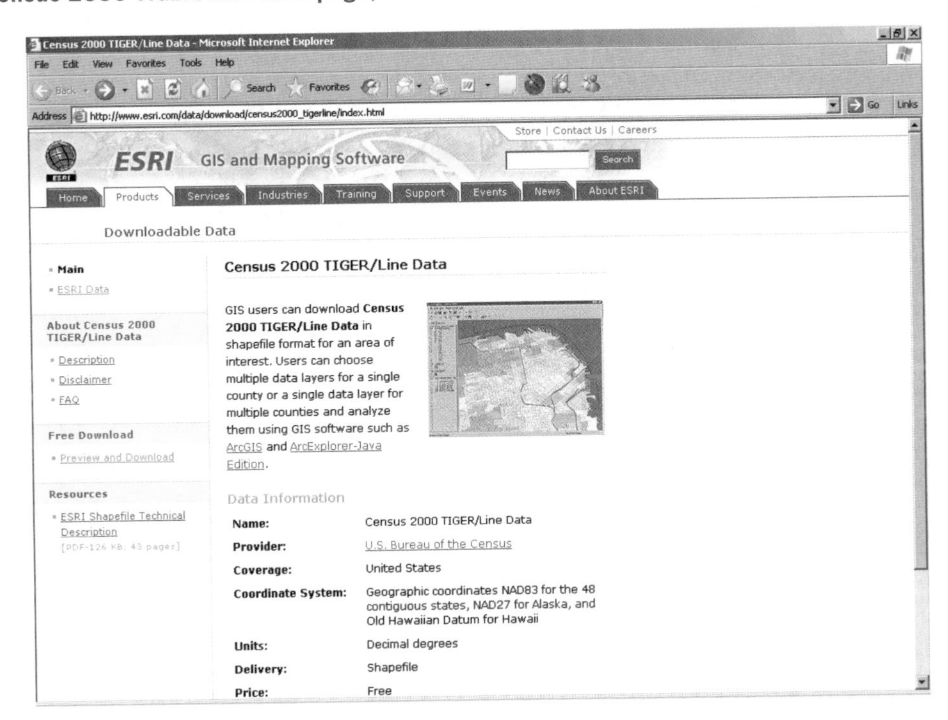

3 **Click the state of Illinois, or click it from the drop-down list and click Submit Selection.**

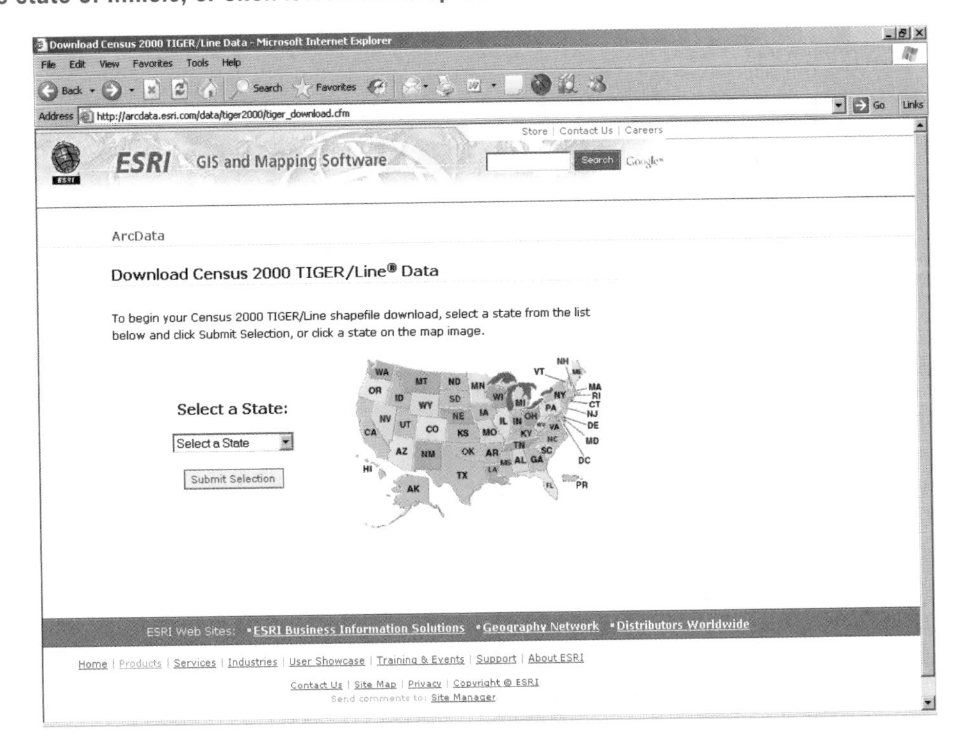

Download census tracts and data for a county

1 **From the Select by County drop-down list, select Cook, then click Submit Selection.**

2 **Check Census Tracts 2000 in the Available data layers list.**

Available data layers	File Size
☐ Block Groups 1990	726.9 KB
☐ Block Groups 2000	698.1 KB
☐ CMSA/MSA Polygons 2000	14.3 KB
☐ Census 2000 Collection Blocks	3.0 MB
☐ Census Blocks 1990	5.9 MB
☐ Census Blocks 2000	3.5 MB
☐ Census Tracts 1990	382.6 KB
☑ Census Tracts 2000	392.2 KB
☐ Congressional Districts - 106th	84.1 KB
☐ Congressional Districts - Current	84.0 KB
☐ County 1990	14.4 KB

Census tracts

3 **Scroll down to see the Available Statewide Layers and check Census Tract Demographics (SF1) from the list.**

Available Statewide Layers	File Size
Census Block Demographics (PL94)	35.6 MB
☐ Census Block Demographics (SF1)	12.6 MB
☐ Census Block Group Demographics (SF1)	738.2 KB
☐ Census County Demographics (PL94)	26.0 KB
☐ Census County Demographics (SF1)	11.4 KB
☐ Census Place Demographics (PL94)	218.4 KB
☐ Census Place Demographics (SF1)	109.4 KB
☐ Census State Demographics (PL94)	1.6 KB
☐ Census State Demographics (SF1)	665.0 bytes
☐ Census Tract Demographics (PL94)	535.7 KB
☑ Census Tract Demographics (SF1)	251.8 KB

Proceed to Download

SF1 data table

4 **Click Proceed to Download.**

You will see a message indicating that your data file is ready.

5 **Click Download File.**

6 **Click Save to save the file to your computer.**

7 **Create a folder called \Gistutorial\UnitedStates\Illinois and then a subfolder called \Gistutorial\UnitedStates\Illinois\CookCounty and save the file there. Close your Web browser.**

Extract files

1 **Use your zip program to extract the zipped files to your \Gistutorial\Illinois\CookCounty folder. (You will need to unzip a total of three files because the file you downloaded contains two zipped files.)**

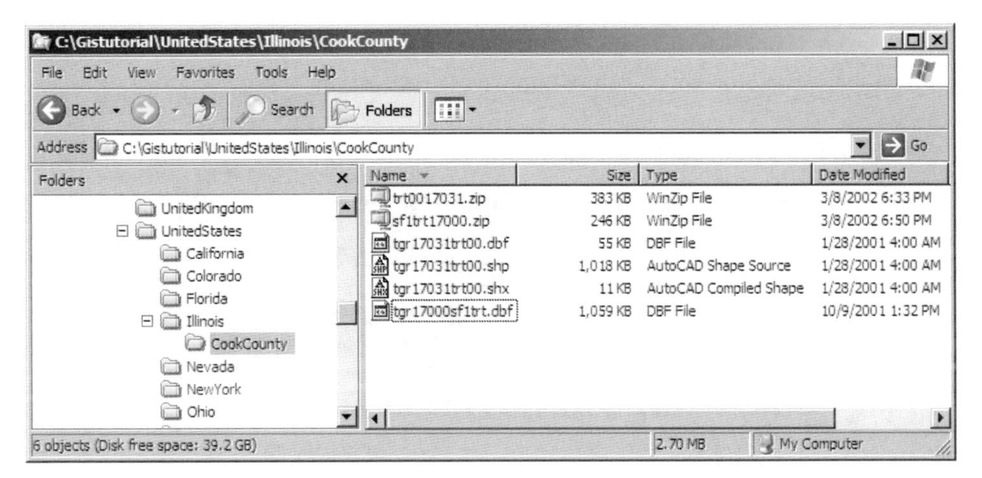

2 **Start ArcMap and add the files to a new, empty map.**

3 **Open the attribute tables for the table and the layer and explore their contents.**

Verify that both the shapefile and table include the matching primary key, STFID, making it easy to join the table to the map.

4 **Save map as \Gistutorial\Tutorial5-2.mxd.**

U.S. Census Bureau

The U.S. Census Bureau Web site is a wonderful resource for census data and TIGER basemaps. Census data can be downloaded from http://census.gov or from the Census Bureau's American FactFinder site *(factfinder.census.gov).*

1 **Start your Web browser and go to *http://www.census.gov.***

2 **From the U.S. Census home page, click the TIGER link in the Geography section.**

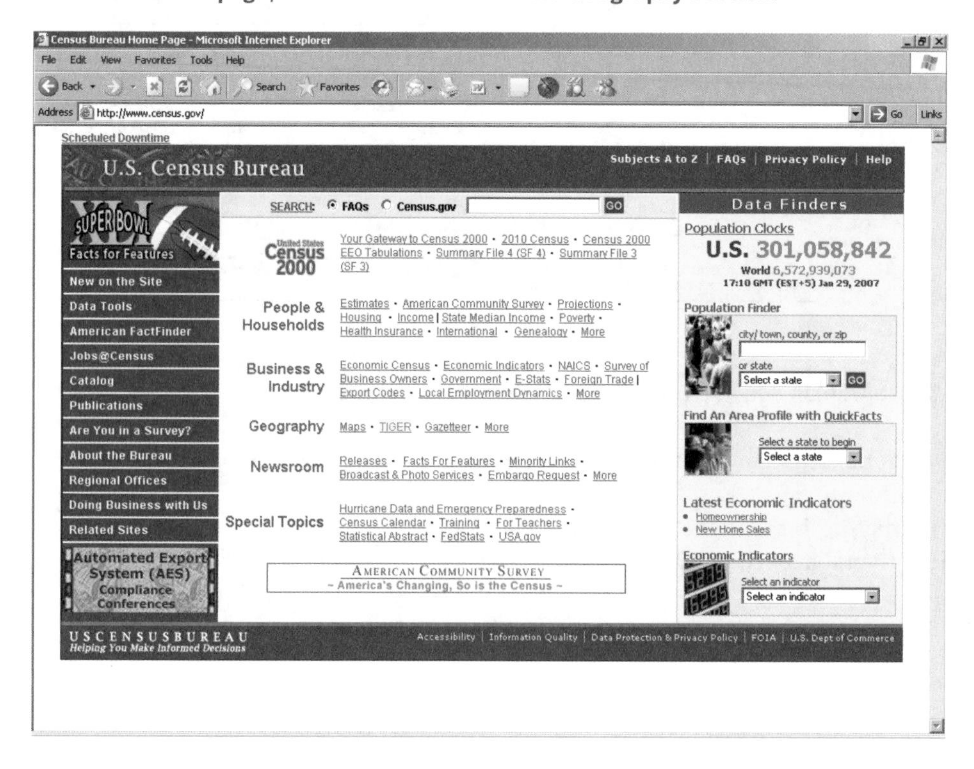

3 **Scroll down and click the Cartographic Boundary Files link.**

4 **Click Download Boundary Files and click the 2000 link for County and County Equivalent Areas.**

5 **Create a new folder, Gistutorial\UnitedStates\Delaware, then scroll down the Web page to the ArcView Shapefile section and download the Delaware zipped file co10_d00_shp.zip to your new folder.**

6 **Expand co10_d00_shp.zip in the Gistutorial\UnitedStates\Delaware folder, make a temporary map document to view the resulting map and attribute table, and close the map document without saving it.**

American FactFinder

The American FactFinder is an especially useful site for downloading census data tables to join to ESRI maps.

1 Use your Web browser to go to *factfinder.census.gov.*

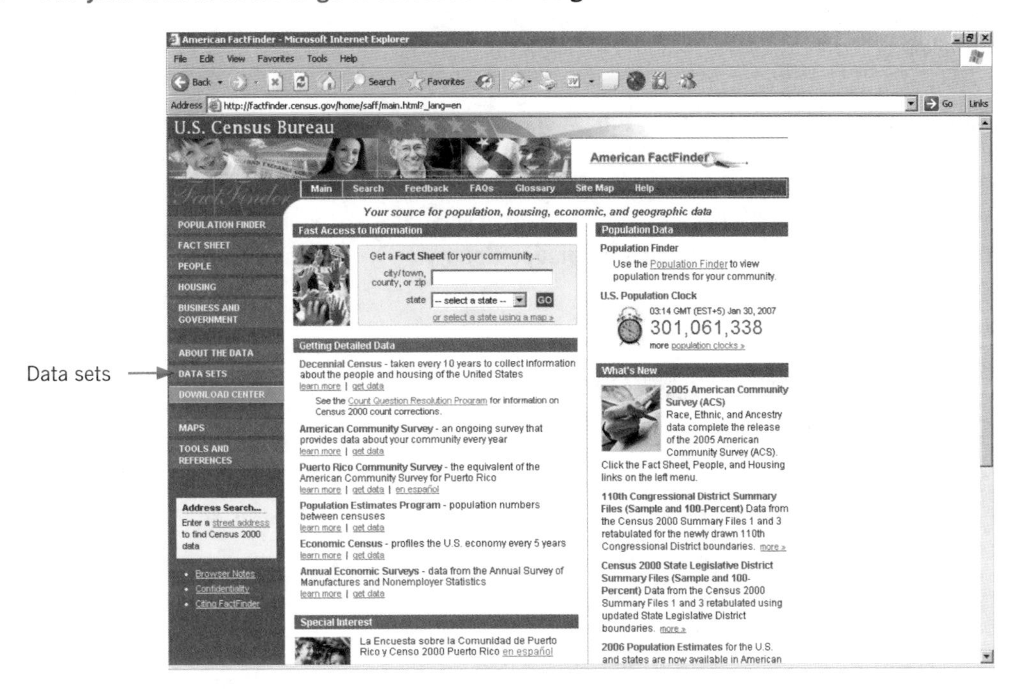

2 Click DOWNLOAD CENTER in the left panel.

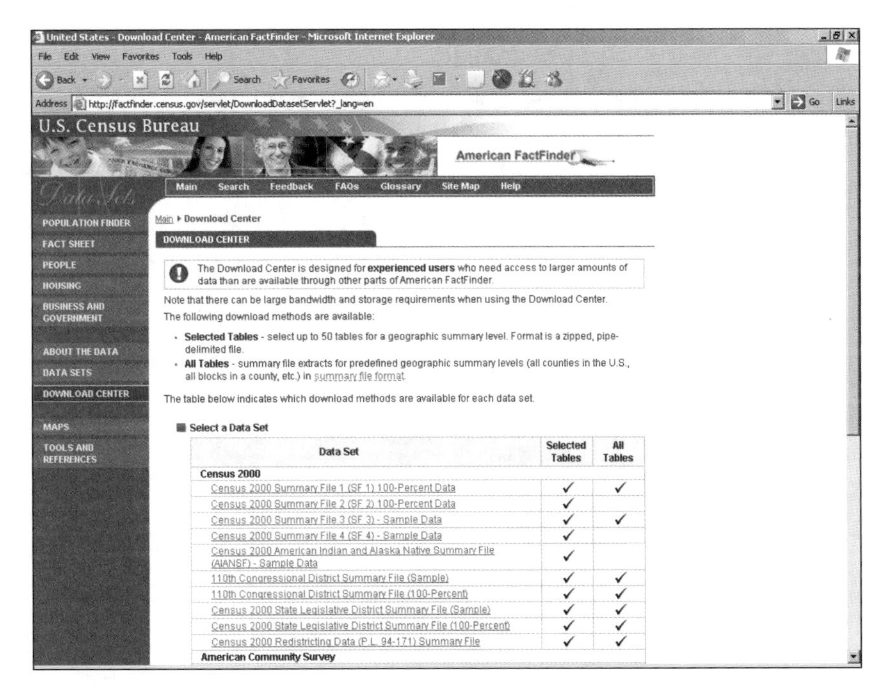

Create and download American FactFinder data tables

The census SF 1, SF 2, SF 3, and SF 4 data tables can be downloaded and joined to existing shapefiles or geodatabase feature class.

1 **From the DOWNLOAD CENTER page, click the Census 2000 Summary File 3 (SF 3) - Sample Data link.**

2 **In the Select a Geographic Summary Level panel, click All Census Tracts in a County (140).**

3 **Click the drop-down list for state and click Arizona.**

4 **Click the county drop-down list and click Maricopa County.**

5 **Click the Selected Detailed Tables radio button.**

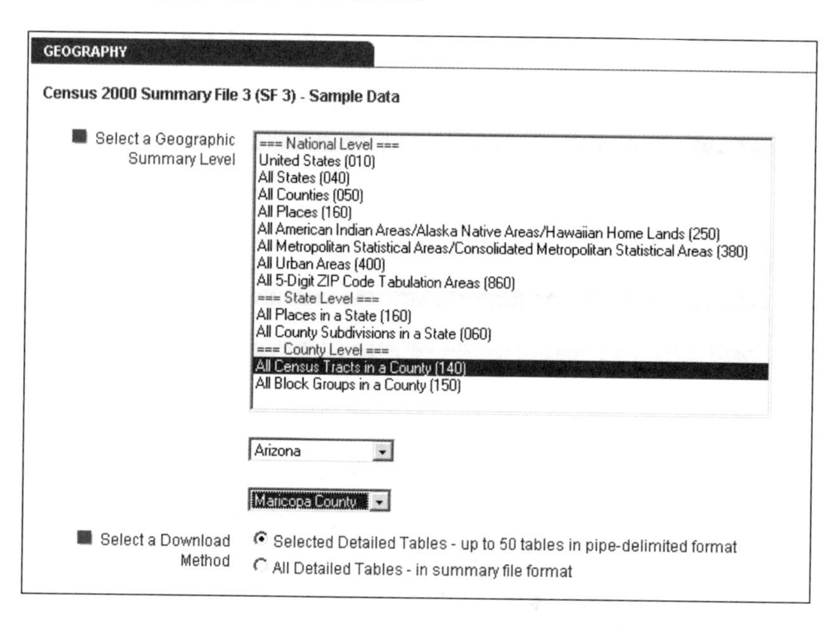

6 **Click Go.**

7 Scroll down, click P30 Means of Transportation to Work for Workers 16+ Years, and click Add.

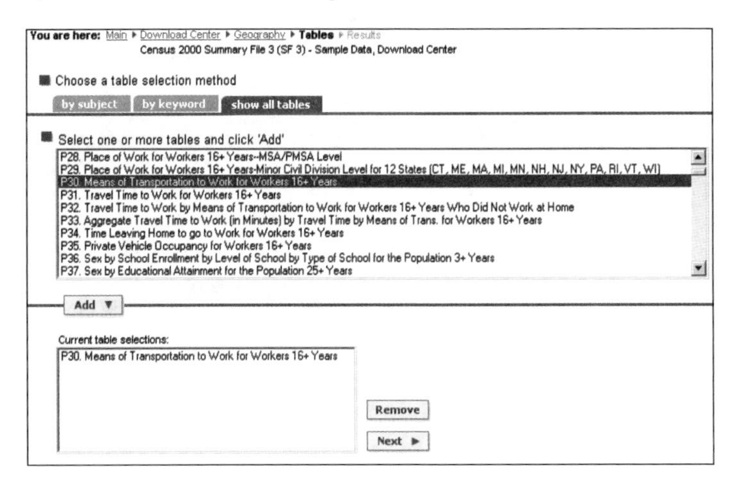

8 Click Next, Start Download.

9 Save the resulting zipped file to \Gistutorial\MaricopaCounty\, then extract the dc_dec_2000_sf3_data1.tx file to the same folder.

This file can be imported into Excel, processed, and eventually joined to the Maricopa County census tract map.

YOUR TURN

Download a second data table for Maricopa County tracts.

Vector spatial data formats

Coverages

A coverage is a type of GIS vector data format used to store geographic features. Coverages typically store one or more feature classes that are topologically and thematically related. For example, in a cadastral dataset it is common for a coverage to store the parcel boundaries as polygons and the lot lines as arcs (lines). You can add coverage data to ArcMap and use it for analysis and presentation, but coverage data cannot be edited with ArcMap.

When browsing data within Windows Explorer, coverages appear as folders containing several files. The graphic below shows four coverages within the EastLiberty folder; the contents of the Building coverage appear in the right-hand side of the window.

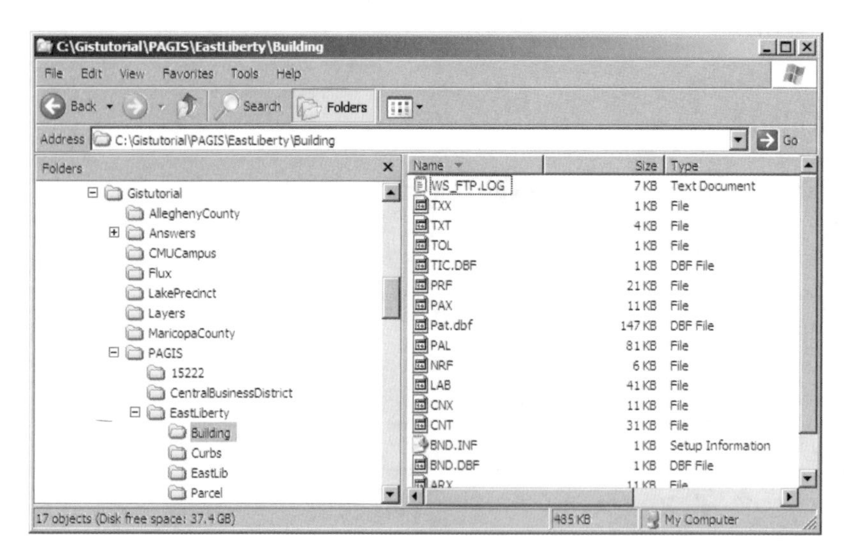

A better way to browse GIS data is with ArcCatalog. In fact, you should use only ArcCatalog to copy, move, or browse your GIS data because that's exactly its purpose—to help you manage your GIS data. The graphic shown (at right) contains an image of the same four coverages shown in the screen capture above, but this time they are being viewed with ArcCatalog. Notice that each coverage name appears beside a yellow icon, and that the feature classes inside the selected Parcel coverage appear in Contents tab of the Preview pane.

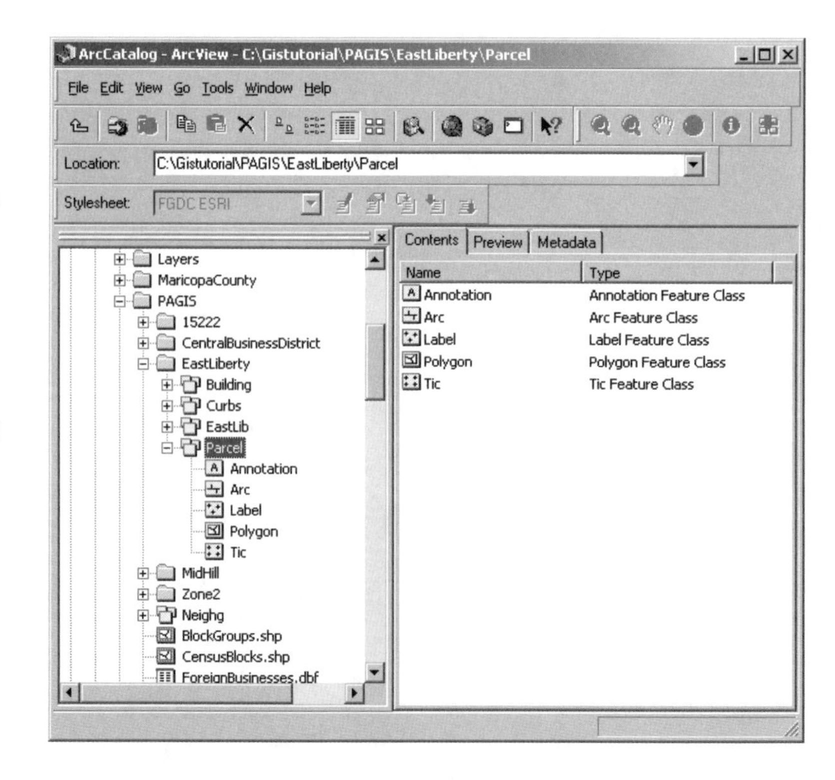

Add a coverage to ArcMap

1 Start ArcMap with a new, empty map, then click the Add Data button.

2 Navigate to **\Gistutorial\PAGIS\EastLiberty** and double-click Building.

3 Click the Polygon feature class, then click Add.

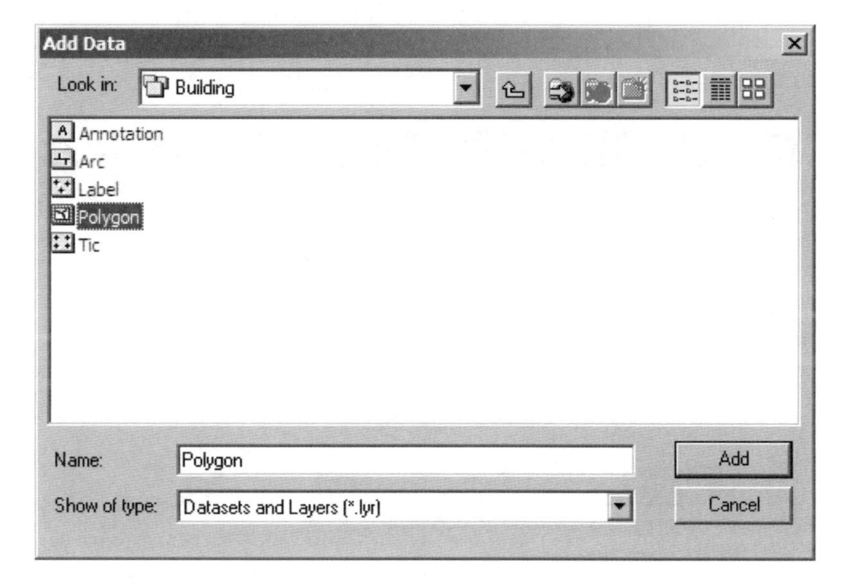

Convert a coverage to a shapefile

Because coverages cannot be edited in ArcMap, they must be exported as shapefiles or geodatabase feature classes before you can edit their attribute tables or geometry.

1 In the table of contents, right-click the Building Polygon layer and click Data, Export Data.

2 Browse to the **\Gistutorial\PAGIS\EastLiberty** folder and name the output shapefile **Building.shp**.

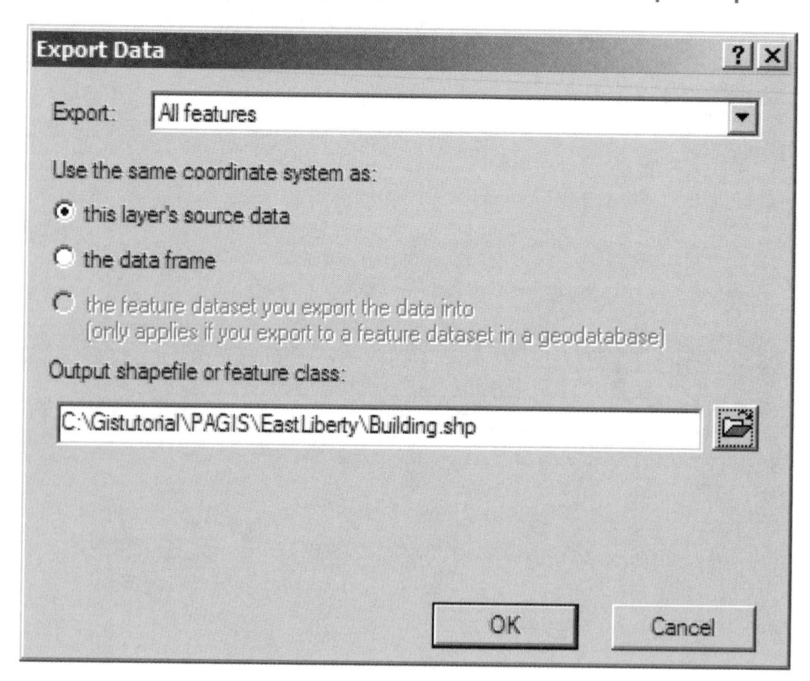

3 Click OK, then click Yes to add the exported data to the map as a layer.

The polygons in the new shapefile, Building.shp, can now be edited. You will learn about editing shapefiles in tutorial 6.

4 Save the map as **\Gistutorial\Tutorial5-3.mxd**.

Shapefiles

Shapefiles are another type of GIS vector data format used to store geographic features. Unlike coverages, which are made up of a collection of feature classes, a shapefile can represent only one feature class. For example, if you wanted to store parcel boundaries and lot lines using shapefiles, you would have to create two shapefiles—one to store the parcel boundaries as polygons and another to store the lots as lines. Shapefiles can be created, edited, and analyzed using ArcView, ArcEditor, or ArcInfo.

Shapefiles consist of at least three files: a .dbf file, a .shx file, and a .shp file. Each of these files is prefixed with the shapefile's name. The .dbf file stores the attributes, the .shp file stores the geometry of the features, and the .shx file stores an index of the spatial geometry. The graphic below shows a shapefile named Parks as it appears in Windows Explorer.

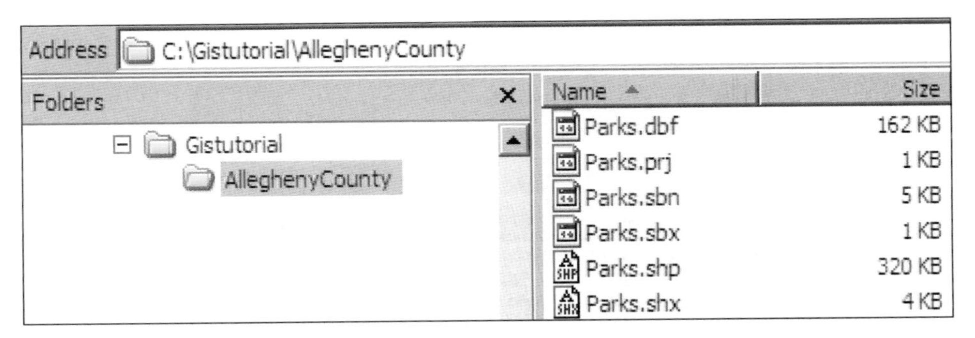

In the graphic below, the Parks shapefile (along with several other shapefiles and datasets) is shown in ArcCatalog. In ArcCatalog, a shapefile appears as one file and is represented with a green icon that indicates whether it is a point, line, or polygon vector map layer.

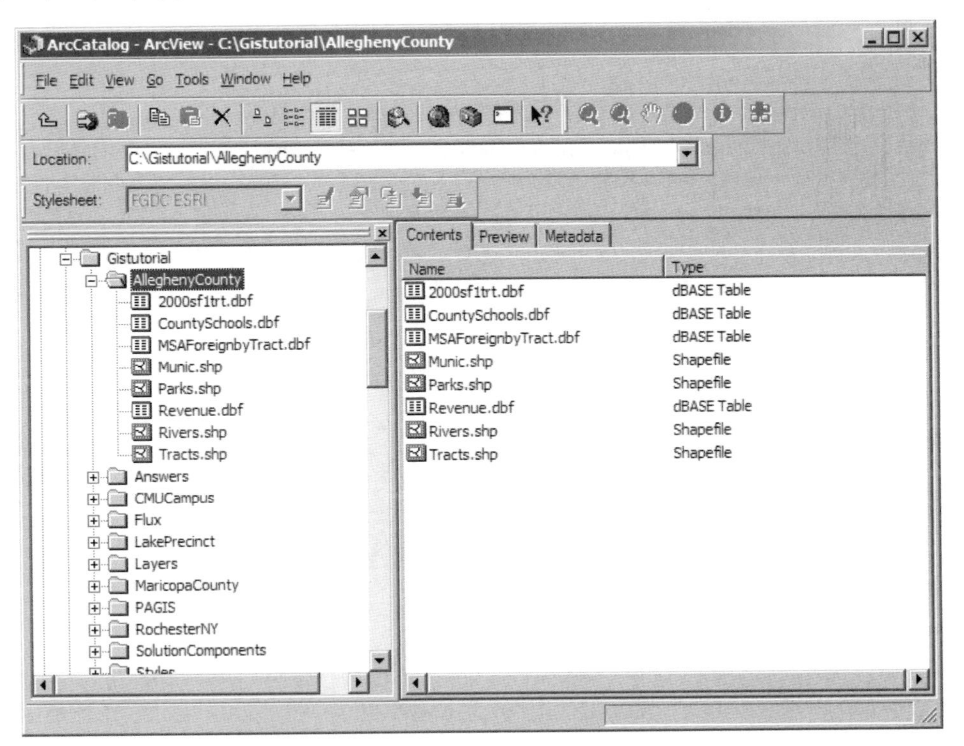

Add a shapefile in ArcMap

1 In ArcMap, create a new, empty map, then click the Add Data button.

2 Browse to the **\Gistutorial\AlleghenyCounty** folder, hold the Ctrl key and click **Munic.shp**, **Parks.shp**, and **Rivers.shp**, then click Add.

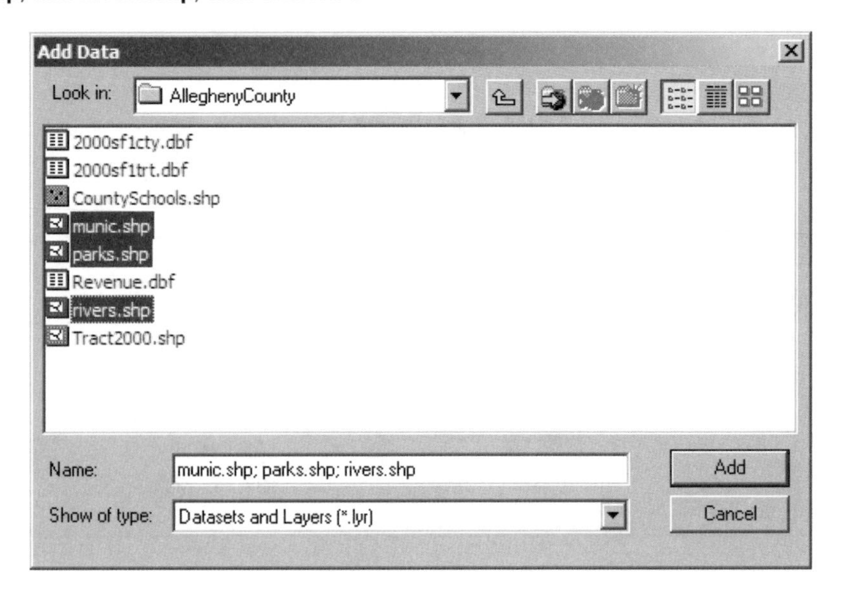

3 Save the map as **Gistutorial\Tutorial5-4mxd.**

Interchange (.e00) files

Many local planning agencies and GIS consultants provide their coverage data as interchange (.e00) files. Interchange files are the preferred method for sharing coverage data because all of the folders and files associated with a coverage are placed into one .e00. With ArcView, an interchange file can be converted back to a coverage by using the Import from Interchange File tool. In this example, you will import ZIP Codes for Arizona that were originally downloaded from the U.S. Census Web site as an .e00 file.

Start ArcCatalog

1 Start ArcCatalog, then click View, Toolbars, ArcView 8x Tools.

2 Click Conversion Tools, Import from Interchange File.

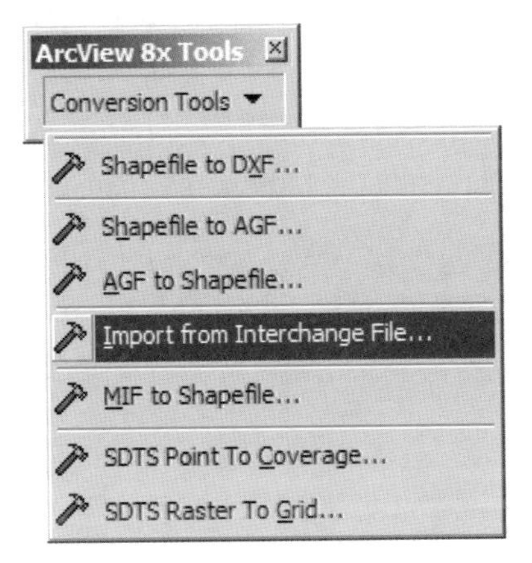

3 For the Input file, browse to **\Gistutorial\UnitedStates\Arizona**, then double-click **Zipcodes.e00**.

4 Name the output dataset **\Gistutorial\UnitedStates\Arizona\Zipcodes** and click OK.

This creates a coverage that can then be added to ArcMap.

5 After the process is complete, use ArcCatalog to verify that the Zipcodes coverage exists and then close ArcCatalog.

Annotation layers

Labels are one option for placing text on an ArcGIS map. Labels are positioned on the map by the software based on a set of labeling properties defined in the Labels tab of the Layer Properties dialog. If you want full control over where labels are placed, you can convert them to annotation.

1 **In a new, empty map in ArcMap, click the Add Data button.**

2 **Browse to \Gistutorial\PAGIS\CentralBusinessDistrict, click the CBDStreets shapefile, and click Add.**

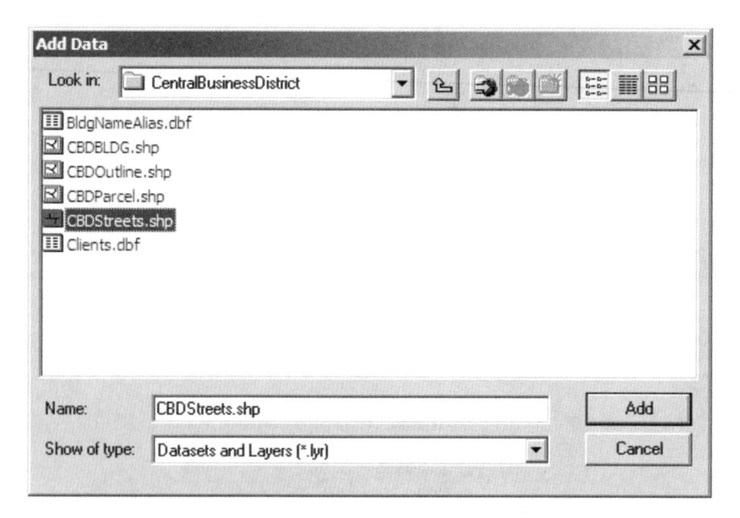

3 **Click the Fixed Zoom In button four times.**

4 **In the table of contents, right-click the CBDStreets layer and click Label Features.**

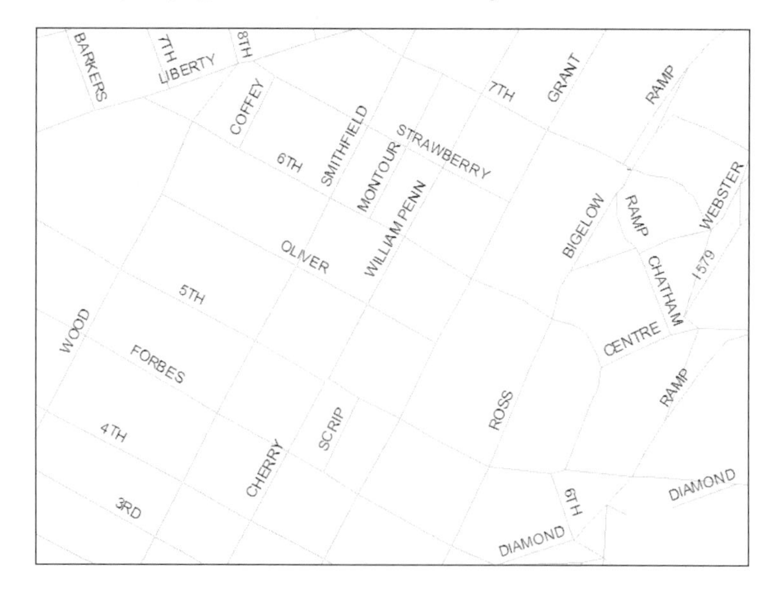

5 **In the table of contents, right-click the CBDStreets layer and click Convert Labels to Annotation.**

6 **In the Store Annotation frame, click the In the map radio button.**

7 **Click in the Annotation Group field and type StreetName.**

8 **Click Convert.**

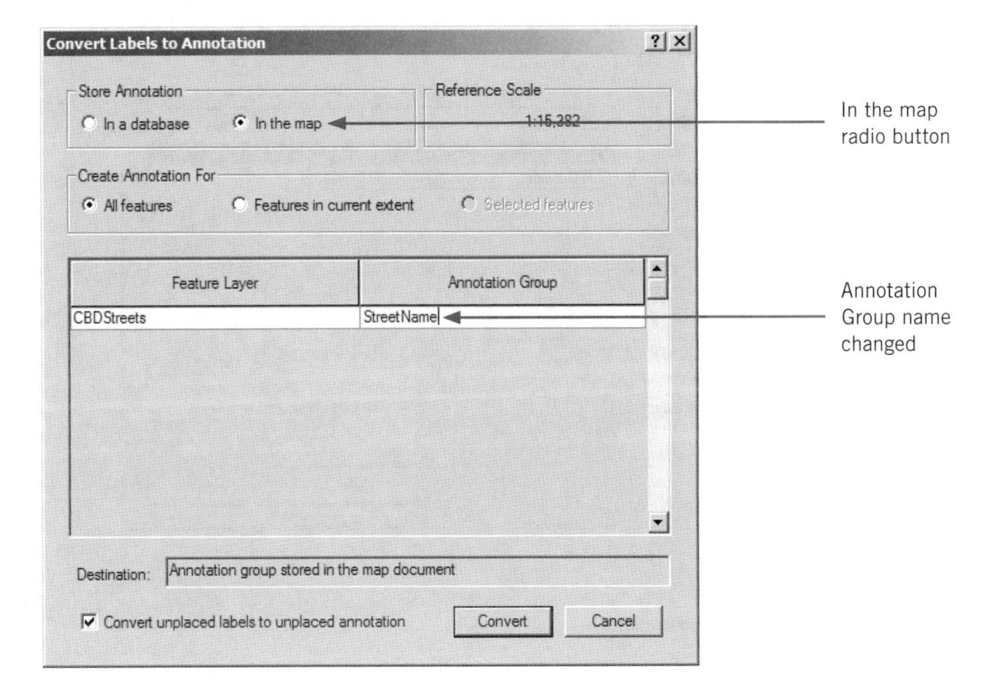

Use annotation layer

After labels have been converted to geodatabase annotation, the annotation class is automatically added to the map. Annotation groups are listed in the Data Frame Properties dialog box on the Annotation Groups tab.

1 **In the table of contents, right-click the Layers data frame and click Properties.**

2 **Click the Annotation Groups tab.**

Annotation labels can be toggled on or off, and their properties can be changed in this dialog box.

3 **Click the StreetName group and click the Properties button.**

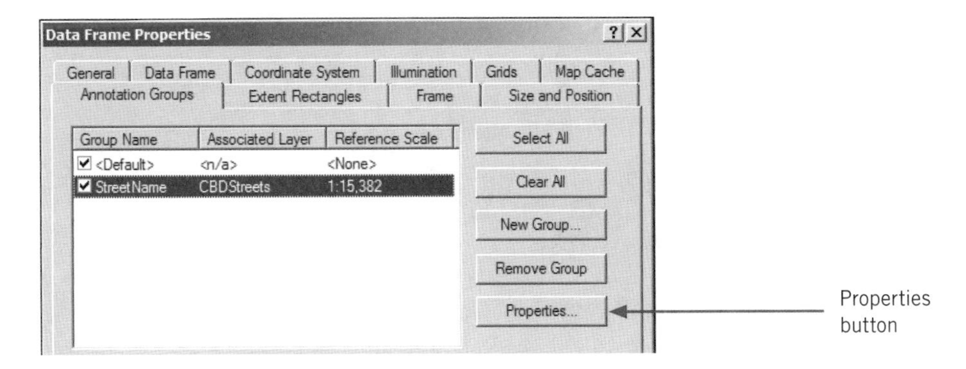

4 Click the Associated Layer drop-down list and click <None>.

This allows you to turn the annotation street names on and off independently from the street centerlines.

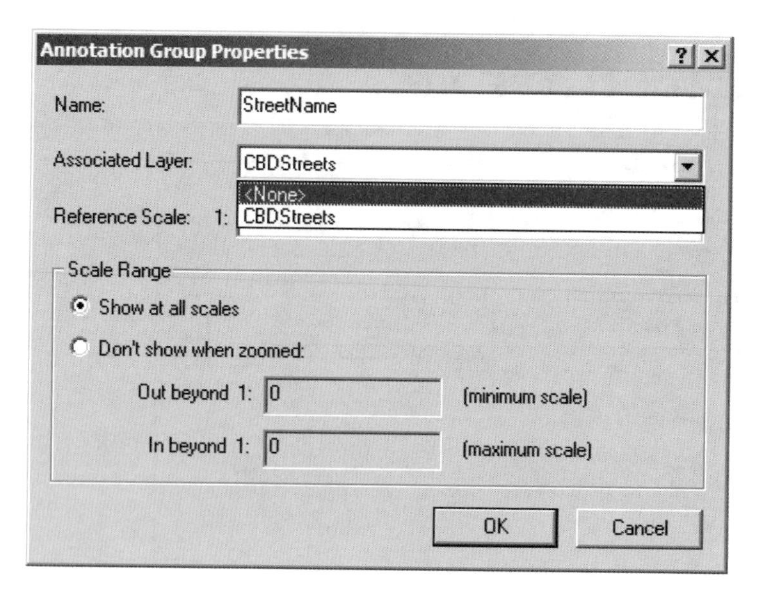

5 Click OK, then click OK again.

6 If necessary, close the Overflow Annotation dialog box.

7 In the table of contents, uncheck the CBDStreets layer to turn it off.

The labels remain on. This is because they are independent from the street layer.

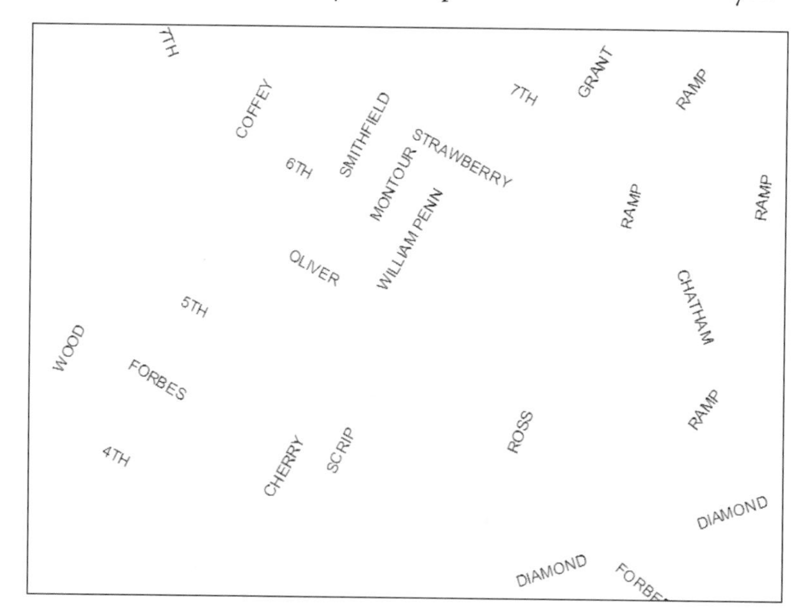

8 Save the map as \Gistutorial\Tutorial5-5.mxd.

CAD files

If an organization has existing CAD (computer-aided design) files, they can be added to ArcMap in their native format. Within ArcMap, CAD files can be viewed but not edited.

ArcMap can add CAD files in one of two formats: as native AutoCAD (.dwg) or as Drawing Exchange Files (.dxf) that can be created from most CAD applications.

When viewed in ArcCatalog, a CAD dataset will appear in the catalog tree with a light blue icon. By selecting the CAD dataset in the catalog tree, you can view its contents in the Contents pane.

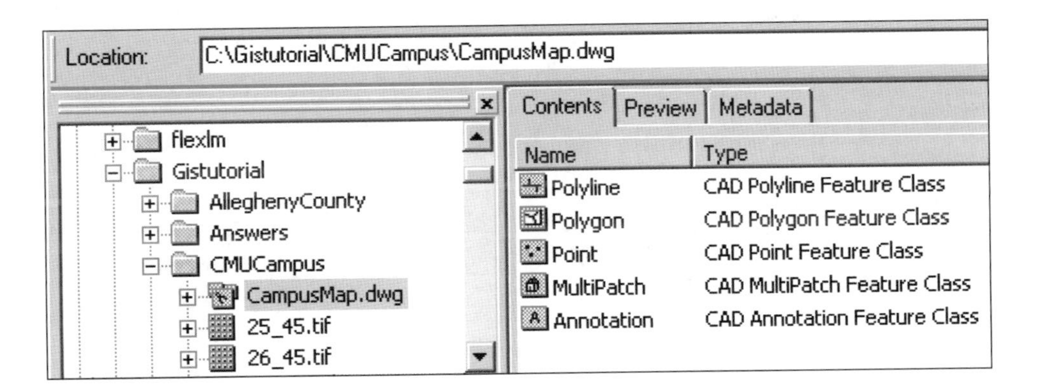

Add a CAD file as a layer for display

1 **Create a new, empty map and click the Add Data button.**

2 **Browse to \Gistutorial\CMUCampus and click the CampusMap.dwg icon.**

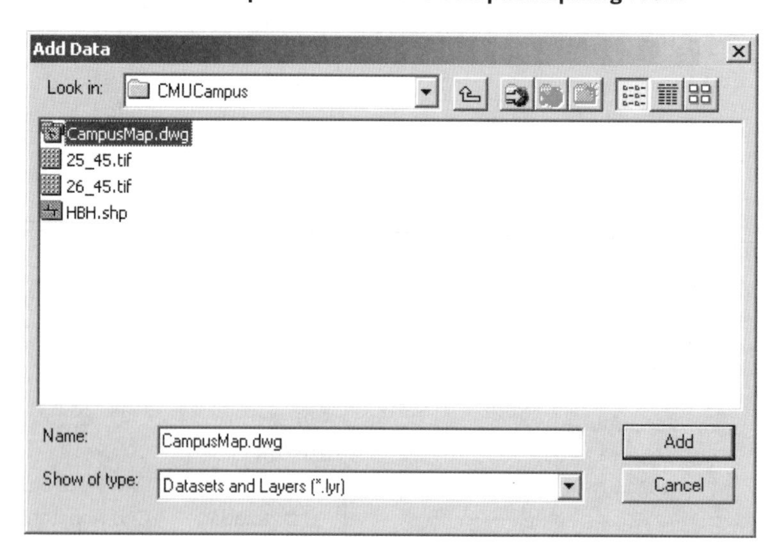

3 **Click the Add button.**

The following map of the Carnegie Mellon University campus appears in ArcMap. It contains many feature types, including lines, polygons, and text. Features on this map cannot be selected or edited. It is for display only. Next, you will add the other CAD file, which allows you to add specific feature types such as lines only.

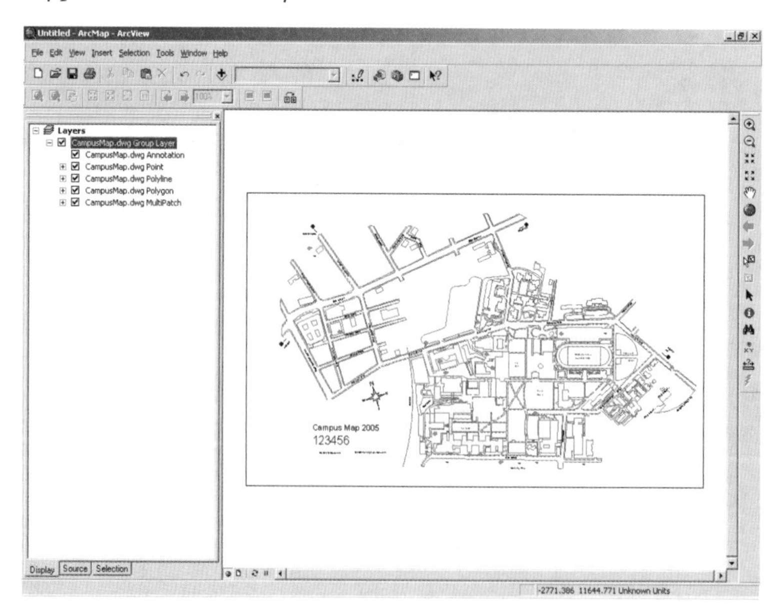

4 **Save the map as \Gistutorial\Tutorial5-6.mxd and leave ArcMap open.**

Add a CAD file as a layer for edit and analysis

1 In the table of contents, remove the **CampusMap.dwg** layer.

2 Click the Add Data button.

3 Navigate to **\Gistutorial\CMUCampus**, double-click the **CampusMap.dwg** with the blue icon.

4 Click the Polyline feature class, then click the Add button.

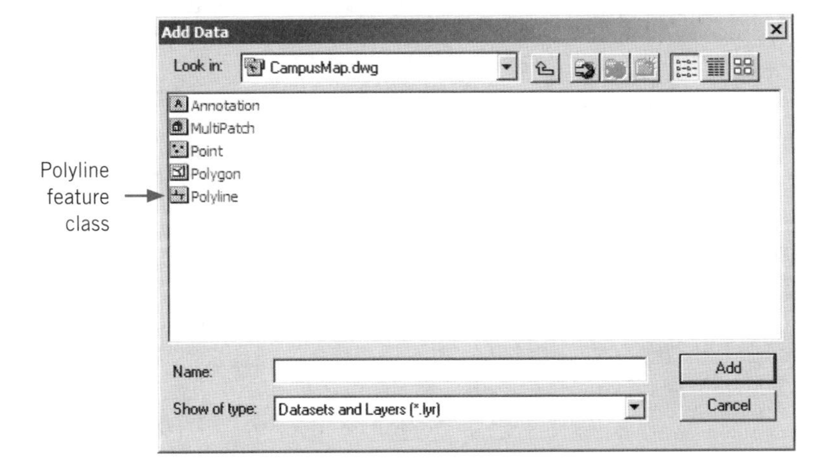

ArcMap adds the polyline feature class from the CAD dataset. The features in this layer can be selected and their properties manipulated, but they cannot be edited.

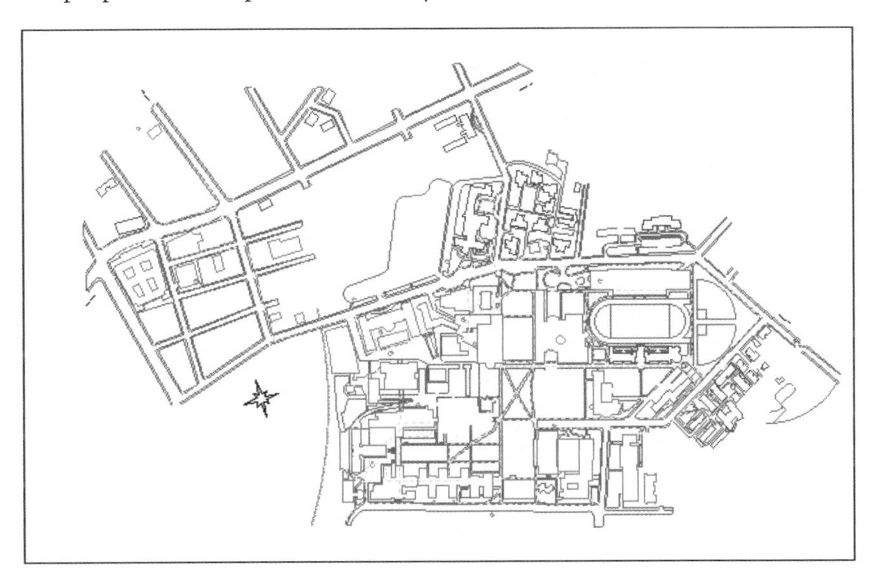

5 In the table of contents, double-click the **CampusMap.dwg** Polyline layer.

6 Click the Drawing Layers tab. Notice that you can turn Layers on and off.

7 Click the Symbology tab, Categories, Unique Values.

8 From the Value Field drop-down list, click Layer, then click the Add All Values button.

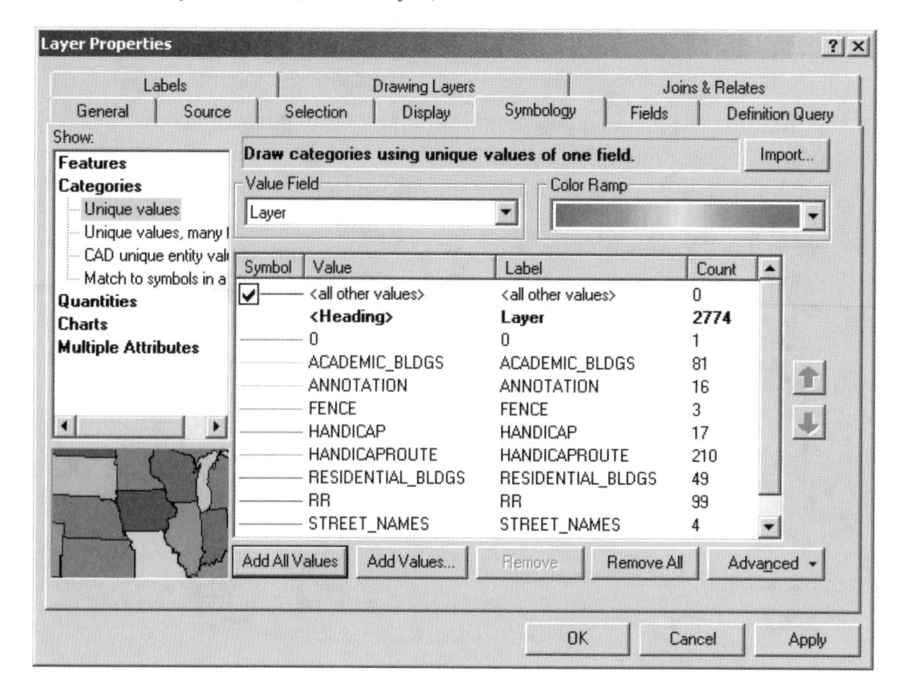

9 Change line colors as desired.

10 Click OK.

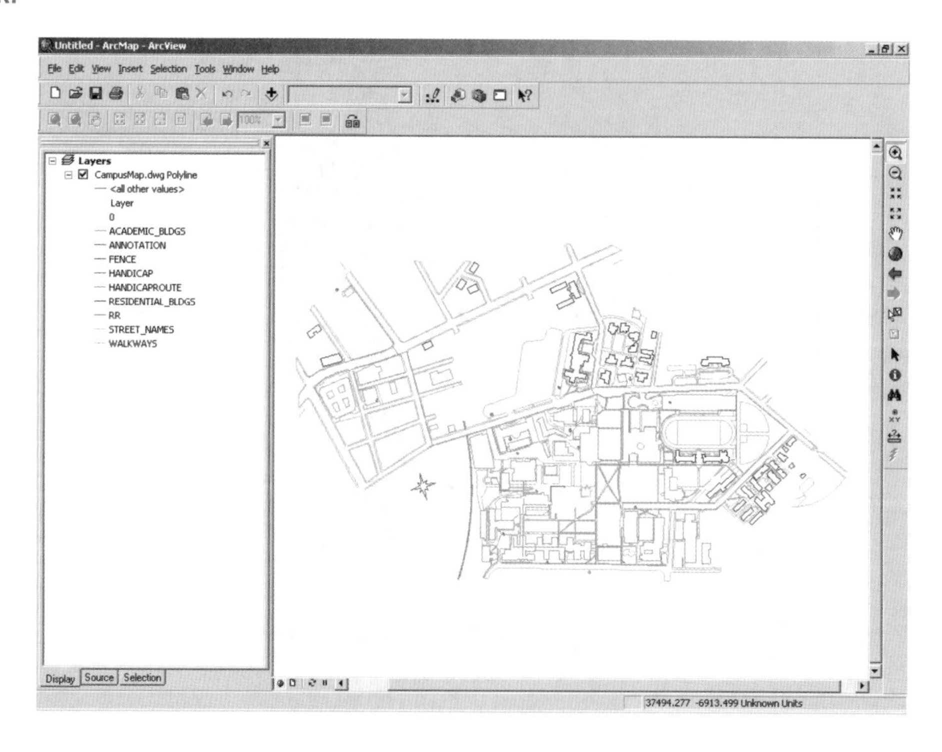

11 Save the map as **\Gistutorial\Tutorial5-7.mxd**.

Export shapefiles to CAD

Sometimes you may need to deliver shapefile data to a person working with CAD software. Using the Export tools in ArcCatalog, you can export shapefiles to .dxf format, which can then be opened by most commercial CAD applications.

Copy the shapefile to the root directory

1 Start **ArcCatalog**.

Note: CAD conversions cannot be done in a folder with more than eight characters in its name.

2 In the catalog tree, browse to **C:\Gistutorial\PAGIS\CentralBusinessDistrict**, right-click **CBDStreets.shp**, click Copy, then browse to **C:** (or a folder such as **C:\TEMP**), right-click, and click Paste.

3 Browse to **C:** (or the folder where you copied your shapefile), right-click **CBDStreets.shp**, then click Export, Shapefile to DXF.

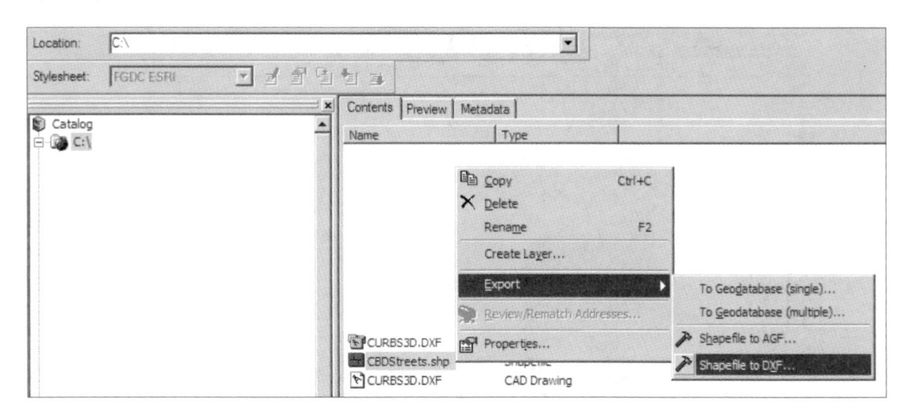

4 From the Decimals drop-down list, click 6 (the number of decimal places used for the coordinates in the output), then type **C:\CBDStreets.dxf** as the Output file name and click OK.

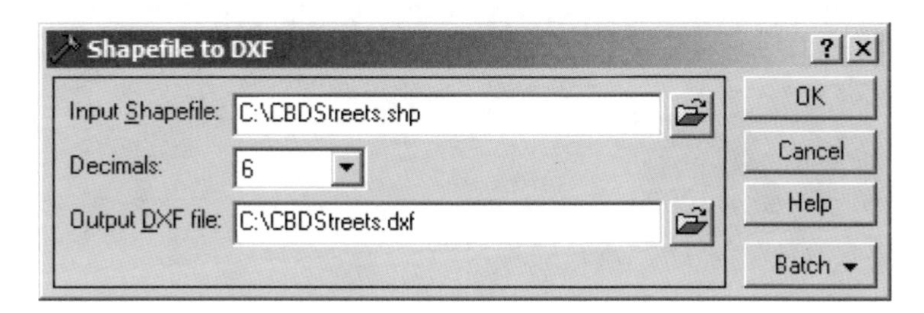

5 Close ArcCatalog.

XY event files

In addition to data sources such as shapefiles, tabular point data that contains x,y coordinates can also be added to a map. For example, you can collect x,y coordinate data using a global positioning system (GPS) device.

To add a table of x,y coordinates to your map, the table must contain two fields—one for the x-coordinates and one for the y-coordinates. The values in the fields may be from any coordinate system such as latitude and longitude.

1 **Create a new, empty map in ArcMap, then click the Add Data button.**

2 **Browse to \Gistutorial\UnitedStates\California, and while holding down the Ctrl key, click CACounties.shp and Earthquakes.dbf, then click Add.**

3 **Right-click Earthquakes and click Open.**

4 **Scroll to the right to see the X and Y fields.**

The x- and y-coordinates here are latitude and longitude values for earthquake locations.

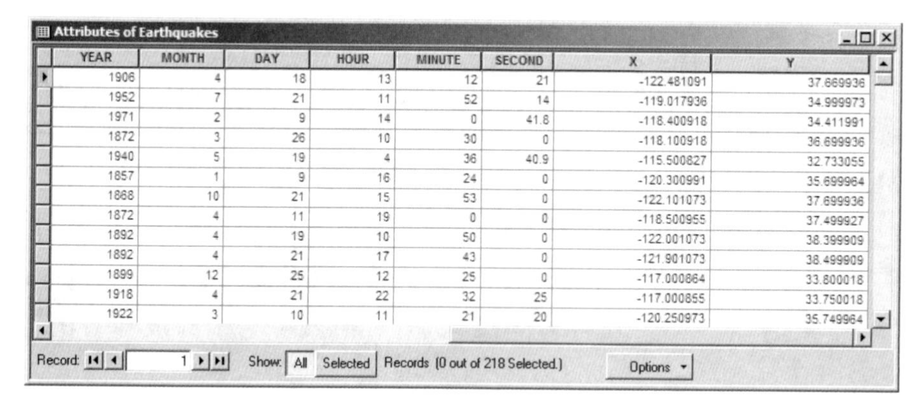

YEAR	MONTH	DAY	HOUR	MINUTE	SECOND	X	Y
1906	4	18	13	12	21	-122.481091	37.669936
1952	7	21	11	52	14	-119.017936	34.999973
1971	2	9	14	0	41.8	-118.400918	34.411991
1872	3	26	10	30	0	-118.100918	36.699936
1940	5	19	4	36	40.9	-115.500827	32.733055
1857	1	9	16	24	0	-120.300991	35.699964
1868	10	21	15	53	0	-122.101073	37.699936
1872	4	11	19	0	0	-118.500955	37.499927
1892	4	19	10	50	0	-122.001073	38.399909
1892	4	21	17	43	0	-121.901073	38.499909
1899	12	25	12	25	0	-117.000864	33.800018
1918	4	21	22	32	25	-117.000855	33.750018
1922	3	10	11	21	20	-120.250973	35.749964

5 **Close the Attributes of Earthquakes table.**

6 **Right-click Earthquakes, click Display XY Data, and choose X and Y for the X and Y fields.**

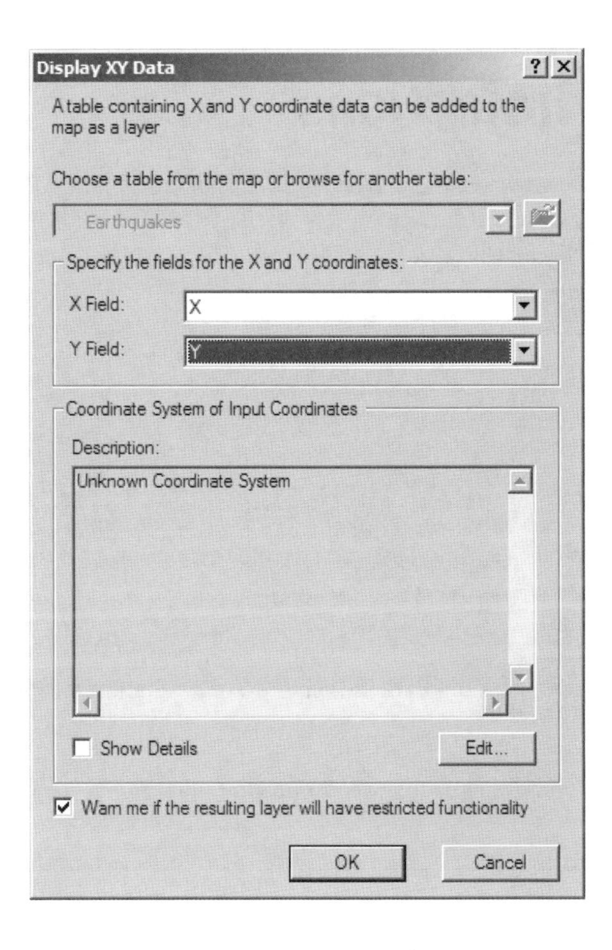

7 Click OK.

This will add point locations for the earthquakes in California as an Event layer. Note that by right-clicking Earthquakes Events and clicking Data, Export Data, you can create a corresponding shapefile.

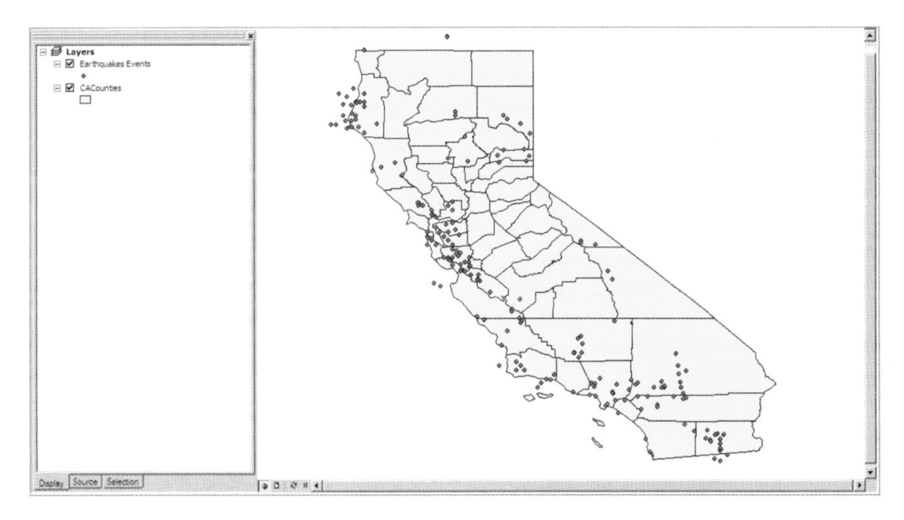

8 Save the map as C:\Gistutorial\Tutorial5-8.mxd.

World and U.S. projections

There are two types of coordinate systems—*geographic and projected.* Geographic coordinate systems use latitude and longitude coordinates on the surface of a sphere while projected coordinate systems use a mathematical conversion to transform latitude and longitude coordinates to a flat surface. Maps obtained from the U.S. Census are typically in geographic coordinates; U.S. maps obtained from local planning departments are typically in projected coordinate systems (e.g., State Plane coordinates).

World projections

Because the earth is spherical and maps are flat, GIS applications require that a mathematical formulation be applied to the earth to represent it on a flat surface. This is called a map projection, and it causes some distortions of distance, area, shape, or direction. ArcMap has over a hundred projections from which you may choose. Typically though, only a few projections are suitable for your data.

1. **In a new, empty map in ArcMap, click the Add Data button, browse to \Gistutorial\World, and add the Country and Ocean shapefiles.**

2. **Change the color of the layers to your liking and move the Country layer to the top of the table of contents.**

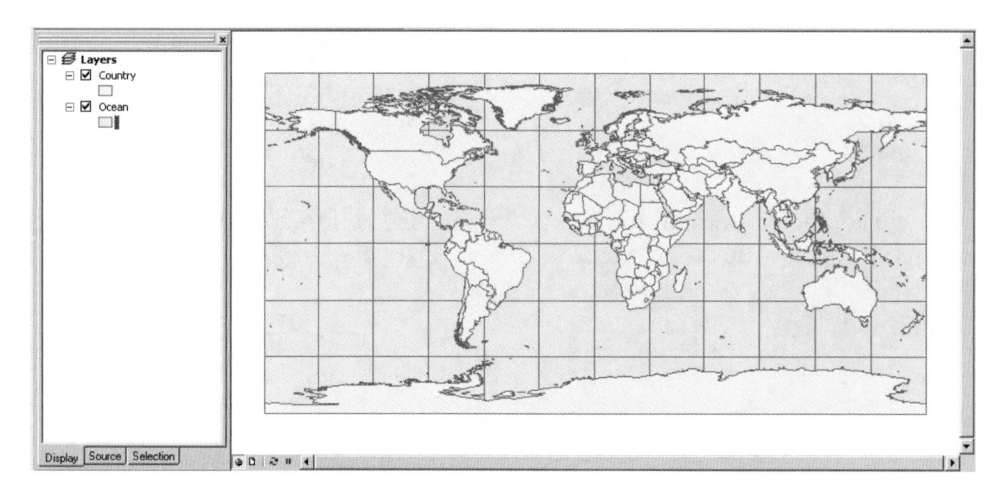

Change the map's projection to Mercator

1 **Right-click the Layers data frame and click Properties.**

2 **Click the Coordinate System tab.**

3 **In the Select a coordinate system box, expand the Predefined folder.**

Select a coordinate system:
Favorites
Predefined
Geographic Coordinate Systems
Projected Coordinate Systems
Layers
<custom>
GCS_WGS_1984

Plus sign clicked

4 **Expand the Projected Coordinate Systems folder, then expand the World folder.**

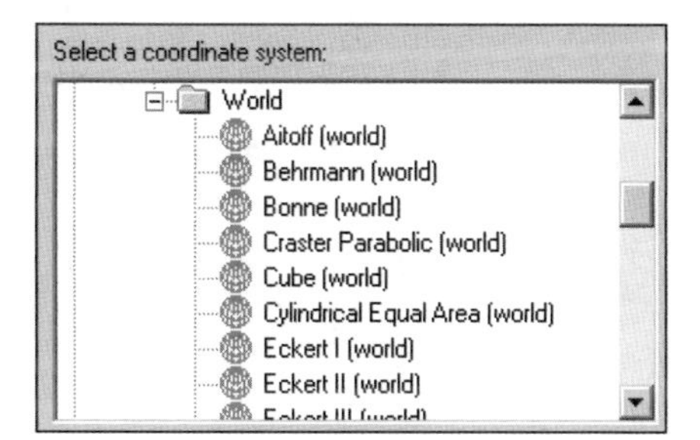

5 **Scroll down the coordinate systems, click Mercator, OK, and the Full Extent button.**

The primary purpose of the Mercator projection is for navigation because straight lines on the projection are accurate compass bearings. This projection greatly distorts areas near the polar regions and distorts distances along all lines except the equator. The Mercator projection is a conformal projection, and thus it preserves small shapes and angular relationships.

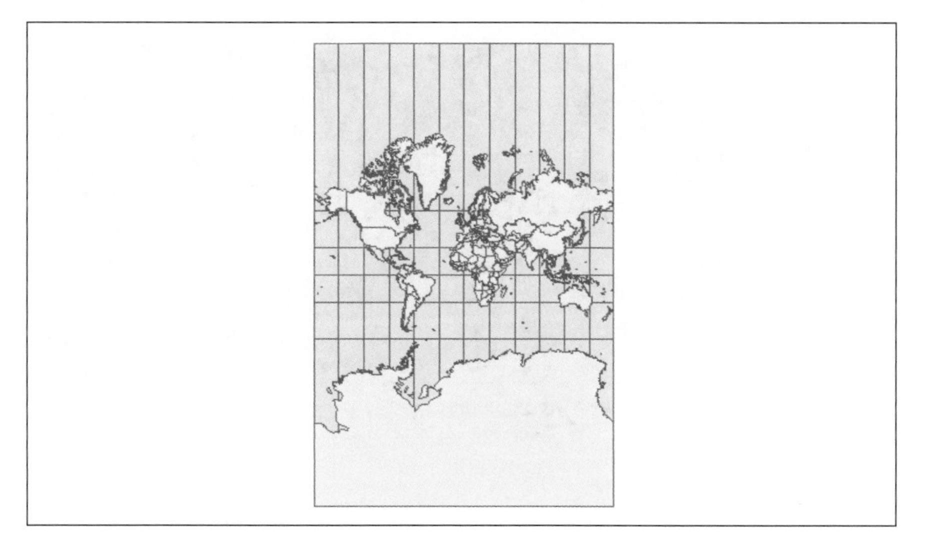

6 **Save the map as \Gistutorial\MercatorProjection.mxd.**

Change the map's projection to Hammer-Aitoff and then to Robinson

1 **Repeat the previous six steps, but this time select the Hammer-Aitoff projection in the fifth step, then save your map as Hammer-Aitoff.mxd.**

This projection is nearly the opposite of the Mercator. The Hammer-Aitoff is good for use on a world map, being an equal-area projection and preserving area. It distorts direction and distance.

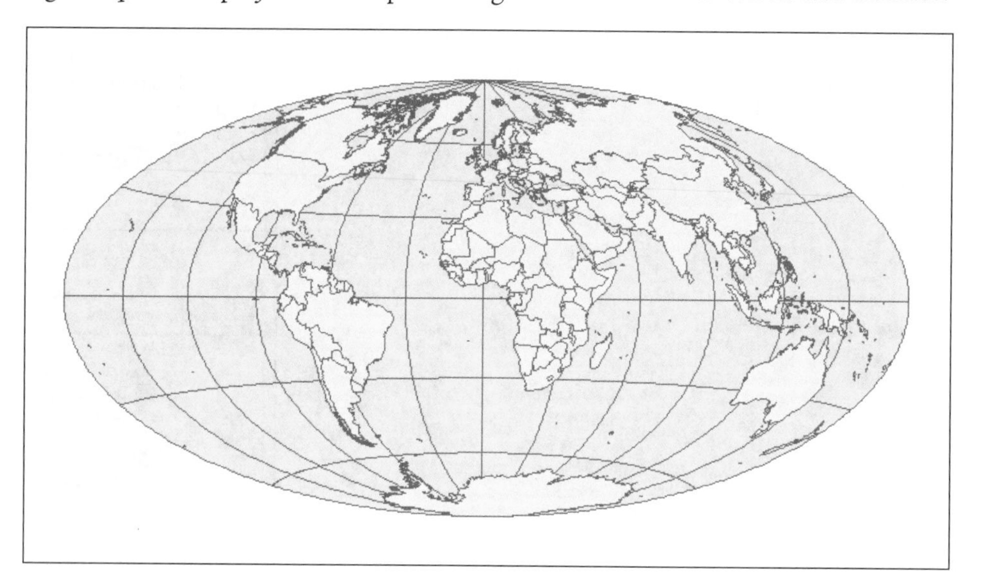

2 **This time apply the Robinson projection to your map, then save the map as Robinson.mxd.**

This projection minimizes distortions of many kinds, striking a balance between conformal and equal-area projections.

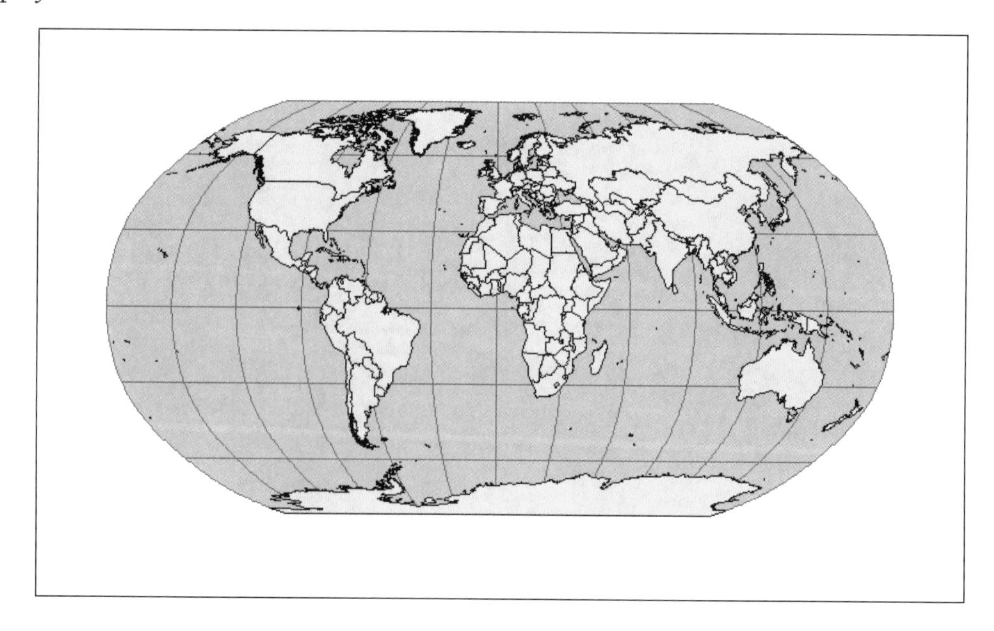

Projections of the USA

In the following sections, you will get experience with projections commonly used for maps of the United States.

Use the Albers equal area conic projection

1 In a new, empty map, click the Add Data button, browse to **\Gistutorial\UnitedStates**, and add **States.shp** to the map.

2 Right-click the Layers data frame, then click Properties and the General tab and change the data frame name to **Albers Equal Area (Coterminous US)**.

3 Click the Coordinate System tab, Predefined, Projected Coordinate Systems, Continental, North America, North America Albers Equal Area Conic.

4 Click OK.

5 Save the map as **\Gistutorial\USProjections.mxd**.

YOUR TURN

Experiment by applying a few other projections to the U.S. map such as North America equidistant. As long as you stay in the correct group—Continental, North America—all of the projections look similar.

State Plane coordinate system

The State Plane coordinate system is a series of projections. It is a coordinate system that divides the fifty U.S. states, Puerto Rico, and the U.S. Virgin Islands into more than 120 numbered sections, referred to as zones, each with its own finely tuned projection. Used mostly by local government agencies such as counties, municipalities, and cities, the State Plane coordinate system was designed for large-scale mapping in the United States. It was developed in the 1930s by the U.S. Coast and Geodetic Survey to provide a common reference system for surveyors and mapmakers. The first step in using a State Plane projection is to look up the correct zone for your area.

Look up a State Plane Zone for a county

1 **Start your Web browser, go to *www.ngs.noaa.gov/TOOLS/spc.html*, and click the Find Zone link.**

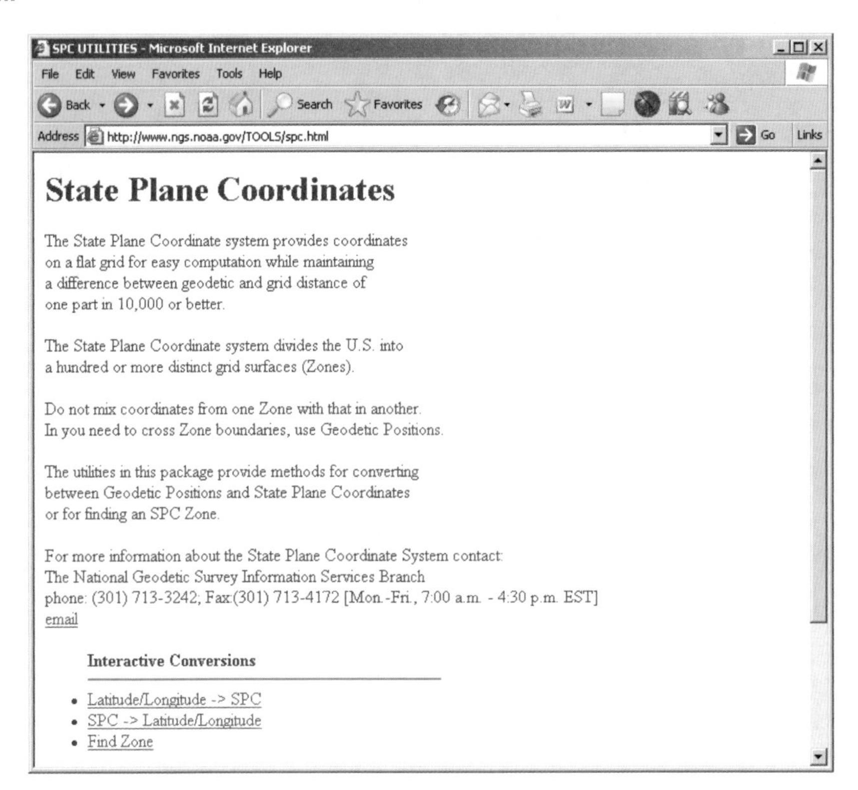

2 **In the resulting Web page, with the By County radio button clicked, click Begin.**

3 **Select Pennsylvania, click Submit, then click Allegheny, Submit.**

The answer is that Pennsylvania's Allegheny County is in State Plane Zone 3702.

4 **Close your browser.**

Add a map without State Plane coordinates

1 **In a new, empty map, click the Add Data button, browse to \Gistutorial\AlleghenyCounty, and add Tracts.shp to the map.**

This dataset was obtained from the U.S. Census Web site for census tracts in Allegheny County, Pennsylvania, for the year 2000. Notice the coordinate readout in the lower right corner of the display. The coordinates are latitude and longitude.

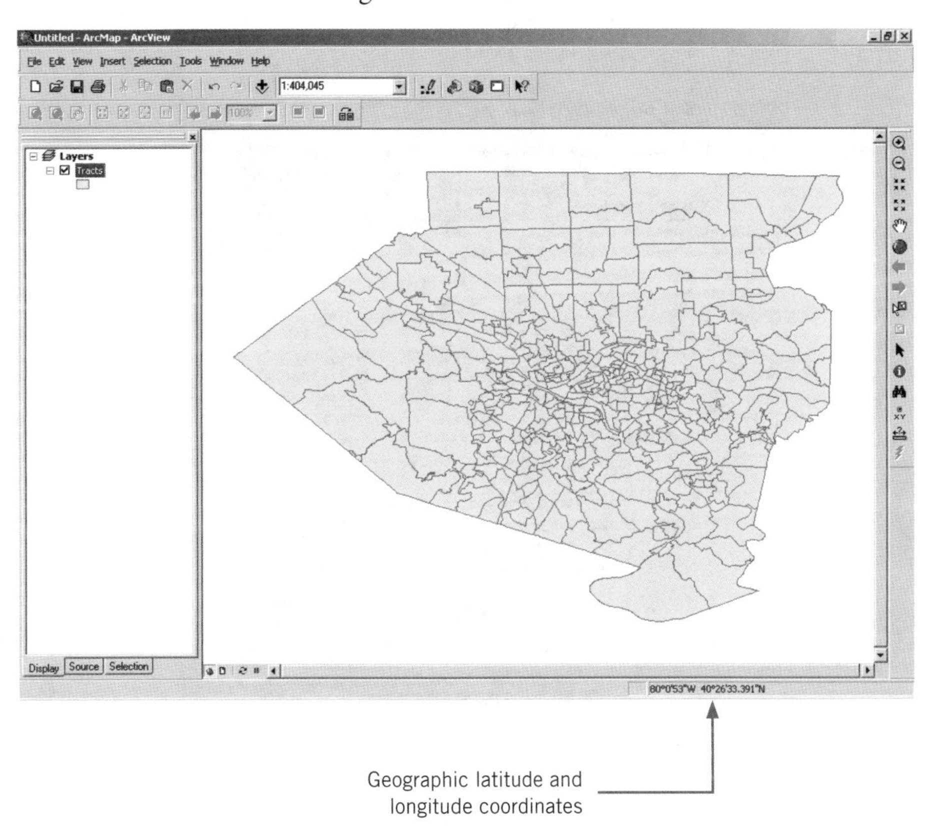

Set the coordinate projection for the map

1 In the table of contents, right-click the Layers data frame and click Properties.

2 Click the Coordinate System tab, and click Predefined, Projected Coordinate Systems, State plane, and NAD 1983 (Feet).

3 Scroll down the list of coordinate systems, hover your cursor over the second Pennsylvania projection (to see the complete projection name, including 3702), and click NAD 1983 StatePlane Pennsylvania South FIPS 3702 (Feet).

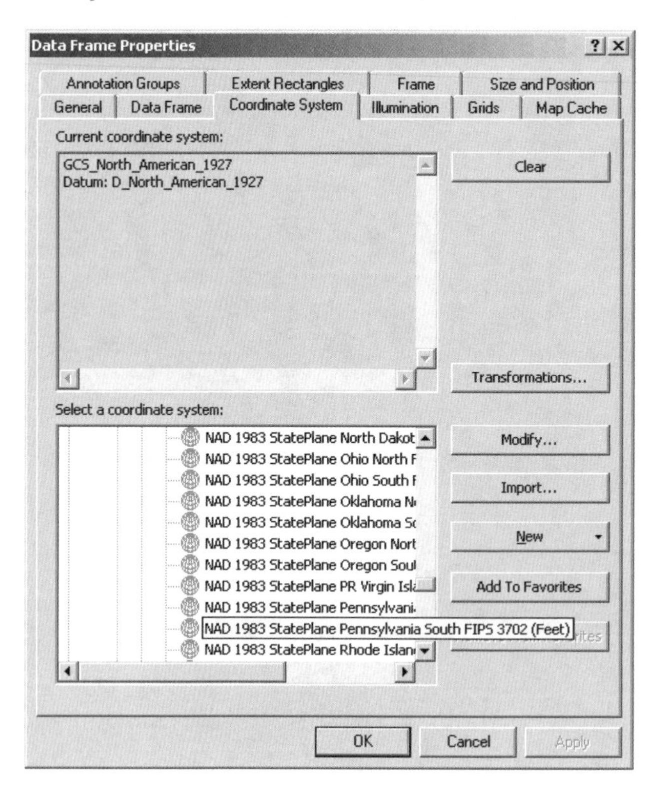

4 Click OK.

The coordinates appearing in the lower right corner of the display now appear in State Plane units. The origin of these coordinates (0,0) is the lower left corner of Pennsylvania.

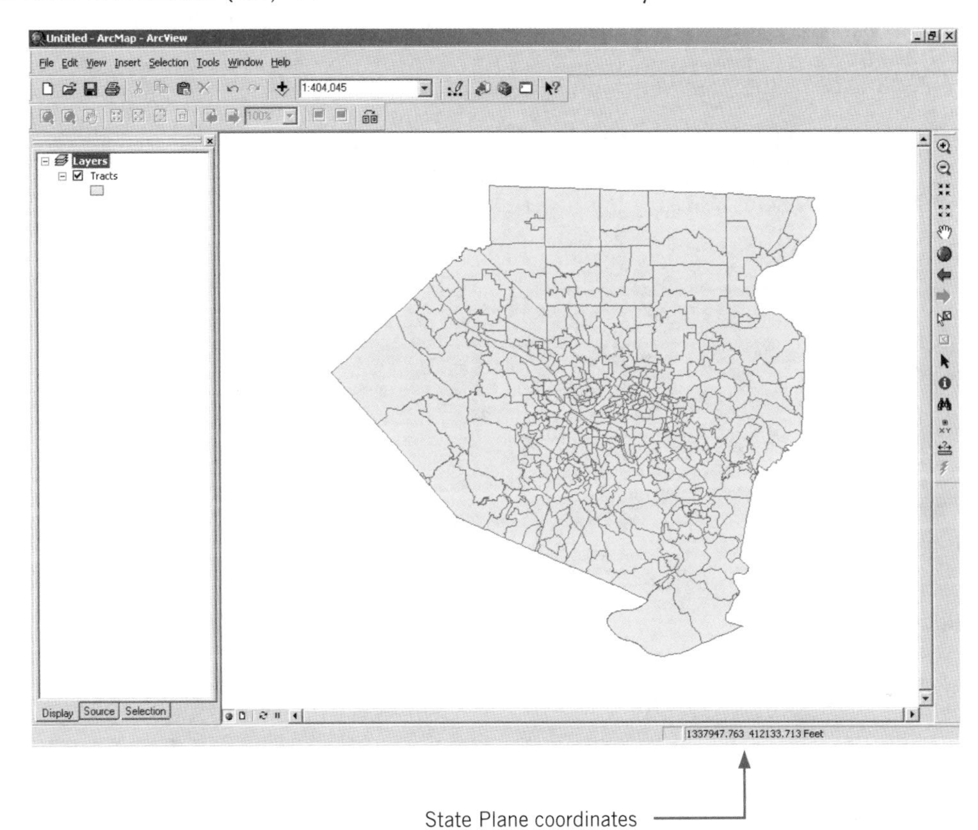

Convert 2000 census tracts to a State Plane shapefile

1 **In the table of contents, right-click the Tracts layer and click Data, Export Data.**

2 **Click Use the same Coordinate System as the data frame.**

3 **Change the name of the Output shapefile to TractsStatePlane.shp and click OK.**

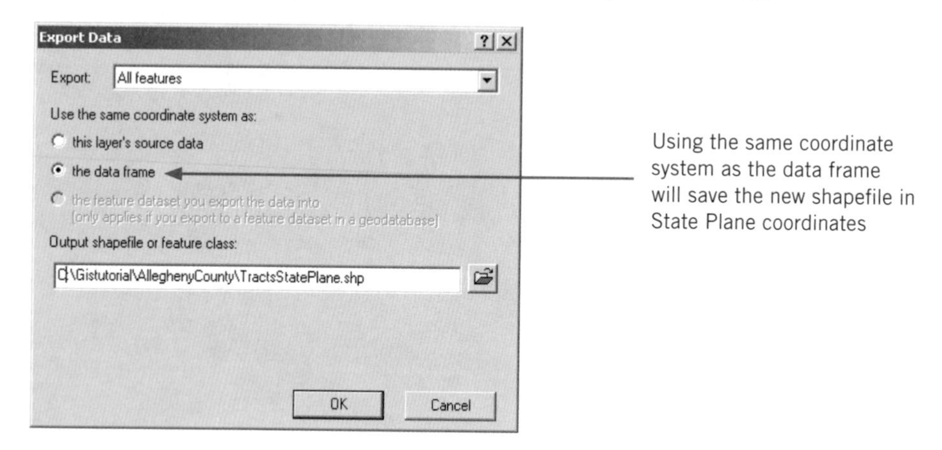

4 **Click Yes to add the exported data to the map.**

5 **Remove the Tracts layer from the map.**

The new shapefile for the census tracts is now permanently in the State Plane coordinate system, Pennsylvania South 1983.

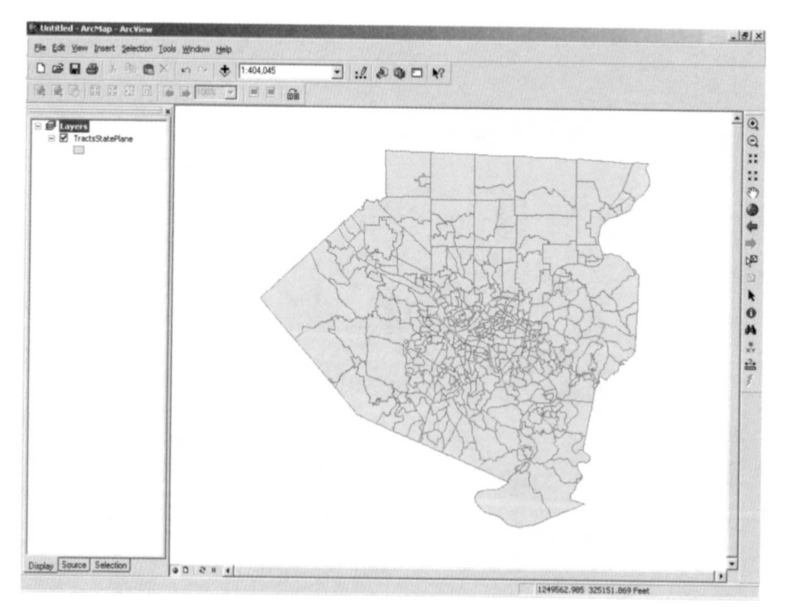

6 **Save the map document as C:\GIStutorial\StatePlaneMap.mxd and leave the document open in ArcMap.**

UTM coordinate system

The Universal Transverse Mercator (UTM) coordinate system was developed by the U.S. military late in the 1940s. It includes 60 longitudinal zones defined by meridians that are 6° wide. ArcGIS has UTM projections available for the northern and southern halves of each zone, divided by the equator. These projects are quite good for areas about the size of a state or smaller.

Look up a UTM Zone

1 **Start your Web browser, go to www.dmap.co.uk/utmworld.htm, and determine the zone for western Pennsylvania.**

You should find that western Pennsylvania is in zone 17 north.

2 **Close your browser.**

Set the coordinate projection for the map

1 **In the StatePlaneMap.mxd table of contents, right-click the Layers data frame, and click Properties.**

2 **Click the Coordinate System tab and click Predefined, Projected Coordinate Systems, Utm, Nad 1983, NAD 1983 UTM Zone 17N, OK.**

The coordinate system and map appearance change accordingly. Notice that the coordinates that read out for your cursor position are now in meters. UTM is a metric system and thus uses meters.

3 **Save the map document as C:\Gistutorial\UTMMap.mxd.**

Stored metadata

Metadata is often described as data about the data. In ArcGIS Desktop, ArcCatalog is used to create and view metadata. When creating metadata, some of the information, such as the spatial extent of the data, can be automatically created by the software, but much of the metadata must be entered manually. ArcMap has a menu item for viewing metadata, which you will use now.

View metadata

1 **Open the StatePlaneMap.mxd map document, click the Add Data button, browse to C:\Gistutorial\UnitedStates\Pennsylvania, add PACounties, and drag the resulting map layer to the bottom of the table of contents.**

You will see that the county boundaries you just added are quite crudely drawn and lack the detail of the tracts and municipality layers. The metadata for PACounties explains why.

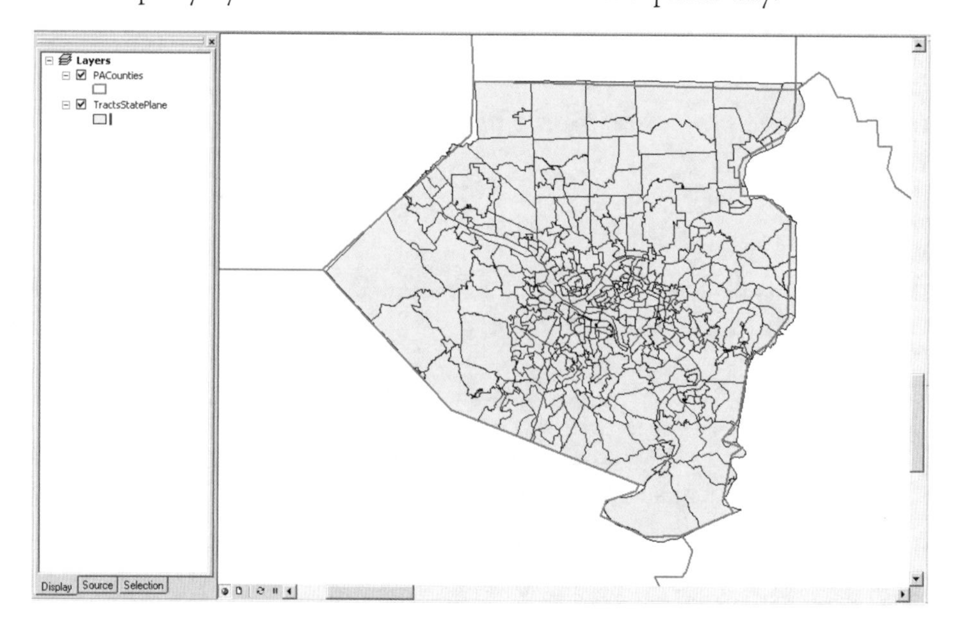

2 **Right-click PACounties and click Data, View Metadata.**

The Metadata window opens with the FGDC ESRI style sheet in use for metadata viewing and the Description tab selected. You can read that the county boundaries have been generalized (had points and lines removed) for speedy processing and viewing at the national level. When zoomed into Allegheny County, the effects of generalization are evident.

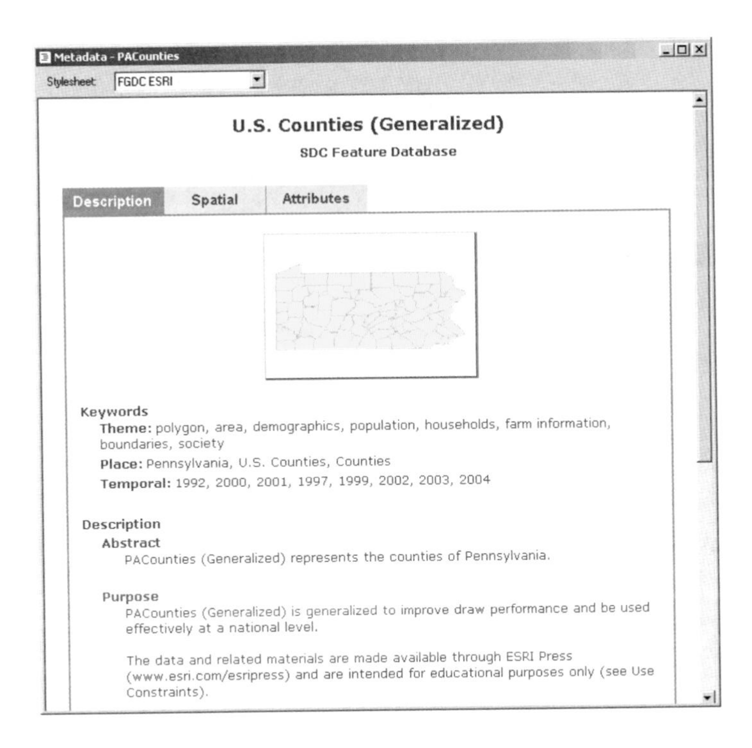

3 Click the Spatial tab.

ArcCatalog automatically generated much of this information, such as the bounding coordinates (or map extent) from the map data.

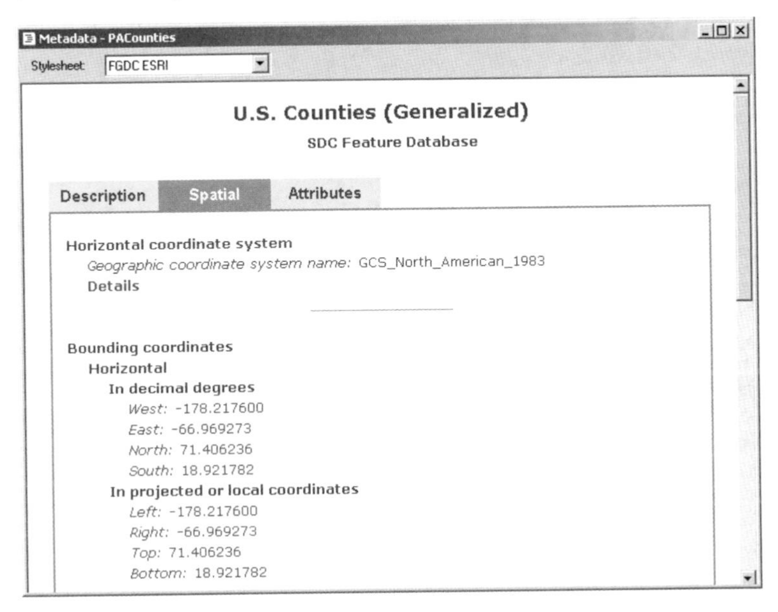

4 Lastly, click the Attributes tab and the STATE_FIPS attribute.

If you click an attribute name, you get metadata on that attribute as seen below for STATE_FIPS.

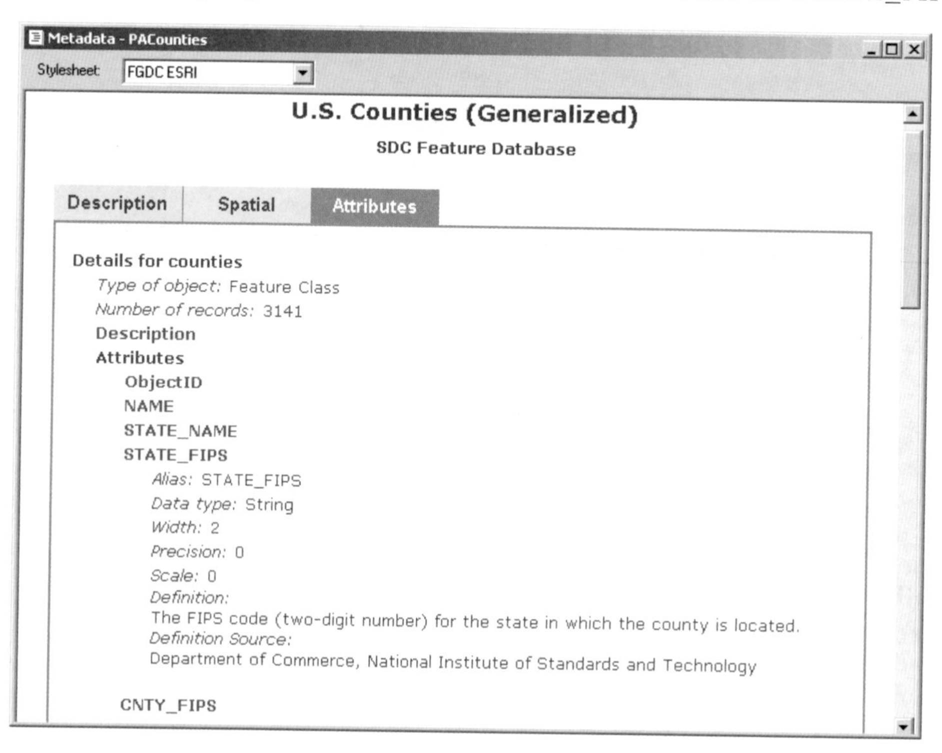

5 Close the Metadata window.

Leave StatePlaneMap.mxd open in ArcMap.

Attribute data

As you have seen in earlier exercises, stand-alone tables can also be added to ArcGIS and manipulated or joined to the attribute tables of other layers in a map. ArcGIS (ArcCatalog and ArcMap) allows direct use of data in a variety of formats, and the program works with them as tables. Formats include text (.txt, .asc, .csv, .tab), dBASE (.dbf), Excel (.xls), and Access (.mdb).

Adding and opening .dbf (dBASE) files

1 **If necessary, start ArcMap. In a new, empty map, click the Add Data button, browse to the \Gistutorial\AlleghenyCounty folder, and add 2000sf1trt.dbf to the map.**

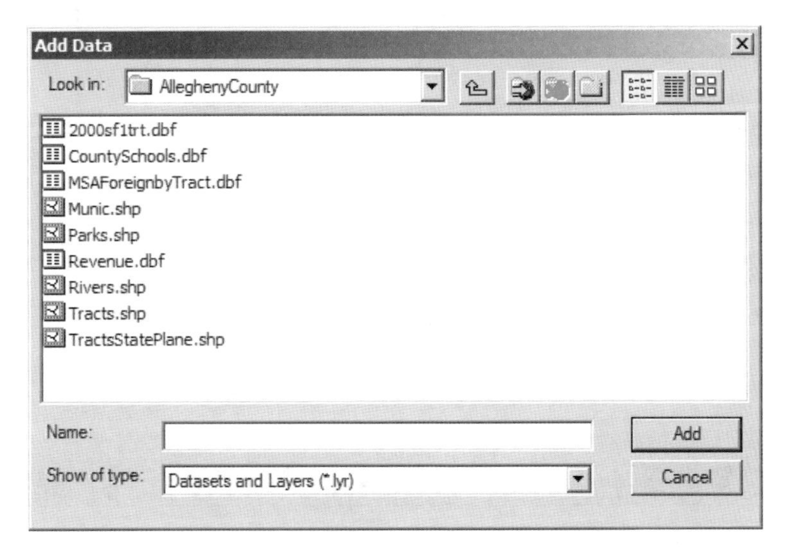

2 **In the table of contents, right-click 2000sf1trt and click Open.**

ArcMap opens the DBF table. This table contains detailed population data for Allegheny County Census Tracts.

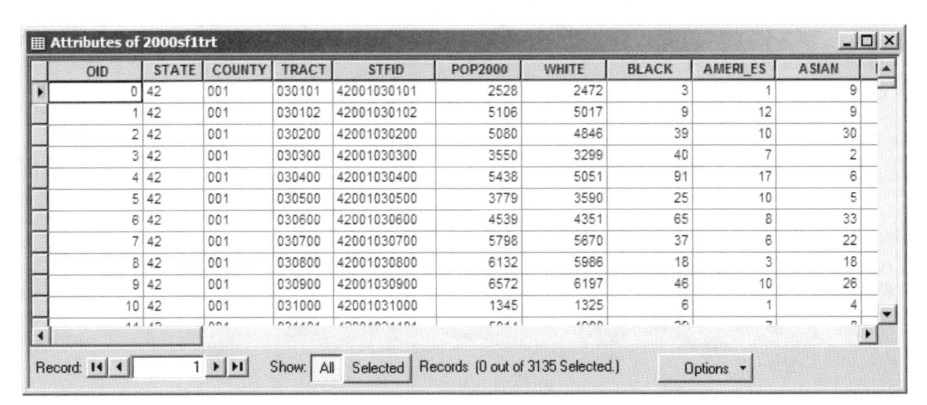
Attributes of 2000sf1trt

OID	STATE	COUNTY	TRACT	STFID	POP2000	WHITE	BLACK	AMERI_ES	ASIAN
0	42	001	030101	42001030101	2528	2472	3	1	9
1	42	001	030102	42001030102	5106	5017	9	12	9
2	42	001	030200	42001030200	5080	4846	39	10	30
3	42	001	030300	42001030300	3550	3299	40	7	2
4	42	001	030400	42001030400	5438	5051	91	17	6
5	42	001	030500	42001030500	3779	3590	25	10	5
6	42	001	030600	42001030600	4539	4351	65	8	33
7	42	001	030700	42001030700	5798	5670	37	6	22
8	42	001	030800	42001030800	6132	5986	18	3	18
9	42	001	030900	42001030900	6572	6197	46	10	26
10	42	001	031000	42001031000	1345	1325	6	1	4

Record: 1 Show: All Selected Records (0 out of 3135 Selected.) Options

Adding and opening comma-delimited .csv files

1 In ArcMap, click the Add Data button, browse to the **\Gistutorial\UnitedStates\Pennsylvania** folder, and add **2000sf1county.csv** to the map.

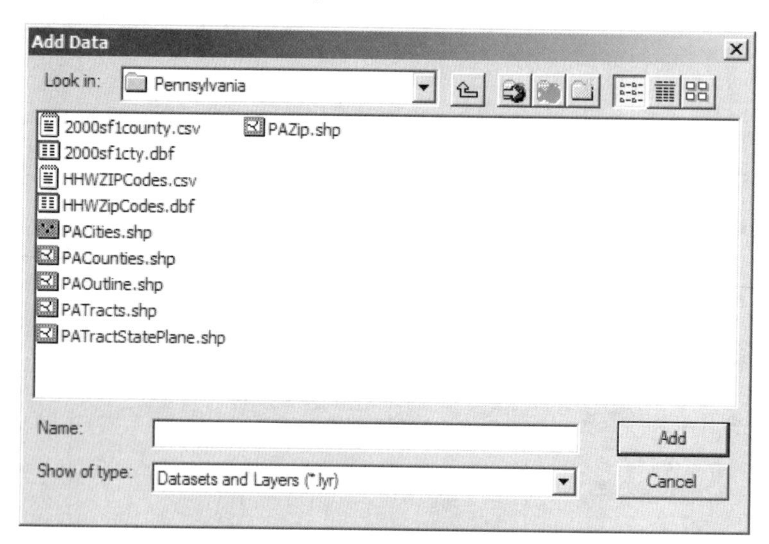

2 In the table of contents, right-click **2000sf1county.csv** and click **Open**.

ArcMap opens the CSV table. This table contains detailed population data for all Pennsylvania counties.

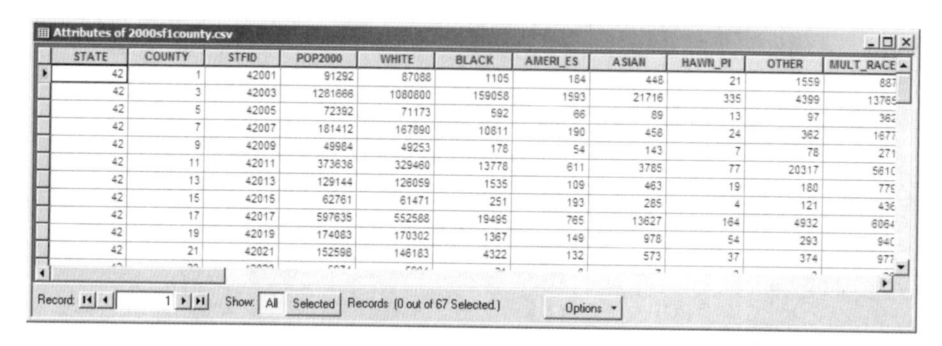

Attributes of 2000sf1county.csv

STATE	COUNTY	STFID	POP2000	WHITE	BLACK	AMERI_ES	ASIAN	HAWN_PI	OTHER	MULT_RACE
42	1	42001	91292	87088	1105	184	448	21	1559	887
42	3	42003	1281666	1080800	159058	1593	21716	335	4399	13765
42	5	42005	72392	71173	592	66	89	13	97	362
42	7	42007	181412	167890	10811	190	458	24	362	1677
42	9	42009	49984	49253	178	54	143	7	78	271
42	11	42011	373638	329460	13778	611	3785	77	20317	5610
42	13	42013	129144	126059	1535	109	463	19	180	779
42	15	42015	62761	61471	251	193	285	4	121	436
42	17	42017	597635	552588	19495	765	13627	164	4932	6064
42	19	42019	174083	170302	1367	149	978	54	293	940
42	21	42021	152598	146183	4322	132	573	37	374	977

Record: 1 Show: All Selected Records (0 out of 67 Selected.) Options

Microsoft Access tables

ArcGIS allows direct import of Microsoft Access tables into ArcMap.

1 **Click the Add Data button, browse to the \Gistutorial\Flux folder, then double-click FLUXEvent.mdb.**

2 **Click the tAttendees table and click Add.**

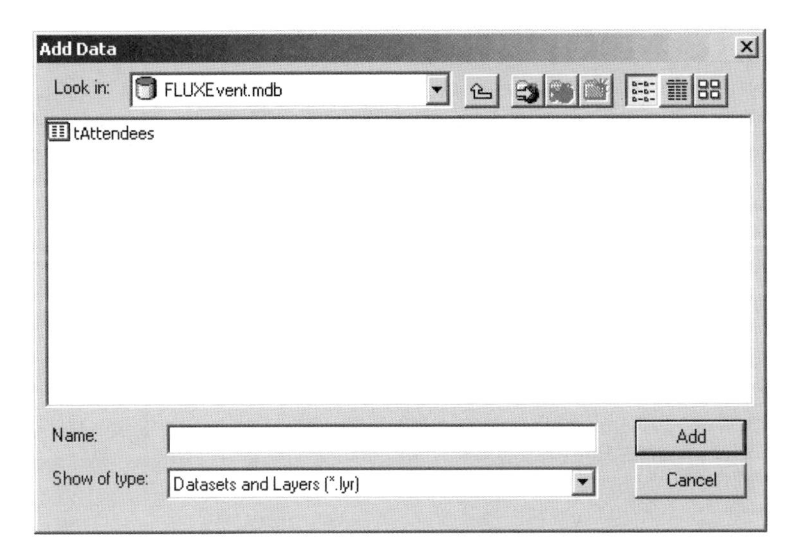

3 **In the table of contents, right-click tAttendees and click Open.**

ArcMap opens the tAttendees table. This table contains all the attendees of a recent FLUX event in Pittsburgh, Pennsylvania. You will geocode these addresses later in this book.

Attributes of tAttendees

ID*	Custom6	Date	Time	Address	City	State	ZIP Code*	Age
1	A	20020629	<Null>	2415 1ST AVE	SACRAMENTO	CA	95818	27
2	A	20020629	<Null>	224 NORTH ST	STERBENVILLE	OH	43952	32
3	A	20020629	<Null>	PO BOX 622 535 4TH ST	MARIANNA	PA	15345	32
4	A	20020629	<Null>	5126 JANIE DRIVE	PITTSBURGH	PA	15227	34
5	A	20020629	<Null>	305 AVENUE A	PITTSBURGH	PA	15221	40
6	A	20020629	<Null>	1431 CRESSON ST	PITTSBURGH	PA	15221	26
7	A	20020629	<Null>	5133 DEARBORN STREET	PITTSBURGH	PA	15224	33
8	A	20020629	<Null>	1122 MORRISON ST	PITTSBURGH	PA	15212	29
9	A	20020629	<Null>	352 FIELDING DRIVE	PITTSBURGH	PA	15235	23
10	A	20020629	<Null>	345 MOORE AVE	PITTSBURGH	PA	15210	21
12	A	20020629	<Null>	588 S AIKEN AVENUE	PITTSBURGH	PA	15232	34

Record: 1 Show: All Selected Records (0 out of 1265 Selected.) Options

4 **Close all the open tables, save the map as \Gistutorial\Table.mxd, and close ArcMap.**

Assignment 5-1

Map of Maricopa County, Arizona, voting districts, schools, and voting-age population using data downloaded from the ESRI Web site

One community that uses GIS for elections is Maricopa County, Arizona, one of the nation's fastest growing communities with 3.2 million residents and 1.5 million registered voters. Maricopa County uses GIS to ensure accurate voting boundaries, maintain voter lists, locate polling places, plan voting precincts, recruit poll workers, and deliver supplies.

In this exercise you will focus on skills needed to download data from an ESRI Web site and prepare the data to use in ArcGIS. For the Maricopa County voting GIS, you will download voting districts, streets, and census blocks for the purpose of building an interactive GIS to be used for selecting schools for use as polling sites. You will also download block-level census data from the ESRI Web site, join this data to the block map, and use it to display the spatial distribution of the voting-age population. Provided is an x,y table that shows schools and their geographical coordinates. Such data can be downloaded from Web sites such as http://www.hometownlocator.com, but using it requires extensive editing and would take too much time for this assignment.

Start with the following:

Shapefiles

From the ESRI Census 2000 TIGER/Line Data Portal, download the following shapefiles and save them in C:\GIStutorial\Answers\Assignment5

- **tgr04013blk00.shp**—Census Block 2000 polygons
- **tgr04013lkA.shp** —Line Features (roads)
- **tgr04013vot00.shp**—Voting Districts 2000 polygons

Data table

From the same ESRI Web site, download the following database and save it in C:\GIStutorial\Answers\Assignment5

- **tgr04000sf1blk.dbf**—Census Block Demographics (SF1)

From C:\Gistutorial\MaricopaCounty, use

- **CountySchools.dbf**—The coordinates have the correct State Plane projection for Maricopa County

Create an interactive GIS

Create a new map document called C:\Gistutorial\Answers\Assignment5\Assignment5-1.mxd. Use scales to display detailed layers when zoomed in to 1:100,000 scale. At that scale, display labels for voting districts, schools, streets. This provides a tool for analyzing potential voting places, voting district by voting district.

Look up the State Plane zone for Maricopa County and use it for your map document's data frame. Add spatial reference data for the ESRI shapefiles: Geographic Coordinate Systems, North America, NAD 1927 (CGQ77).prj. When you add the x,y data, edit the Coordinate System of Input Coordinates to use the correct State Plane coordinates. Add a field to block census data: Voters = [POP2000] – [AGE_UNDER5] – [AGE_5_17].

For very small-grain spatial data, such as provided by census blocks, a good approach is to use small, square point markers of the same size and with a monochromatic color ramp. Symbolize blocks using graduated symbols for the Voters attribute and use a "trick" to make all symbols the same size. Use size from 4 to 4 to get same size and then double-click each symbol to change color for the monochromatic ramp. Set the background color to No Color. The benefit of the "trick" is that ArcMap uses point markers instead of choropleth maps for the blocks.

Zoom in to a populated area of your map document with a map scale of 1:30,000. Create an 8.5 × 11-inch landscape layout with map, legend, and title. Export the layout as C:\Gistutorial\Answers\Assignment5\Assignment5-1.jpg. Create a Word document, saved as C:\Gistutorial\Answers\Assignment5\Assignment5-1.doc, that has a title, your name, your map layout image, and a paragraph suggesting schools to be used as polling places for observed voting districts.

Assignment 5-2

Maps of Florida counties and census data using data downloaded from the Census Bureau Web site

In this exercise you will focus on skills needed to download data from the U.S. Census Bureau's Web sites and prepare the data for use in ArcGIS. You will download a U.S. Census Cartographic Boundary file of Florida counties as an ARC/INFO interchange (.e00) file. After downloading the .e00 file, you will transform it to an ARC/INFO coverage using the import function in ArcCatalog and finally transform it to a shapefile. The Cartographic Boundary files do not contain census data, so, in addition, you need to download a census data table of interest, prepare it for use in a GIS, and join it to the county boundary file. Finally, you symbolize a choropleth map.

Start with the following:

ArcInfo Export File

- **co12_d00.e00**—ArcInfo export file of Florida counties downloaded from *http://www.census.gov/geo/www/cob/co2000.html* and saved to C:\Gistutorial\Answers\Assignment5\.

Census table files

- **dc_dec_2000_sf3_u_data1.txt**—downloaded from *http://factfinder.census.gov's* Download Center using settings: Census 2000 Summary File 3 (SF 3) - Sample Data, All Counties, P65 Retirement Income in 1999 for Households and saved to C:\Gistutorial\Answers\Assignment5\.

Prepare data for use

For the census data table, open Excel, click File, Open, browse to C:\Gistutorial\Answers\Assignment5\, set Files of type to All Files (*.*), and select dc_dec_2000_sf3_u_data1.txt. In the Import Wizard use the following steps:

Step 1. Select Delimited radio button, Start import at row 1 and click Next.

Step 2. Deselect any delimiter checked, type | (character on keyboard as upper case "\") in the Other field, and click Next.

Step 3. Click Finish.

Delete attributes (columns) GEO_ID, SUMLEVEL. Rename attributes as follows:

GEO_ID2 - GEOID2

GEO_NAME - GEONAME

P065001 - HH

P065002 - HHRETIRE

P065003 - HHNORETIRE

Delete Row 2 (with attribute descriptions). Delete non-Florida rows. Save the table as Retired.dbf and close Excel.

For the Florida counties map layer, transform the co12_d00.e00 file to coverage C:\Gistutorial\Answers\Assignment5\counties\.

Add the county coverage to ArcMap, right-click the coverage, and use the Data option to transform the coverage to shapefile C:\Gistutorial\Answers\Assignment5\counties.shp. Check to make sure that shapefile has spatial reference data (it does).

Open Attributes of Counties.shp in ArcMap, create a new field called GEOID2 with Long Integer data type. Use the field calculator in ArcMap to calculate GEOID2 = [STATE]*1000 + [COUNTY].

Create a choropleth map showing percentages of households with retirement income

Create a new map document called C:\Gistutorial\Answers\Assignment5\Assignment5-2.mxd. Use a UTM projection for Layers data frame. Look up the correct UTM zone. Add Retired.dbs and join it to Counties.

Symbolize a choropleth using HHRETIRE normalized with HH, 5 equal-width intervals using round numbers (such as 15%, 20%, etc.) and county name labels.

Create an 8.5 x 11-inch landscape layout with map, legend, and title. Export the layout as C:\Gistutorial\Assignment\Assignment5\Assignment5-2.jpg. Create a Word document, saved as C:\Gistutorial\Assignment5\Assignment5-2.doc, that has a title, your name, your map layout image, and a paragraph describing the pattern of households with retired income in Florida.

What to turn in

If you are working in a classroom setting with an instructor, you may be required to submit the exercises you created in tutorial 5. Below are the files you are required to turn in. Be sure to use a compression program such as PKZIP or WinZip to include all files as one .zip document for review and grading. Include your name and assignment number in the .zip document (YourNameAssn5.zip).

ArcMap documents

C:\Gistutorial\Answers\Assignment5\Assignment5-1.mxd
C:\Gistutorial\Answers\Assignment5\Assignment5-2.mxd

Shapefiles and databases

C:\Gistutorial\Answers\Assignment5\tgr04013blk00.shp
C:\Gistutorial\Answers\Assignment5\tgr04013lkA.shp
C:\Gistutorial\Answers\Assignment5\tgr04013vot00.shp
C:\Gistutorial\Answers\Assignment5\tgr04000sf1blk.dbf
C:\Gistutorial\Answers\Assignment5\Retired.dbf
C:\Gistutorial\Answers\Assignment5\counties.shp

Microsoft Word documents

C:\Gistutorial\Answers\Assignment5\Assignment5-1.doc
C:\Gistutorial\Answers\Assignment5\Assignment5-2.doc

OBJECTIVES

Digitize and edit a polygon layer
Digitize a point layer
Digitize a line layer
Snap features

GIS Tutorial 6

Digitizing

Sometimes it is necessary to add spatial features such as points, lines, or polygons to existing map layers. This tutorial introduces you to editing spatial data. Within the following exercises you will learn how to digitize new vector features and add attribute data to a table. You will also adjust vector data spatially to make it align with a georeferenced aerial photo.

Create a new polygon shapefile

There are a few different ways to create shapefiles with ArcGIS Desktop. A common method is to use ArcCatalog.

Create a new polygon shapefile

1 Start ArcCatalog.

2 In the catalog tree, browse to **\Gistutorial\PAGIS** and click the **MidHill** folder.

This is the folder in which the new shapefile will be created.

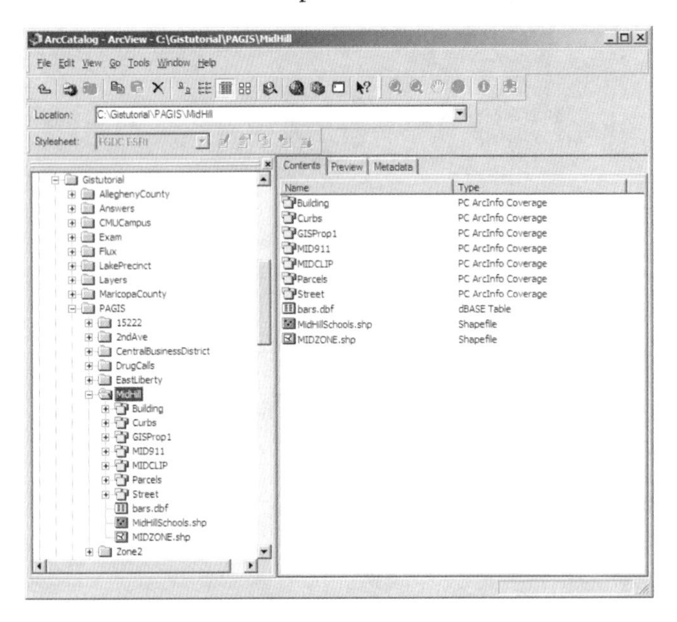

3 Click File, New, Shapefile.

4 In the Name field, type **CommercialZone**.

5 From the Feature Type drop-down list, choose Polygon.

6 Click Edit, Select, Projected Coordinate Systems, Add, State Plane, Add, NAD 1927, Add, NAD 1927 StatePlane Pennsylvania South FIPS 3702.prj, Add, OK.

7 Click OK.

8 Close ArcCatalog.

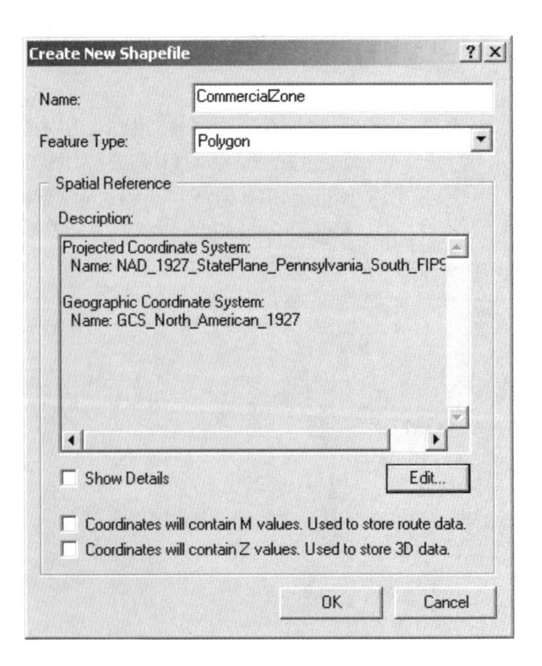

Start ArcMap: Open a map document

1 **From the Windows taskbar, click Start, All Programs, ArcGIS, ArcMap.**

2 **Click the An existing map radio button in the ArcMap dialog box.**

3 **Click OK.**

4 **Browse to the drive on which the Gistutorial folder has been installed, select Tutorial6-1.mxd, and click Open.**

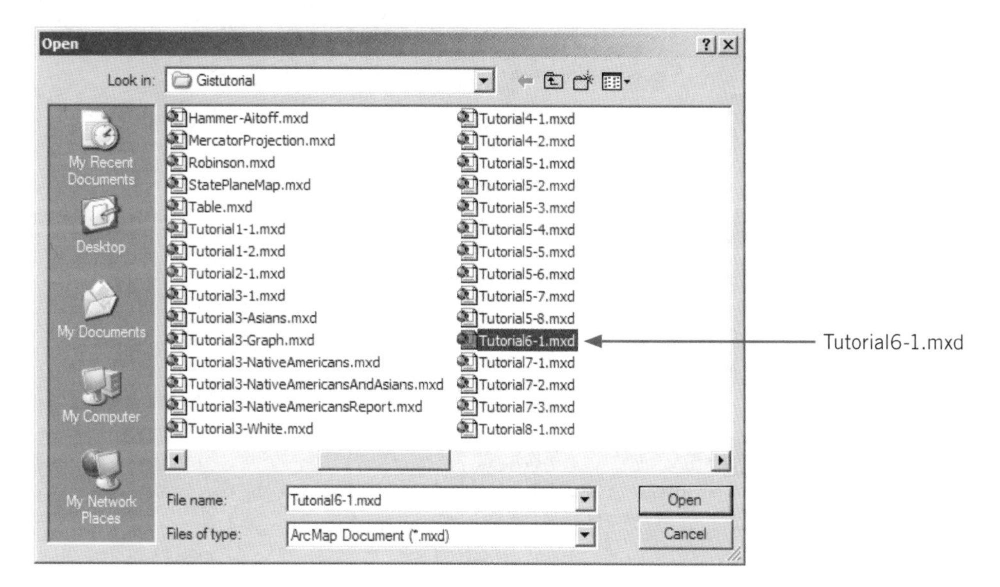

The Tutorial6-1 map document opens in ArcMap, showing a map of the Middle Hill Neighborhood of Pittsburgh, Pennsylvania.

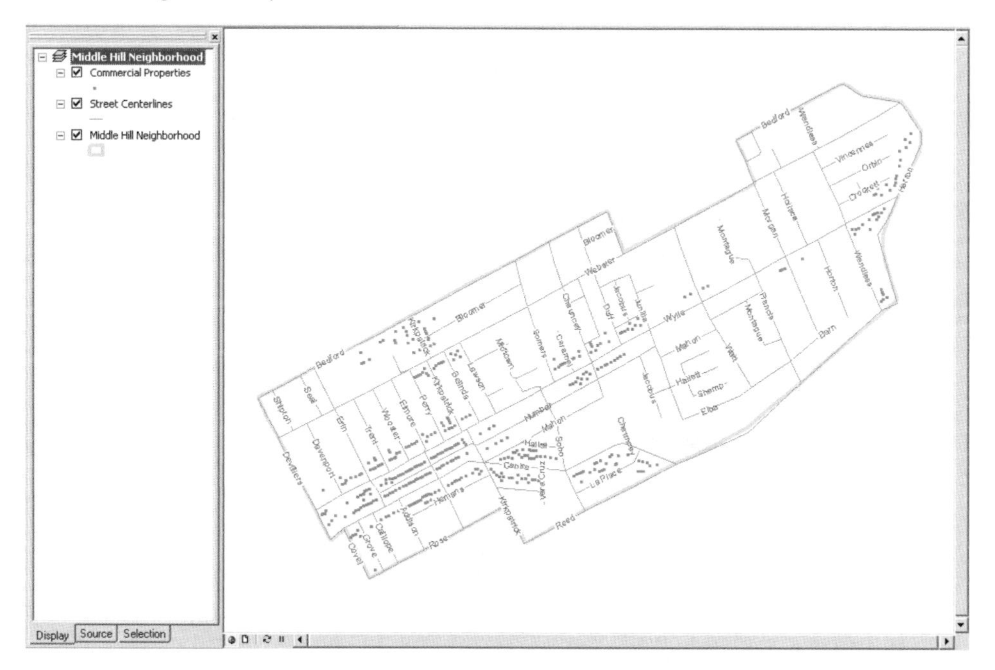

Add the shapefile to a map

1 Click the Add Data button.

2 Browse to **\Gistutorial\PAGIS\MidHill** in the Add Data dialog box.

3 Click the **CommercialZone.shp** shapefile.

4 Click Add.

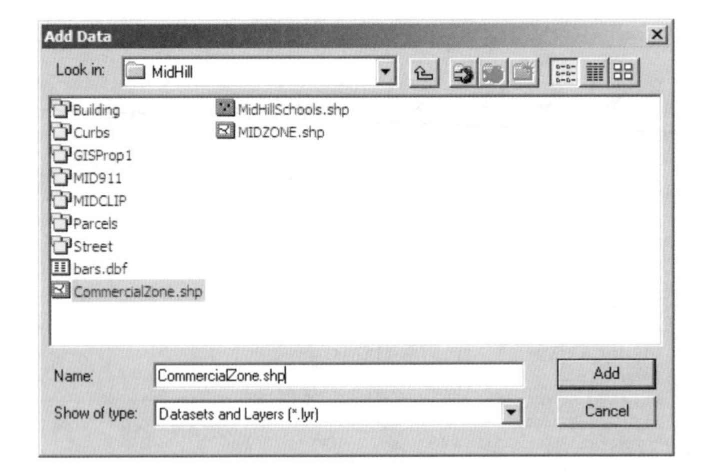

The CommercialZone shapefile is added to the map, although there are no features in the shapefile yet.

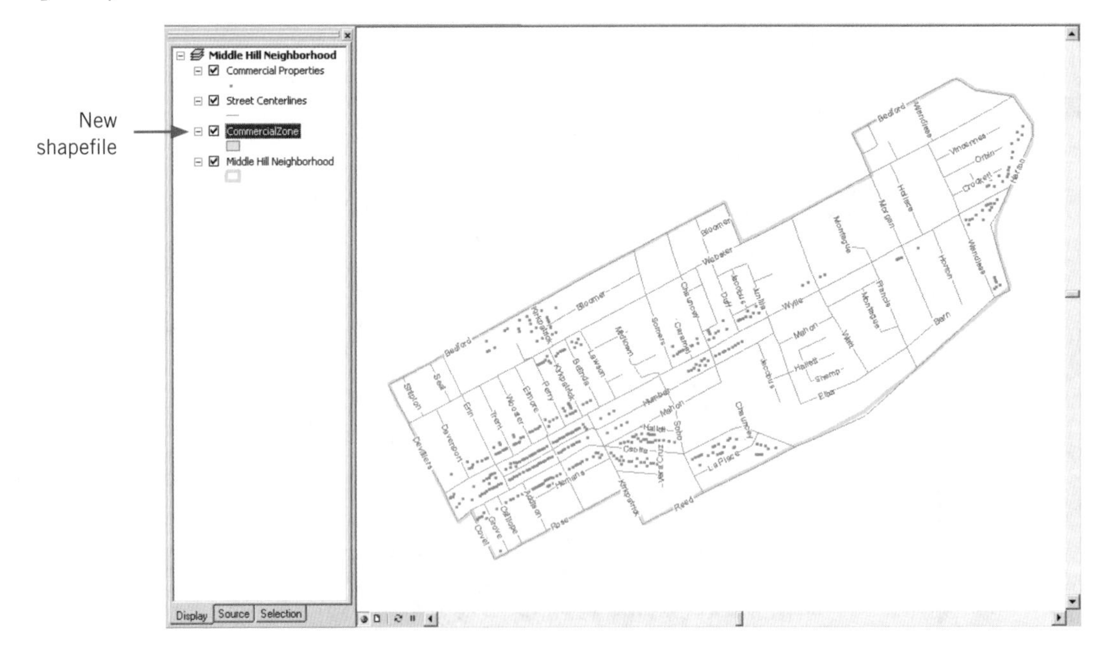

Digitize and edit the polygon layer

Open the Editor toolbar

1 Click Tools, Editor toolbar.

The Editor toolbar appears. You can move it or dock it anywhere in ArcMap.

2 On the Editor toolbar, click Editor, Start Editing.

3 If necessary, click \Gistutorial\PAGIS\MidHill as the folder from which to edit the shapefile, then click OK.

The dialog box asking for the folder from which to edit the shapefile may not appear. If it doesn't, be sure to have the correct layer selected, as shown below.

On the Editor toolbar, the Target layer should be CommericalZone and the Task should be Create New Feature.

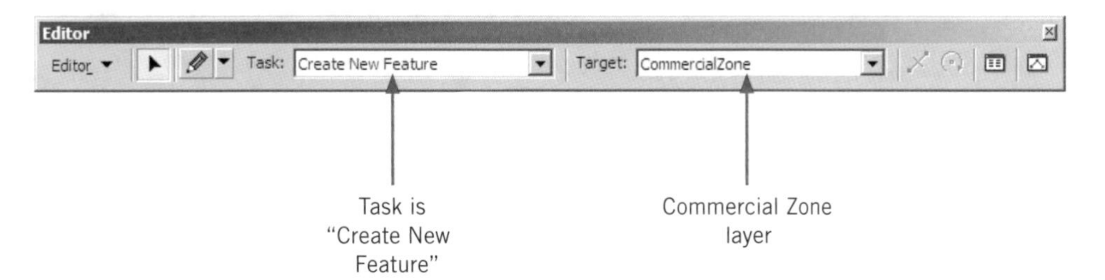

Practice digitizing a polygon

1. At the bottom of the table of contents, click the Selection tab. Make sure that only CommercialZone is checked, then click the Display tab.
2. From the Editor toolbar, click the Sketch tool. 
3. Position the crosshair cursor anywhere on the map and click the left mouse button to place a vertex.
4. Move your mouse and click a series of vertices one at a time to form a polygon (do not double-click!).
5. Double-click the last vertex, placing it just before the first vertex that you entered.

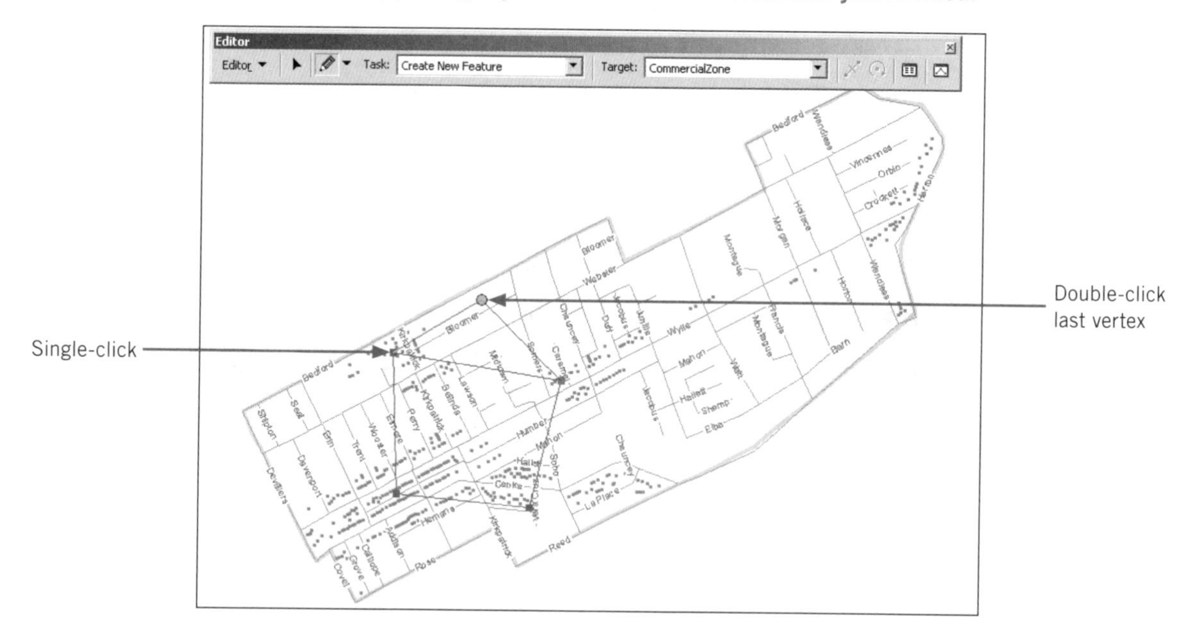

Practice editing a polygon

1 **From the Editor toolbar, click the Edit tool.** 

2 **Click anywhere inside your new polygon and then grab the polygon by clicking and holding anywhere on its boundary.**

The cursor becomes a four-sided arrow when you are ready to grab the polygon.

3 **Drag the polygon a small distance and release.**

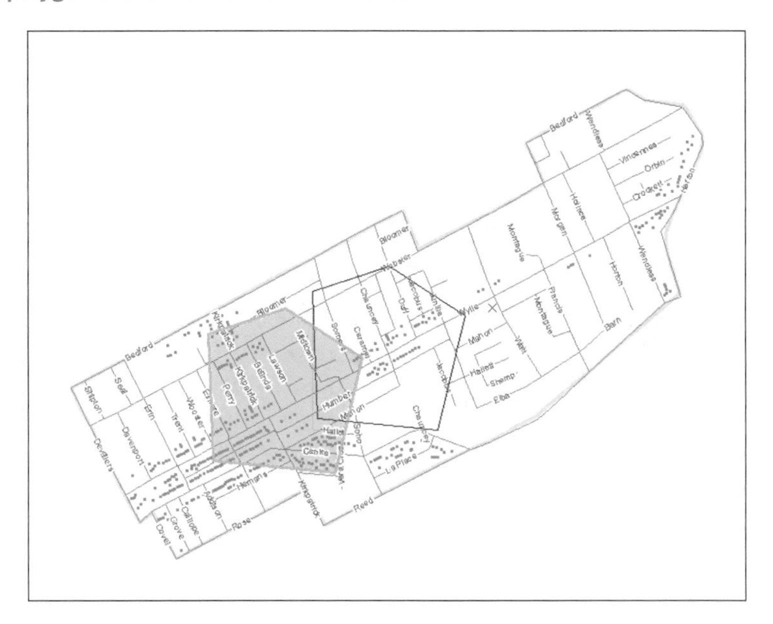

4 **Double-click the outline of the new polygon.**

Grab handles appear at the location of each vertex in the polygon. Next, you will see that you can edit the shape of a feature by moving a vertex.

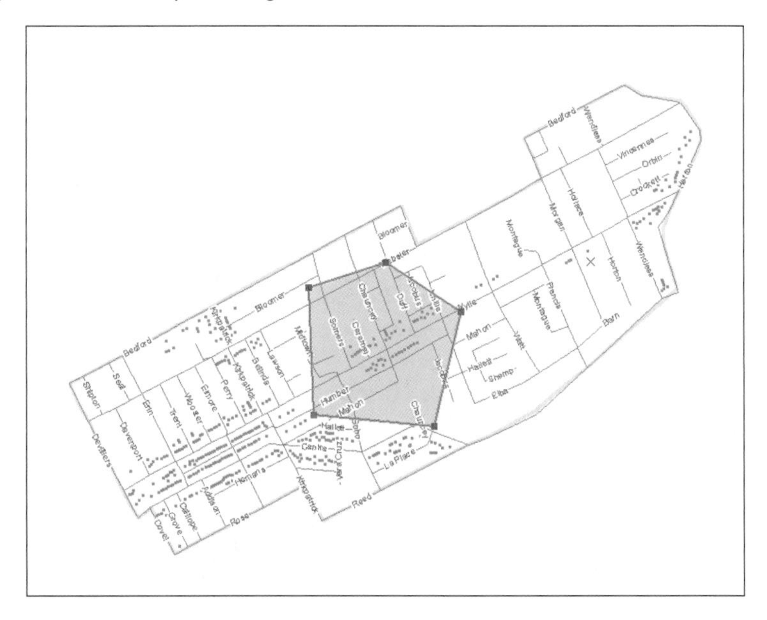

5 **Position the cursor over one of the vertices.**

6 **Click and drag the vertex somewhere nearby and release.**

The polygon's shape changes correspondingly.

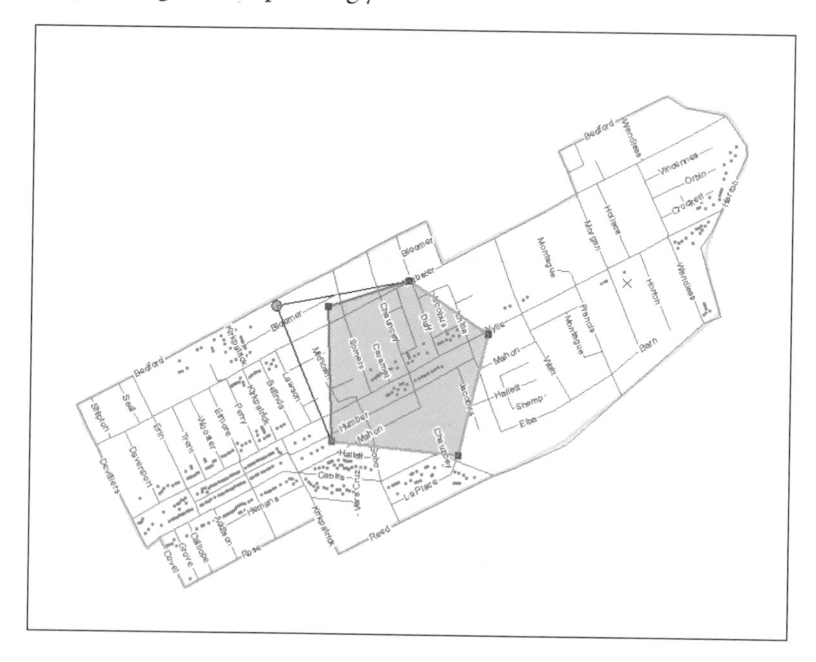

7 **Click anywhere on the map or polygon to confirm the new shape.**

8 **Click the polygon shape again.**

9 **Press the Delete key to erase the polygon.**

The polygon is erased. Next you will practice editing digitized polygons and learn how to add, delete, and move vertices.

Vertices

Move vertex points

It's good practice to learn how to work with vertices. Over the next several steps, you will move, add, and delete vertices from a new polygon.

1 **Zoom to a small area, click the Sketch tool,**  **and draw another new polygon feature.**

2 **Click the Edit tool.** 

3 **Double-click the new polygon.**

Square markers appear on the polygon at its vertex locations.

4 **Position the cursor over one of the vertices.**

5 **Click and drag the vertex somewhere nearby and release.**

The polygon's shape changes correspondingly.

6 **Click anywhere on the map or polygon to confirm the new shape.**

Add vertex points

1 **Double-click inside the polygon to make the vertex locations visible.**

2 **Move the mouse along the line between two vertices.**

3 **With your cursor directly over the line, right-click and choose Insert Vertex.**

A new vertex is added at the location of the cursor. This vertex can now be moved to change the polygon's shape.

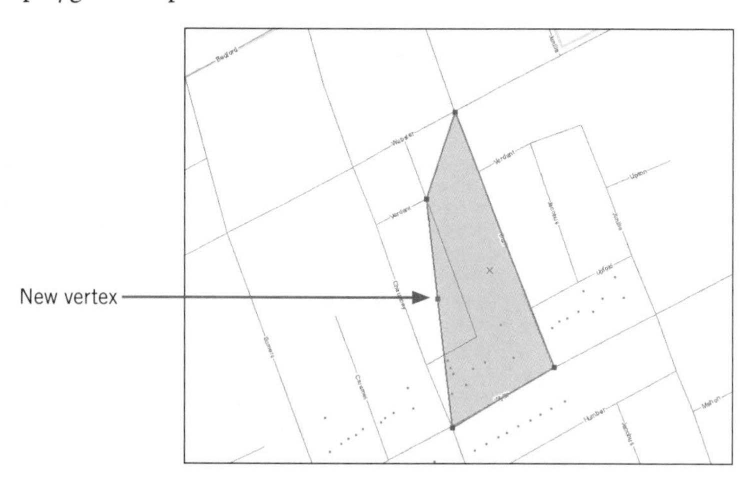

4 **Position the cursor over the new vertex, then click and drag the vertex to a new position and release.**

5 **Click anywhere on the map to confirm the new shape.**

Delete vertex points

1 **Double-click inside the new polygon.**

2 **Place your mouse cursor over any vertex in the new polygon.**

3 **Right-click and choose Delete Vertex.**

4 **Click the map.**

YOUR TURN

Practice changing the shape of the new polygon by moving, adding, and deleting vertices.

Drawing and editing tips

Advanced Edit tools

There are several advanced editing tools. Here you will try the Generalize and Smooth tools, which both affect the shape of digitized polygons.

1 **Click Editor, More Editing Tools, and Advanced Editing.**

2 **On the Editor toolbar, click the Sketch tool and digitize an arbitrary polygon with at least 25 vertices, some close together.**

Leave your new polygon selected. If it is not selected, click inside it with the Edit tool.

3 **On the Advanced Editing toolbar, click the Generalize tool, type a Maximum allowable offset of 100, and click OK.**

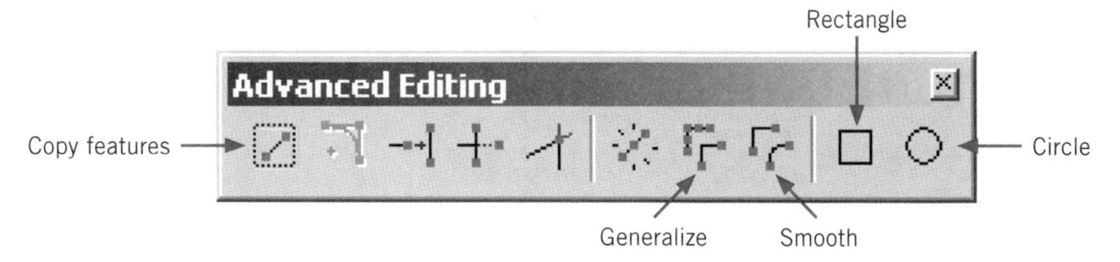

The result is a polygon with fewer vertices, no two of which have a line segment between them less than 100 feet. You can click the Undo button to try a different offset.

4 **Next, click the Edit tool and click inside the same polygon, then click the Smooth tool, type a maximum allowable offset of 10, and click OK.**

This adds many shape vertices to create smooth curves between the polygon's vertices.

5 **Close the Advanced Editing toolbar.**

6 **Click the Edit tool, click inside the polygon, and press the Delete key.**

Specify a segment length

1 **On the Editor toolbar, click the Sketch tool and digitize a point.**

2 **Right-click and click Length (or press Ctrl + L).**

3 **Type segment length of 250, press Enter, rotate the 250-foot line to the desired orientation, and click.**

The line is constrained to be 250 feet long.

4 **Digitize a few more points with specified segment lengths, then double-click to finish the polygon.**

5 **Delete the polygon.**

Edit tasks

The Task drop-down list on the Editor toolbar contains a set of editing tasks that affects how the Sketch tool functions.

1 **Select a polygon in the CommercialZone layer.**

2 **From the Editor toolbar, click the Tasks drop-down list.**

The editing tasks are organized into four groups. Most of the tasks are used in conjunction with the Sketch tool. For example, to cut a polygon into two new polygons, select the polygon you want to cut, choose the Cut Polygon Features task, then use the Sketch tool to define the line along which you want to cut the selected polygon. For more information on how to use the editing tasks, refer to the ArcGIS Desktop Help.

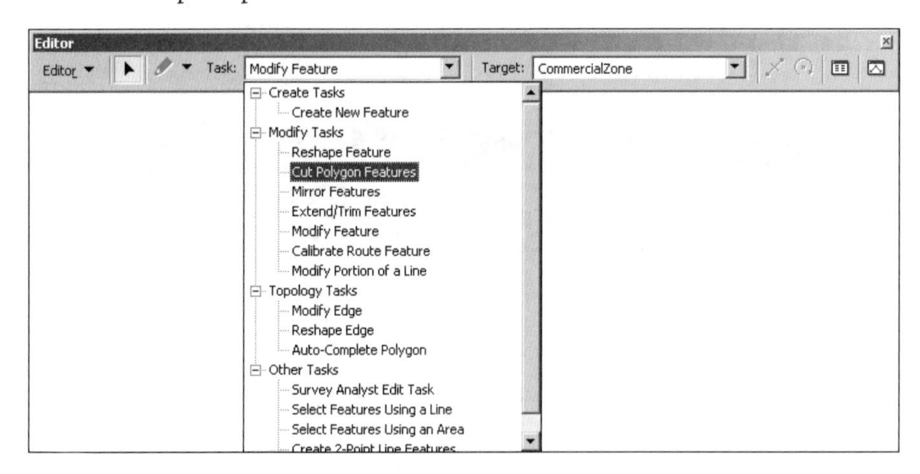

3 **Experiment with the Cut Polygon Features tool.**

4 **Close the Advanced Editing toolbar.**

Adding other graphics

In addition to the editing tools, ArcMap also has a set of tools on the Drawing toolbar used to draw graphics. Objects drawn with the drawing tools are stored in the map document as graphics, not as features in a feature class.

1 **Click View, Toolbars, Draw.**

2 **From the Drawing toolbar, click New Rectangle and draw a rectangle on the map.**

3 **Click any of the drawing tools and draw additional graphics.**

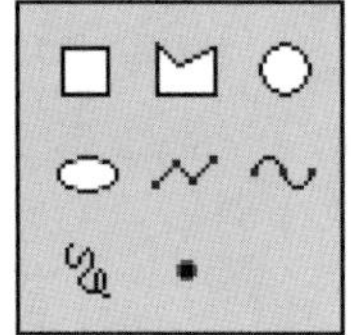

Digitize the commercial zone polygons

1 **Delete any of the practice polygons and graphics you created in the previous steps.**

2 **Zoom to the cluster of commercial centroids at the top left of the map.**

3 **Click the Create New Feature tool and the Sketch tool and roughly digitize the first polygon, seen below, by clicking one vertex at a time and double-clicking to finish.**

Where possible, use street centerlines as a guide for digitizing your arcs.

4 **Click Editor, Stop Editing to close the edit session.**

5 **Click Yes to save your edits.**

YOUR TURN

Click the Full Extent button. Repeat step 2 to roughly digitize polygons around the remaining seven commercial polygons. After completing each polygon, click Editor, Save Edits. When you complete the final polygon, close your edit session.

Note: You must complete this Your Turn *before moving on in the exercises.*

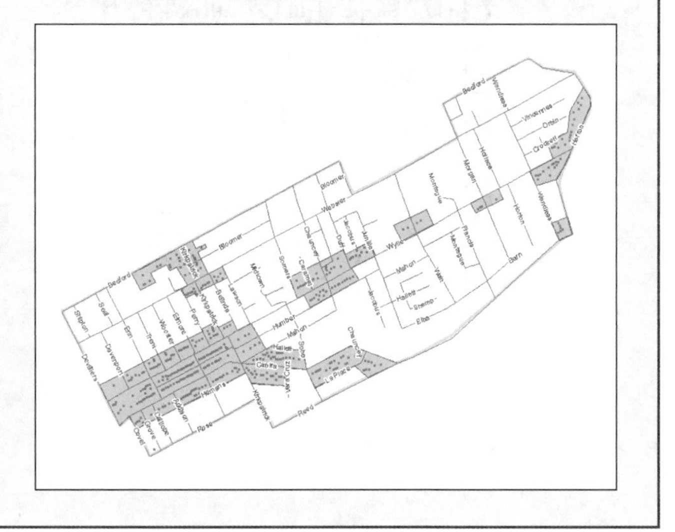

Add feature attribute data

Now that you have digitized the commercial polygons, you will assign zone numbers to them.

1 **Right-click the CommercialZone layer in the table of contents.**

2 **Click Open Attribute Table.**

The Attributes of CommercialZone table has three columns named FID, Shape, and ID, all created by ArcMap.

3 **In the Attributes of CommercialZone, click Options, Add Field.**

If you are unable to select this option, go back to the Editor toolbar, click Stop Editing, and then try again to add a field.

4 **In the Name field, type ZoneNumber, leave the type as Short Integer, then click OK.**

5 **Click Editor, Start Editing, click in the top cell of the ZoneNumber field, type 1, and press Enter.**

6 **In sequential order, continue numbering the remaining cells in the ZoneNumber field.**

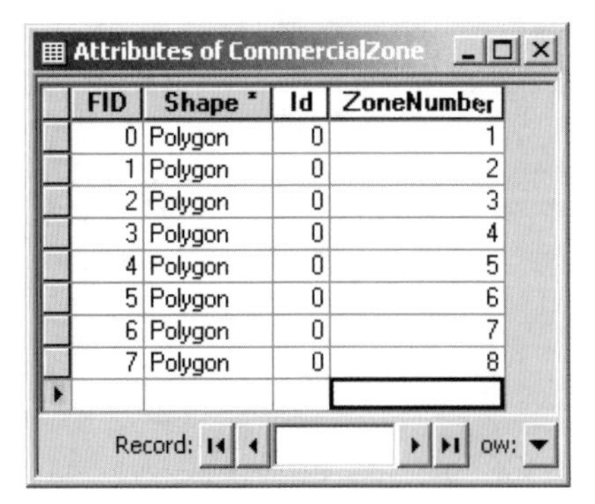
Attributes of CommercialZone

FID	Shape *	Id	ZoneNumber
0	Polygon	0	1
1	Polygon	0	2
2	Polygon	0	3
3	Polygon	0	4
4	Polygon	0	5
5	Polygon	0	6
6	Polygon	0	7
7	Polygon	0	8

Record: ow:

7 **Click Editor, Stop Editing.**

8 **Click Yes to save your edits.**

Label the commercial zones

1 Turn off the Commercial Properties layer.

2 In the table of contents, right-click the CommercialZone layer and click Properties.

3 Click the Labels tab.

4 In the Text String frame, choose ZoneNumber from the Label Field drop-down list.

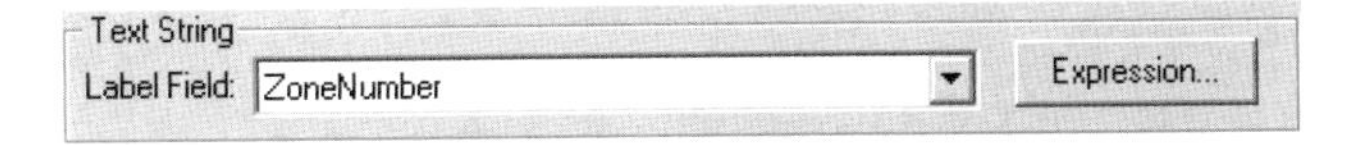

5 In the Text Symbol frame, set the font's type to Arial, the size to 12, and the style to Bold.

6 Click OK.

7 Right-click the CommercialZone layer and click Label Features.

Your label numbers may not match those below, depending on your order of digitizing.

8 Save the map document.

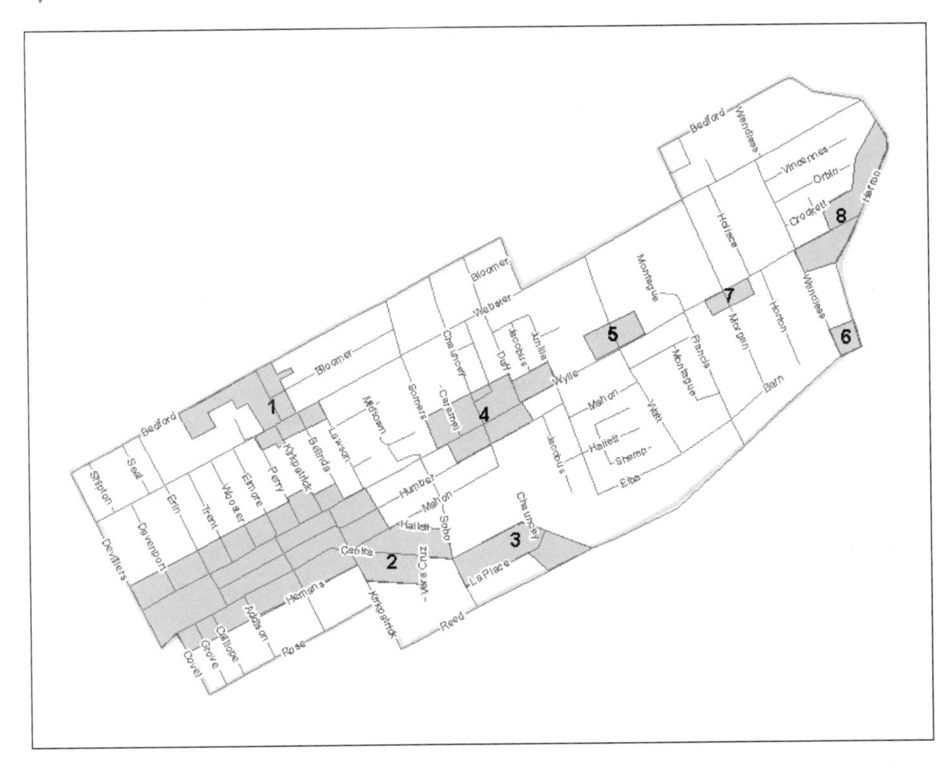

Digitize a point layer

Many local agencies use GIS for homeland security. Two GIS layers common to emergency preparedness applications are evacuation routes and shelter facilities. In this exercise, you will digitize shelter locations as points.

Create a point layer for evacuation shelters

1 Start ArcCatalog.

2 In the catalog tree, browse to **\Gistutorial\PAGIS**, then click the **MidHill** folder.

3 Click File, New, Shapefile.

4 In the Name field, type **EvacShelter**.

5 From the Feature Type drop-down list, choose Point.

6 Click Edit, Import, browse to the **\Gistutorial\PAGIS\Midhill** folder, click **Streets.shp**, Add, OK.

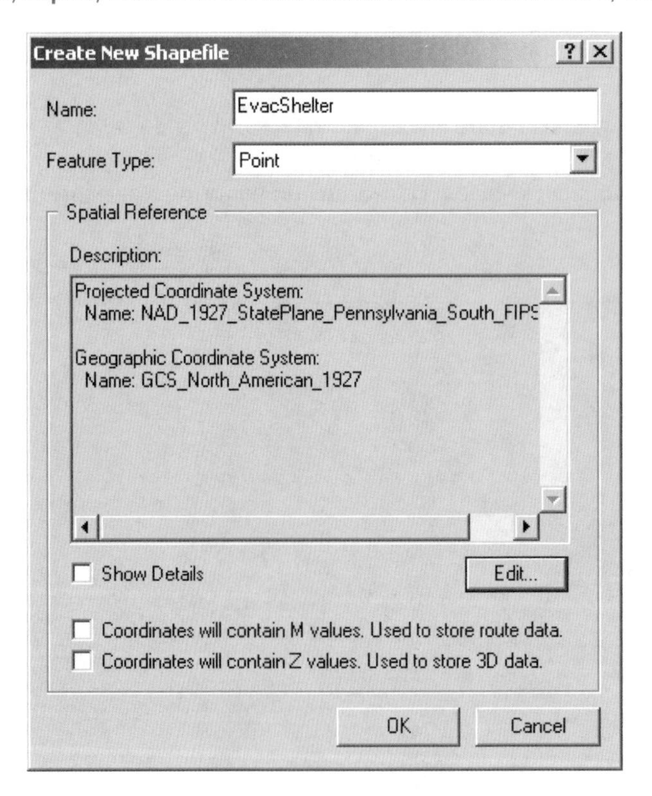

7 Click OK.

8 Close ArcCatalog.

Add evacuation shelter points

1 Open **Tutorial6-2.mxd** from the **\Gistutorial** folder.

2 Click the Add Data button, navigate to **\Gistutorial\MidHill**, and add **EvacShelter.shp** to the map.

3 In the table of contents, check EvacShelters.shp, Building Rooftops, Street Centerlines, and Middle Hill Neighborhood to turn them on. Turn off all other layers in the map.

4 In the table of contents, click the legend symbol for the EvacShelters layer.

5 Change the symbol to Square 2, the color to Big Sky Blue, and the size to 10.

6 Click OK.

7 From the Editor toolbar, click Editor, Start Editing.

8 If prompted, click **\Gistutorial\PAGIS\MidHill** as the folder from which to edit the shapefile.

Be sure the Target layer is EvacShelter.

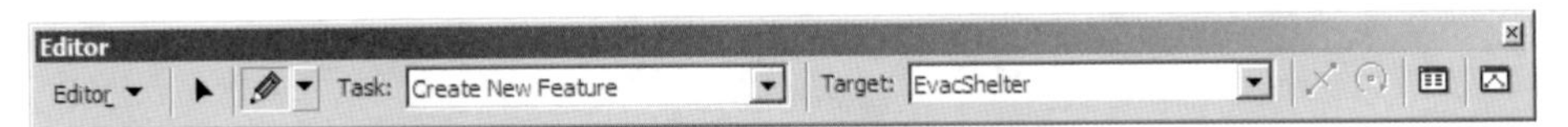

9 Click the Sketch tool. 

10 The blue squares in the map below represent the shelter locations. Using the Sketch tool, click the corresponding locations in your map to add the shelter points to the EvacShelters shapefile.

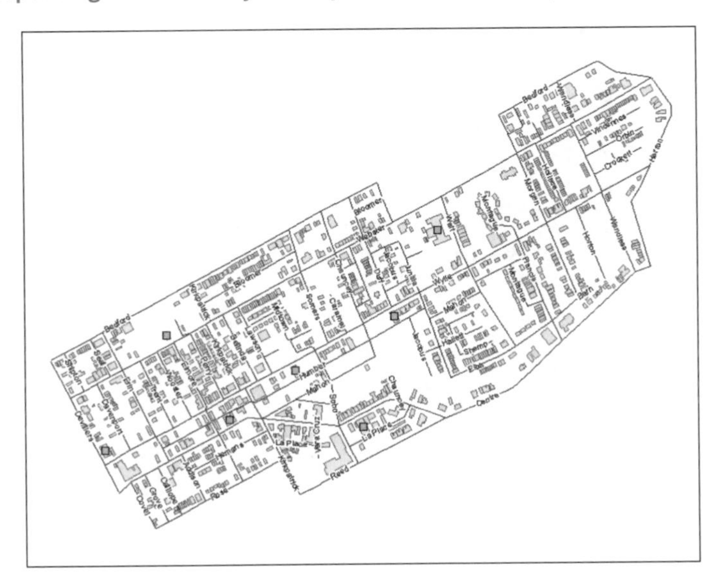

11 When you are finished adding the evacuation shelter points, click Editor, Stop Editing.

12 Click Yes to save edits to **EvacShelter.shp**.

Add a name field to the EvacShelter attribute table

1 In the table of contents, right-click the **EvacShelter.shp** layer.

2 Click Open Attribute Table.

3 In the Attributes of EvacShelter, click Options, Add Field.

4 In the Name field, type **Name**. From the Type drop-down list choose Text, then click OK.

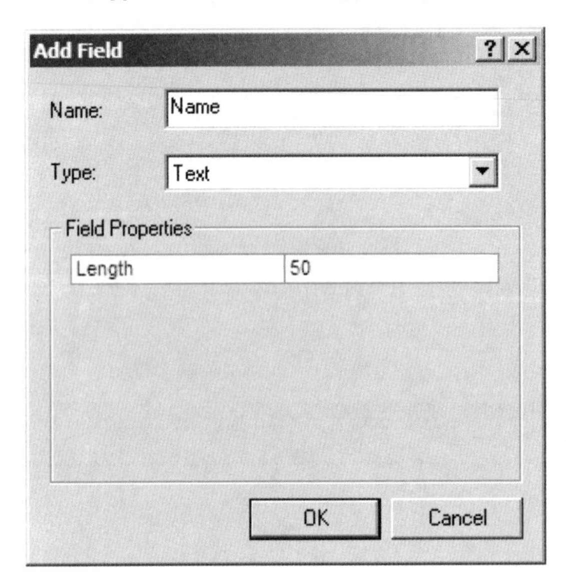

Add name attributes to the EvacShelter records

1 **From the Editor toolbar, click Editor, Start Editing.**

2 **If prompted, click \Gistutorial\PAGIS\MidHill as the folder from which to edit the shapefile.**

If the dialog box asking for the layer does not appear, make sure that the EvacShelter is the Target layer on the Editor toolbar.

3 **In the Attributes of EvacShelter table, click the small gray box to the left of the first record in the table.**

This will highlight the record in the table and the related feature in the map.

Attributes of EvacShelter

FID	Shape*	Id	Name
4	Point	0	
5	Point	0	
6	Point	0	

Record: 1 Show: All Selected Records (1 out of 9 Selected.) Options

4 **Using the map and table shown below as your key, add the Id and Name attributes to the selected record, then repeat the process to add the Id and Name attributes to the remaining records.**

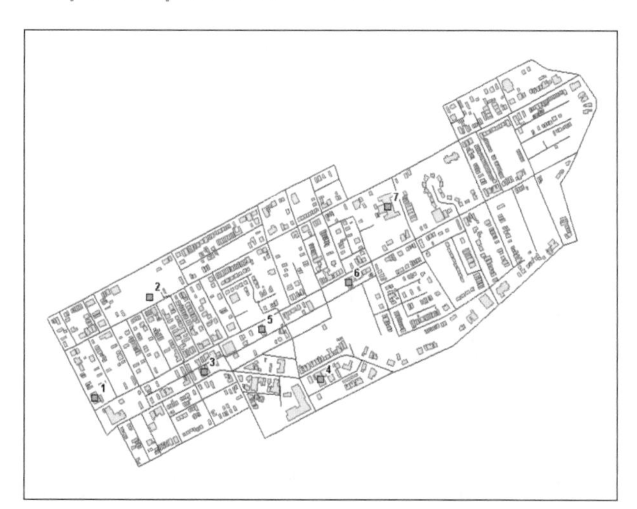

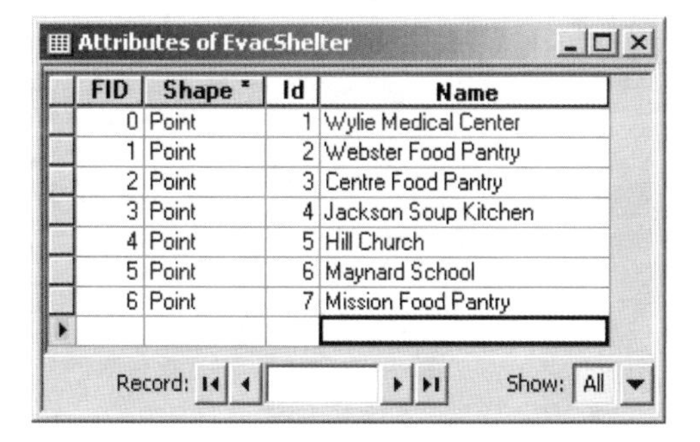

Attributes of EvacShelter

FID	Shape *	Id	Name
0	Point	1	Wylie Medical Center
1	Point	2	Webster Food Pantry
2	Point	3	Centre Food Pantry
3	Point	4	Jackson Soup Kitchen
4	Point	5	Hill Church
5	Point	6	Maynard School
6	Point	7	Mission Food Pantry

Record: Show: All

5 **When you are finished adding the attributes, from the Edit menu choose Stop Editing, then click Yes to save your edits. Leave ArcMap open.**

Digitize a line layer

An evacuation route can now be created to and from the evacuation shelters. This is accomplished by digitizing a line feature.

Create a line shapefile for an evacuation route

1 Start ArcCatalog.

2 In the catalog tree, browse to **\Gistutorial\PAGIS**, then click the **MidHill** folder.

3 Click File, New, Shapefile.

4 In the Name field, type **EvacRoute**.

5 From the Feature Type drop-down list, choose Polyline.

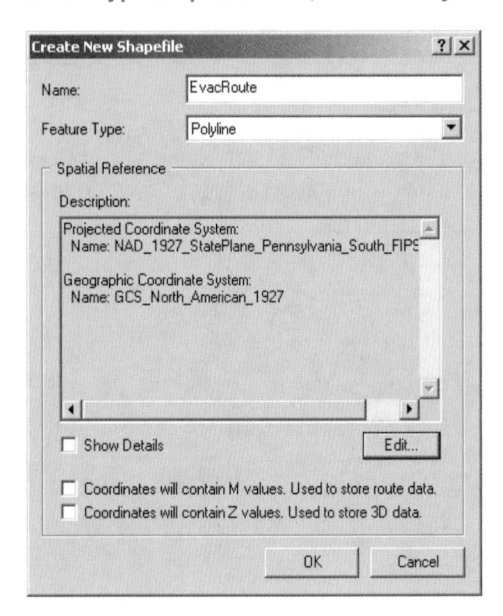

6 Click Edit, Import, browse to the **\Gistutorial\PAGIS\MidHill** folder, click **Streets.shp**, Add, OK.

7 Click OK.

8 Close ArcCatalog.

Change the line symbol for the evacuation route

1 In ArcMap, add **EvacRoute.shp** to the map.

2 In the table of contents, click EvacRoute's line symbol to open the Symbol Selector.

3 In the Symbol Selector, scroll far down the list of line symbols to find and select the Arrow at End symbol.

4 Change the Color to Mars Red and Width to 2.

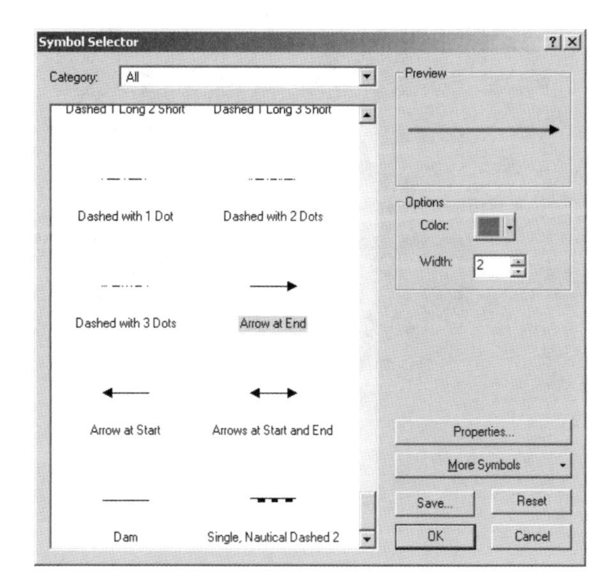

5 Click OK.

Prepare area for digitizing and start editing

1 Click the Zoom In button,  **then use it to zoom to the western half of the Middle Hill neighborhood.**

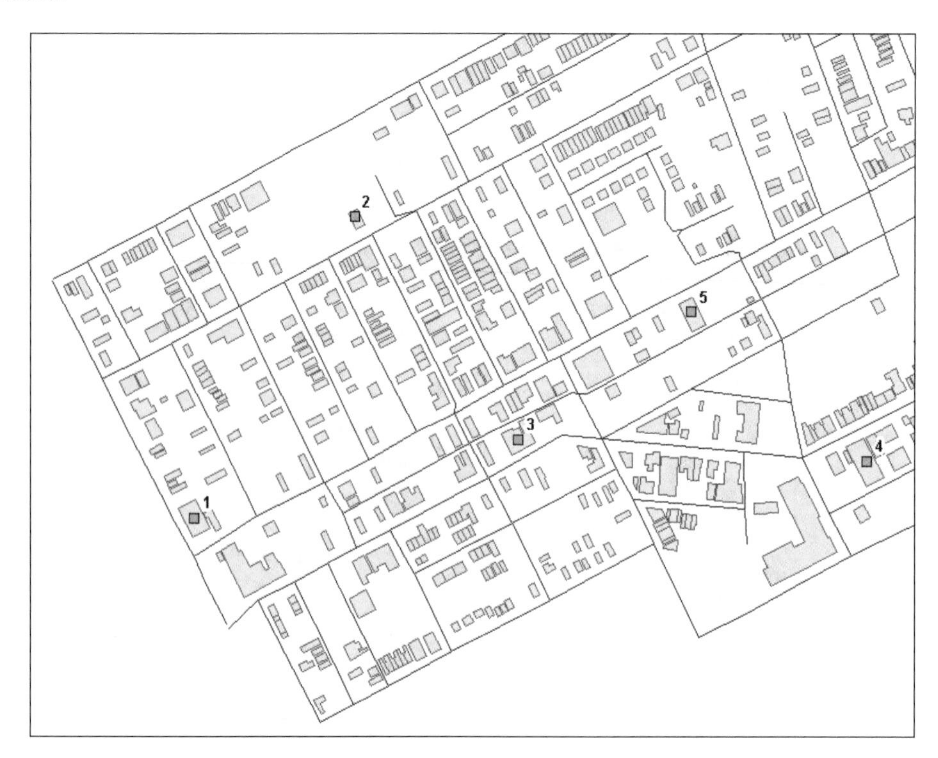

2 From the Editor toolbar, click Editor, Start Editing.

3 If the Start Editing dialog box appears, click the \Gistutorial\PAGIS\MidHill folder that will allow you to edit the EvacRoute layer, then click OK.

Make sure the Target layer is EvacRoute and the Task is Create New Feature.

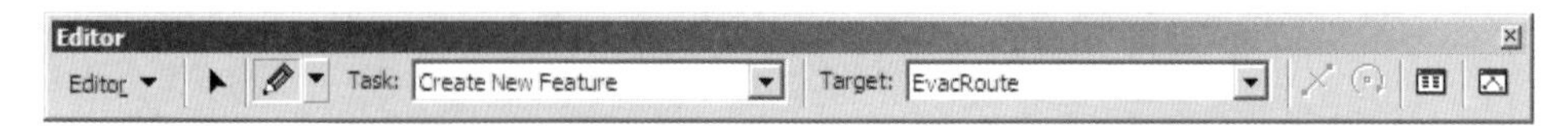

Digitize by snapping to features

When digitizing, you can snap the vertices you're adding to existing features. Snapping is used to reduce digitizing errors by ensuring that the features which must be connected actually *are* connected. Within ArcMap you can snap to vertices, or endpoints, anywhere along a line or polygon boundary (called edge snapping), or to the midpoint of a line.

1 Click the Sketch tool. 

2 Click Evacuation Shelter 1 to add the route's starting point.

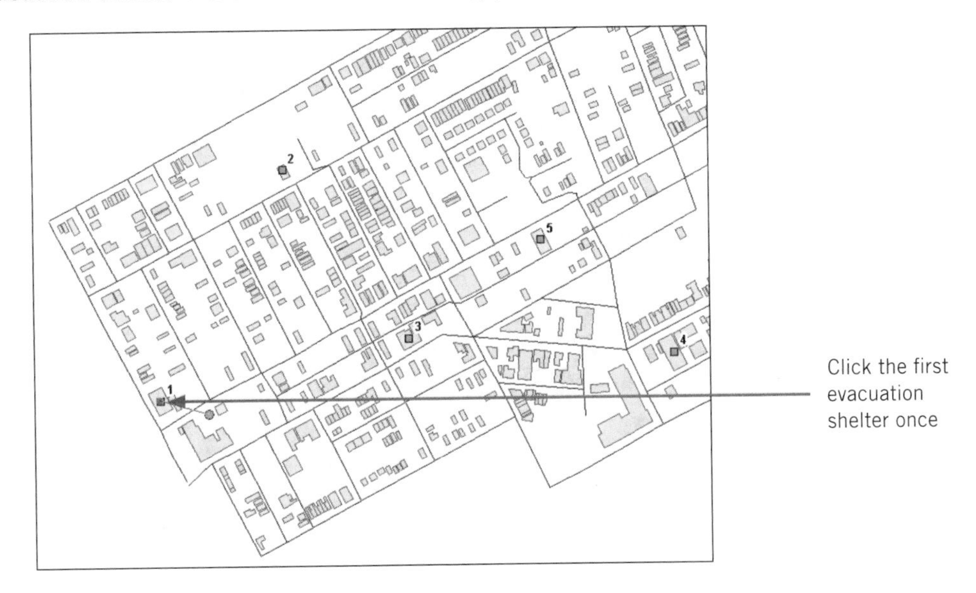

3 Move the cursor to the nearest street, right-click, and click Perpendicular from the context menu.

4 Click Center Avenue once to place a vertex there.

ArcMap will force the line to be perpendicular with the street centerline.

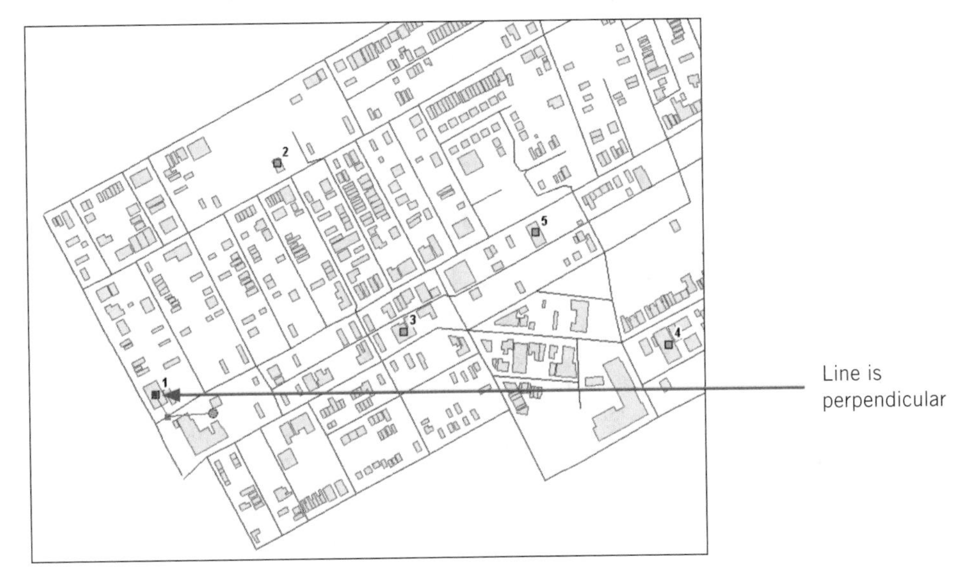

5 Move the cursor to the first intersection on the street centerline, right-click, and click Snap to Feature, Endpoint.

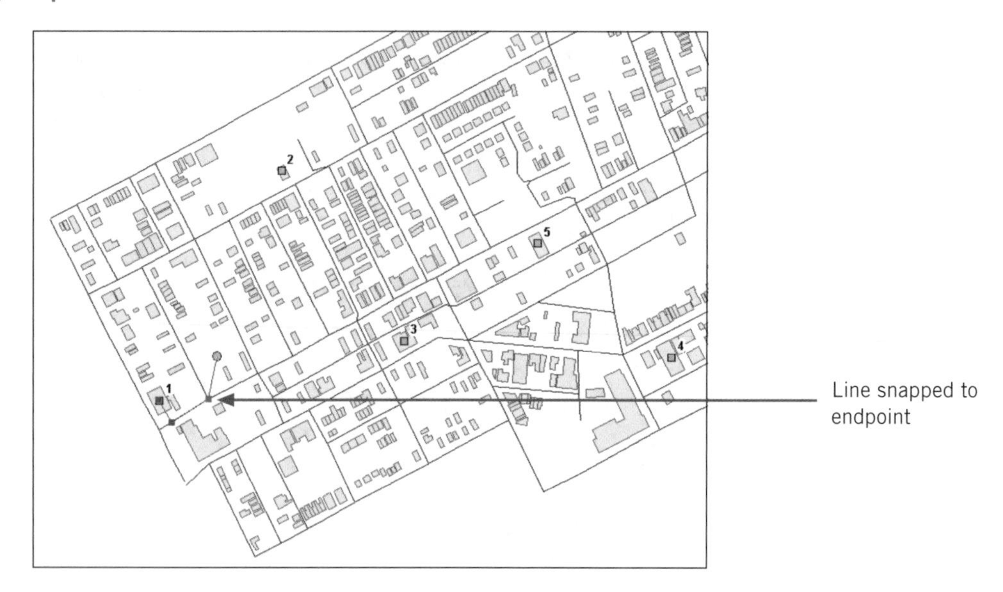

6 Continue snapping to street intersections along the evacuation route shown below. Double-click to finish the route at Evacuation Shelter 2. (If necessary, use other snapping options from the Snap to Feature context menu.)

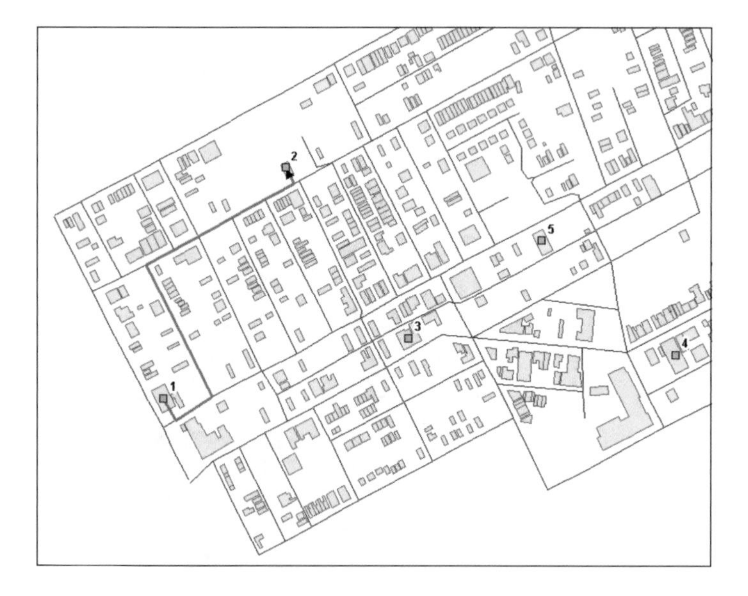

Save your edits and the map

1 From the Editor toolbar, click Editor, Stop Editing.

2 Click Yes to save your edits.

3 Save your map document.

YOUR TURN

Use the Sketch tool to digitize the remaining line segments connecting the evacuation shelters.

Use snapping tools for Endpoints, Midpoints, Edges, and Vertices.

Use the function key shortcuts for faster digitizing.

Use the Undo button if you make a mistake.

Label the evacuation shelters by name instead of number.

Save the edits and map document as Tutorial6-2 when you are finished.

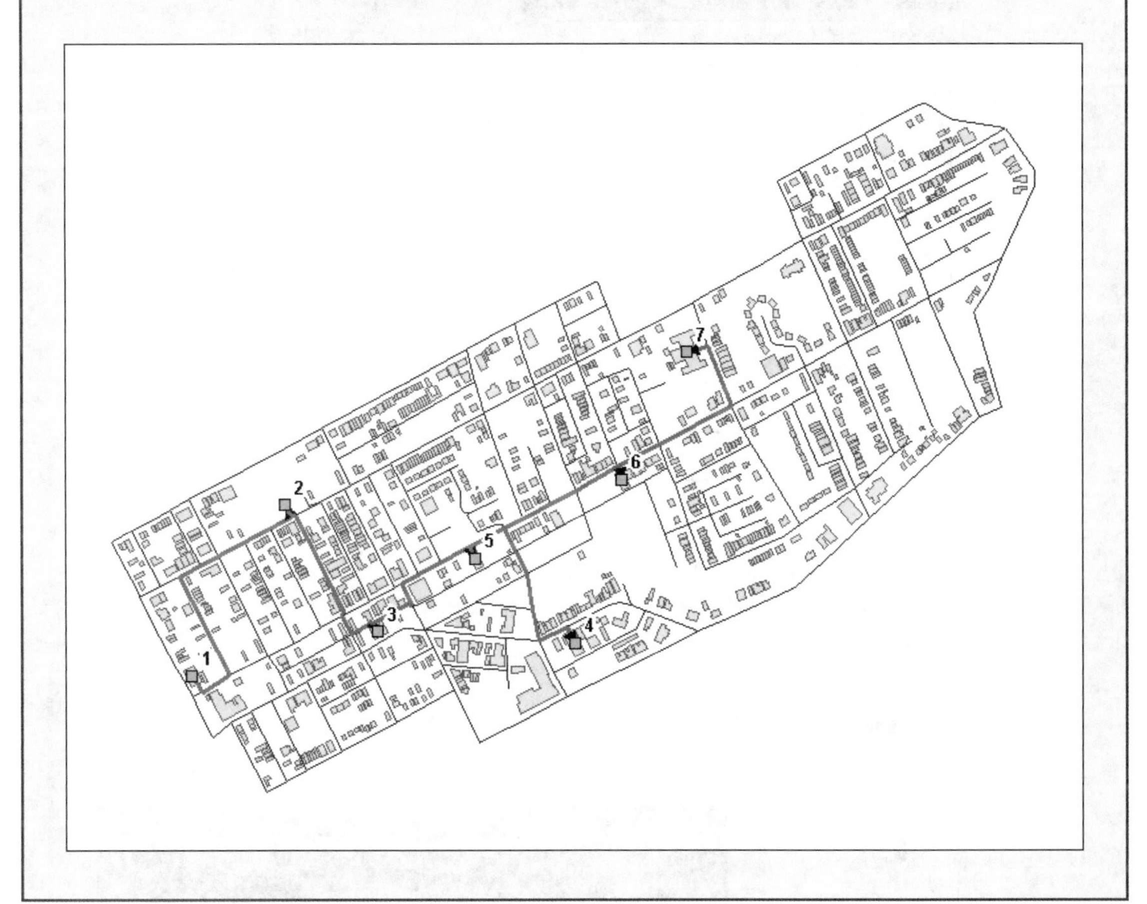

Spatial adjustment

The ArcMap spatial adjustment tools are used to transform, rubber sheet, and edge match features in a shapefile or geodatabase feature class. In this exercise, you will transform an outline of a building so that it correctly overlays an aerial photograph.

Add aerial photos to a map

1 In a new, empty map, click the Add Data button.

2 Browse to the **\Gistutorial\CMUCampus** folder and add **25_45.tif** and **26_45.tif** to the map.

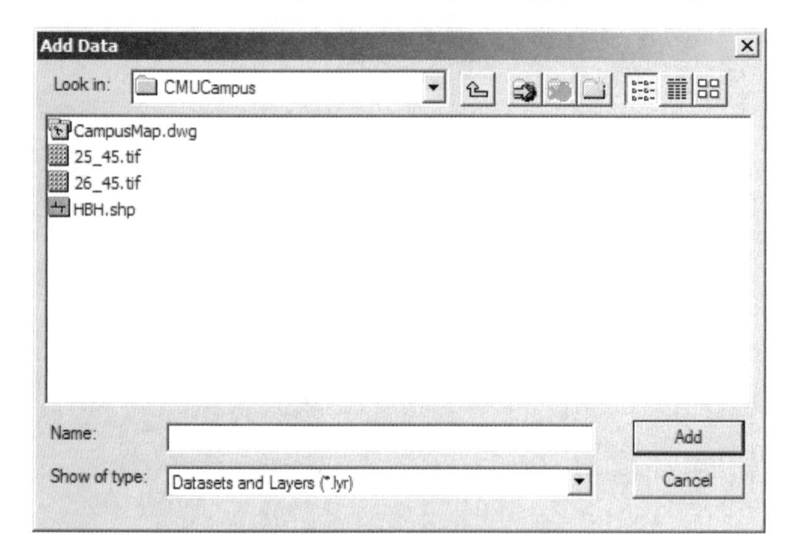

The two images you just added to ArcMap were obtained from the Southwestern Pennsylvania Commission *(spcregion.org)* 2000–2001 Aerial Photography program. Both aerial photos are georeferenced to State Plane South NAD 1983.

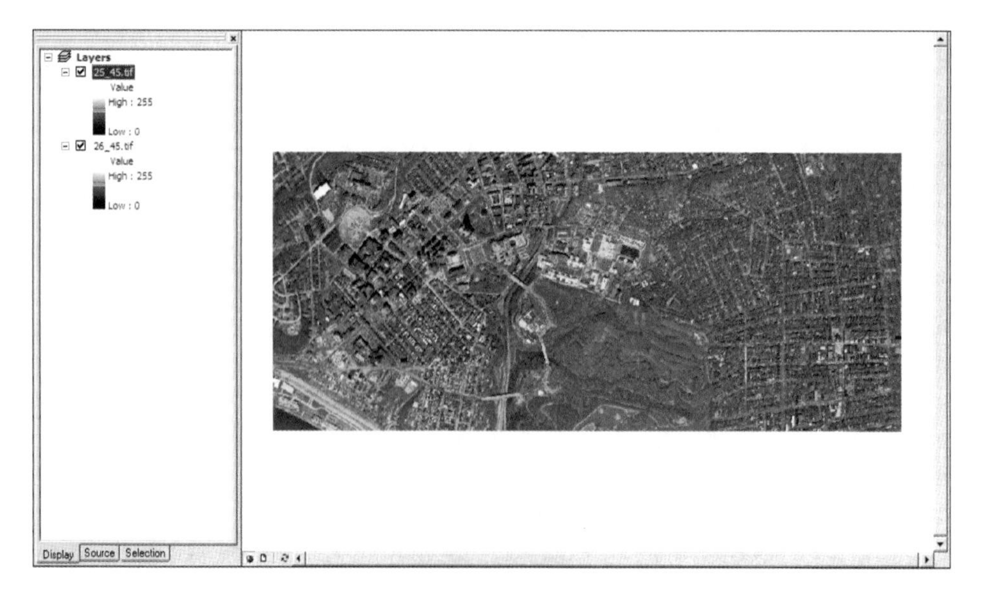

Adjust the transparency values of the aerial photos

1 In the table of contents, right-click **25_45.tif** and click Properties.

2 Click the Display tab.

3 In the Transparency field, type **20**.

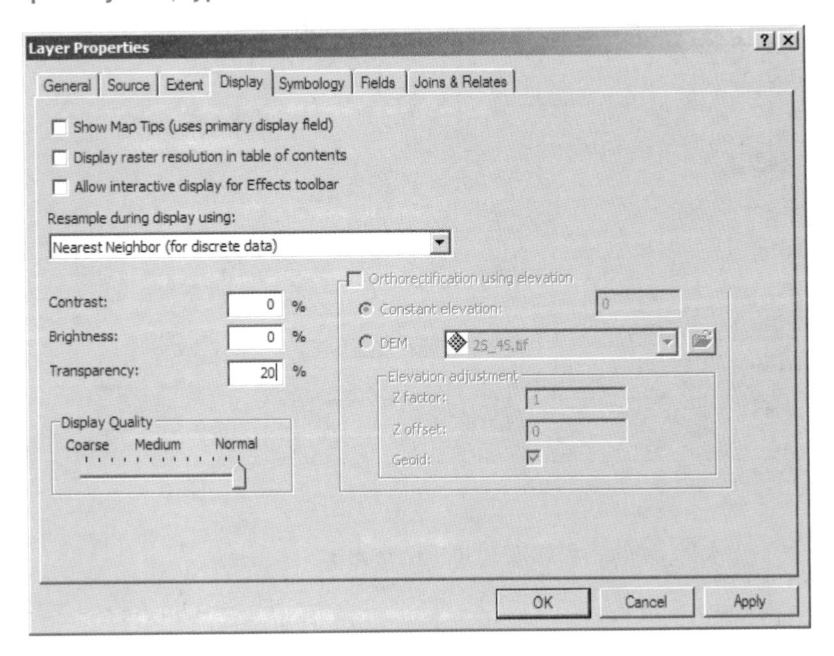

4 Using the same method outlined in the previous three steps, set the transparency value of the 26_45.tif layer to 20%.

5 In the map display, zoom to the Hamburg Hall building of the Carnegie Mellon University (CMU) campus as shown below.

Add a building outline

1 **Click the Add Data button.**

2 **Browse to the \Gistutorial\CMUCampus folder and add HBH.shp to the map.**

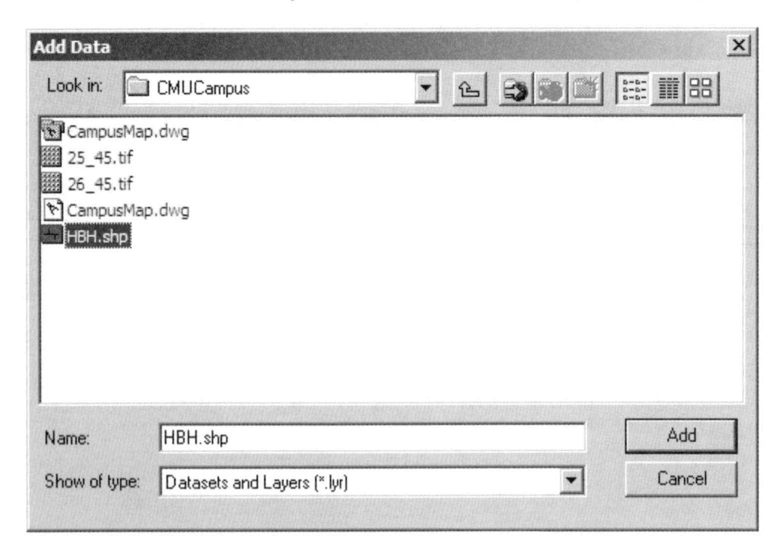

3 **In the table of contents, rename the HBH layer Hamburg Hall.**

4 **Change Hamburg Hall's symbol color to Mars Red and its symbol width to 1.5.**

The Hamburg Hall layer originated as a CAD drawing from the university's facilities management department. As you can probably tell by looking at the map, the Hamburg Hall layer is not in proper alignment or scale with the buildings shown on the aerial photos. In the next steps, you will adjust the Hamburg Hall layer so that it properly aligns with the aerial photo.

Move the building

1 From the Editor toolbar, click Editor, Start Editing, then in the Starting to Edit In a Different Coordinate System window click Start Editing.

2 Make sure that Hamburg Hall is the target and click the Edit tool. 

3 Click the outline of the Hamburg Hall building.

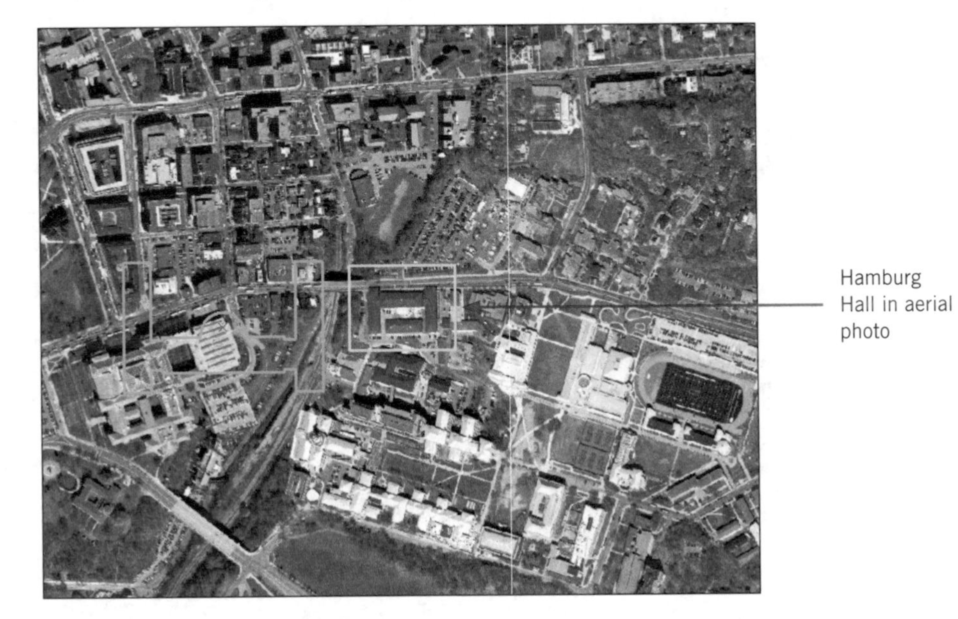

4 Place your mouse cursor directly over the outline of the Hamburg Hall feature so that the cursor icon changes to a four-headed arrow.

5 Click and drag the Hamburg Hall feature to the location shown above.

6 Zoom in to the building feature, as shown below.

Rotate the building

1. Click the Edit tool and double-click the outline of the Hamburg Hall shapefile to activate grab handles.
2. From the Editor toolbar, click the Rotate tool. 
3. Click the lower-right grab handle of Hamburg Hall and, while holding down the mouse button, rotate it 180 degrees, as shown below.

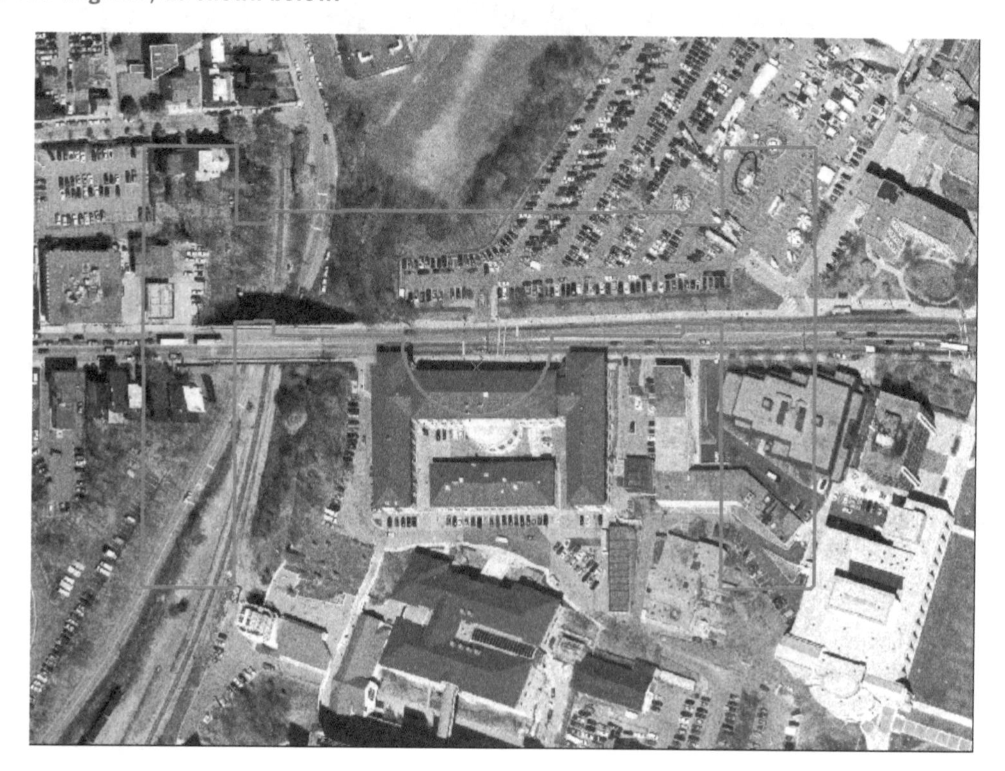

Transform the building to the aerial photo

Add displacement links

1 **Click Editor, More Editing Tools, and Spatial Adjustment.**

This opens the Spatial Adjustment toolbar.

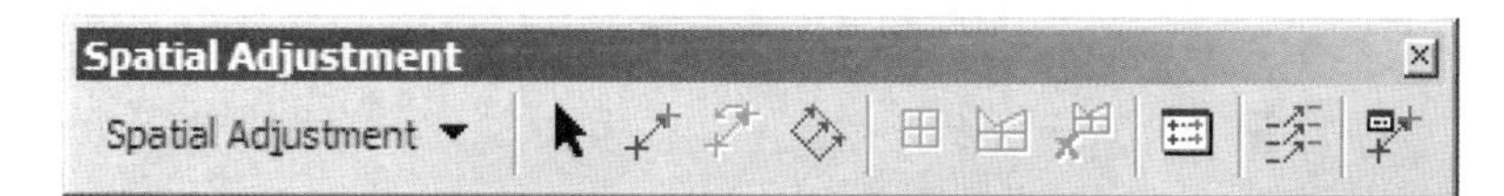

2 **Click Spatial Adjustment, Adjustment Methods, Transformation - Similarity.**

3 **Click the New Displacement Link button.** 

4 **Click the upper left corner of the Hamburg Hall building feature.**

5 **Click the corresponding location on the aerial photo.**

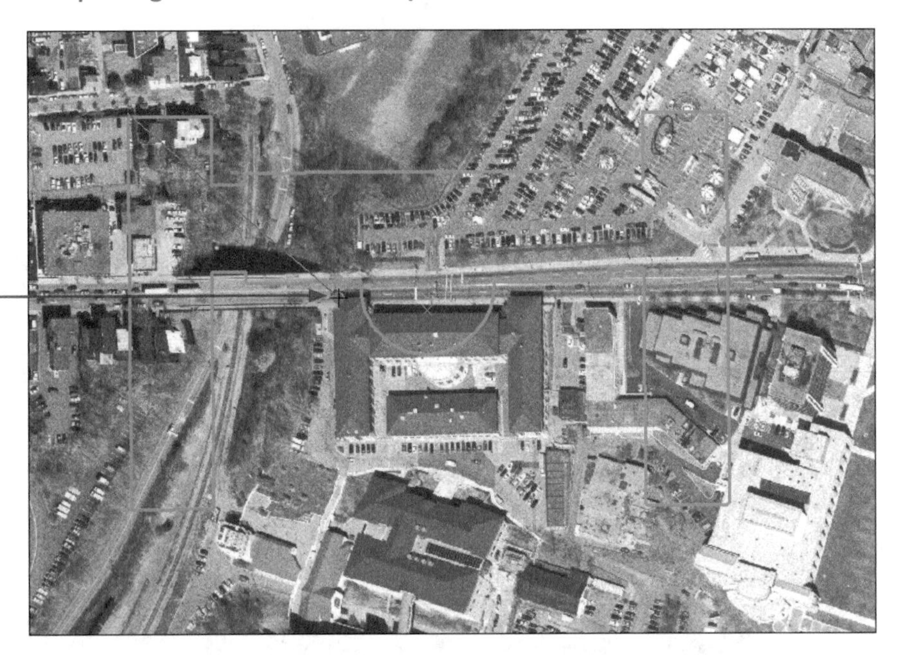

6 **Continue adding displacement links to the building feature and the aerial photo, as shown below.**

Edit displacement links

If you select the wrong position on the building or map, you can use the edit displacement tools to adjust your picks.

1 From the Spatial Adjustments toolbar, click the Select Elements button.

2 Click one of the displacement links.

3 Click the Modify Links button.

4 Click and drag the link to a new position.

5 Drag the link back to its original location, if necessary.

Adjust the building

1 From the Spatial Adjustment toolbar, click Spatial Adjustment, Adjust.

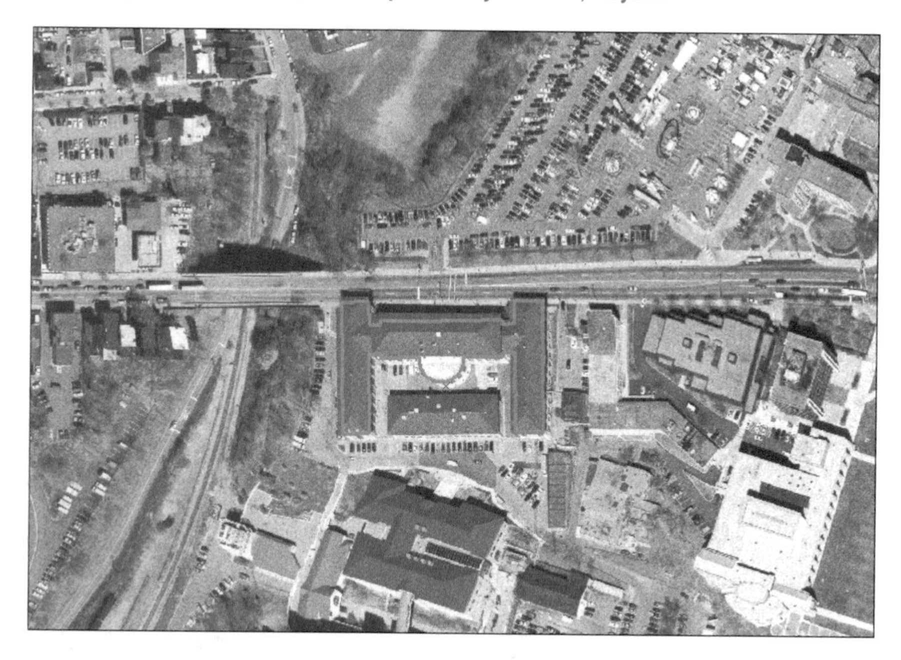

Based on the displacement links you set, the geometry of the Hamburg Hall feature is scaled down in size and moved to match the geometry of the feature in the aerial photo. If the resulting match is not very good, select the Hamburg Hall feature, redefine new displacement links, and run the Adjust command again.

2 Stop editing and save your edits.

3 Save your map as **\Gistutorial\Tutorial6-3.mxd**.

Assignment 6-1

Digitizing police beats

Community-oriented police officers are responsible for preventing crime and solving underlying community problems related to crime. Among other activities, these officers walk "beats," which are small networks of streets that define areas. Often the beats are designed in cooperation with community leaders who help set policing priorities. Beats change as problems are solved and priorities change. Hence, it is good to have the capability to digitize and modify police beats.

In this exercise, you will digitize two new polygon police beats for Pittsburgh's Zone2 Police District based on street centerlines that make up these beats.

Start with the following:

- **C:\Gistutorial\PAGIS\Zone2\StreetsZone2.shp**—TIGER streets for Zone 2 Police District.
- **C:\Gistutorial\PAGIS\Zone2\Zone2.shp**—polygon layer for boundary of Zone2 Police District.

Create a police beat map

In ArcCatalog, create two new polygon shapefiles called Beat1.shp and Beat2.shp in \Gistutorial\Answers\Assignment6. Import spatial reference data from Zone2.shp. See the guidelines on the next page for what streets should make up the beats. To each beat shapefile add a new attribute called Name with text data type. Enter values Beat #1 and Beat #2 for this attribute in the two respective shapefiles. Use the results to label the beats.

In ArcMap, create a new map document called C:\Gistutorial\Answers\Assignment6\Assignment6-1.mxd with a layout showing a map with an overview of Police Zone 2 and the newly digitized beats and maps zoomed into beats 1 and 2. Show the beats with thick line widths and bright, distinctive colors, and streets as lighter "ground" features. In the overview map, label the beats "Beat #1" and "Beat #2" and label the streets in the detailed maps. Include a scale bar in feet. See hints for digitizing.

Export two JPEG files of the map zoomed to each beat called Beat1.jpg and Beat2.jpg to C:\Gistutorial\Answers\Assignment6. Insert the two images in a Word document saved as C:\Gistutorial\Answers\Assignment6\Assignment6-1.doc.

Street centerline guides for Beat #1

There are twenty-one street segments making up beat #1. Digitize the boundary made up of the outermost streets. It is roughly a rectangle.

- 1 through 199—17th St (four segments)
- 1 through 99—18th St (two segments)
- 1 through 199—19th St (one segment)
- 1 through 199—20th St (four segments)
- 1 through 99—Colville St (one segment)
- 1700 through 1999—Liberty Ave (one segment)
- 1700 through 1999—Penn Ave (three segments)
- 1700 through 1999—Smallman St (four segments)
- 1700 through 1999—Spring Way (one segment)

Street centerline guides for Beat #2

There are twenty street segments making up beat #2. Digitize similar to Beat #1.

- 100 through 299—7th St (two segments)
- 1 through 299—8th St (three segments)
- 100 through 299—9th St (four segments)
- 800 through 899—Exchange Way (one segment)
- 700 through 899—Ft Duquesne Blvd (three segments)
- 700 through 899—Liberty Ave (three segments)
- 100 through 199—Maddock Pl (one segment)
- 700 through 899—Penn Ave (three segments)

Selecting streets hint

Open the feature attribute table for the streets. Move the table so you can see both the table and the streets on the map. Move the NAME field to the left of L_F_ADD, select both columns, and sort ascending. Make multiple selections for a given beat in the table by simultaneously holding down the Ctrl key and clicking rows corresponding to the beat's street segments. The streets layer is a TIGER file map with TIGER-style address number data, so look for street number ranges in the following fields: L_F_ADD, L_T_ADD, R_F_ADD, and R_T_ADD.

Snapping hint

You want your new beat polygons to match the street segments exactly. To do this, you need to use the ArcMap snapping functions. Use the snapping shortcut functions for Endpoint [Ctrl + F5] and Finish Sketch [F2] to snap to existing lines.

Assignment 6-2

Using GIS to track campus information

GIS is a good tool to create "way finding" information maps. These maps can be used in many organizations that have large campuses or complicated buildings (for example, airports, hospitals, office parks, colleges, and universities).

In this exercise, you will create a map of Carnegie Mellon's campus by spatially adjusting buildings to an aerial photo map of the campus. You will also digitize layers showing bus stop and parking lot locations.

Start with the following:

- **C:\Gistutorial\CMUCampus\25_45.tif and 26_45.tif**—digital orthographics of CMU Campus provided by the Southwestern Pennsylvania Commission in the Pennsylvania State Plane South NAD 1983 projection.
- **C:\Gistutorial\CMUCampus\CampusMap.dwg**—CAD drawing of CMU Campus provided by the CMU facilities management department.
- **C:\Gistutorial\PAGIS\PghStreets.shp**—street centerlines useful for locating bus stops.

Add to the campus map

In ArcCatalog, create a new polygon shapefile called CMUParking.shp and a new point shapefile called CMUBusStops.shp saved in C:\Gistutorial\Solutions\Assignment6\. Import spatial reference data from one of the TIF raster map layers.

In ArcMap, create a new map document called C:\Gistutorial\Answers\Assignment6\Assignment6-2.mxd with a layout that shows the aerial photos of the CMU campus and the academic buildings spatially adjusted to match the buildings in the aerial photo. Digitize new shapefiles showing where parking lots and bus stops are located. Show the parking lots as semi-transparent polygons so you can see the parking lots in the aerial photo. Set the transparency of the aerial photos to 20 or 30 percent (using the Display tab of the Layer Properties sheet) to better see the new shapefiles.

Create an 8½ × 11-inch landscape layout with a title, map, and legend. Export the finished layout to a JPEG file saved as C:\Gistutorial\Answers\Assignment6\Assignment6-2.jpg. Insert the map image in a Word document with landscape page setup and save the result as C:\Gistutorial\Answers\Assignment6-2.doc.

CAD drawing and spatial adjustment hints

- Add the CMU Campus CAD drawing as polyline features. In the Add Data dialog box, double-click CampusMap.dwg, Polyline. In the table of contents, right-click CampusMap.dwg Polyline and click Properties. Click the Drawing Layers tab and leave only Academic_Bldgs turned on. Export the buildings as a new shapefile called AcademicBldgs.shp to \Gistutorial\Answers\Assignment6\.
- Use ArcMap's editing tools to move the buildings closer to the aerial image. Then use spatial adjustment tools and zooming functions to adjust the buildings to the aerial photo. Continue using editing tools, such as Move and Rotate, to adjust the buildings according to the aerial photo below.
- It is tricky to transform the academic buildings to the raster map. If you get an approximate transformation that has some mismatches, that is acceptable. Do the following steps: Zoom to the full extent to see the raster images and academic buildings. Using the Edit tool, Select all of the academic buildings in AcademicBldgs.shp. Drag the buildings to be adjacent to the raster images. Select buildings on the four corners of campus in the shapefile and points on them to match up with the raster map image. On the Spatial Adjustment toolbar, use the New Displacement Link to roughly draw four lines from the buildings layer to the raster image. Zoom in to a point on the buildings map, click the corresponding link with the Select Elements tool on the Spatial Adjustment tool bar, click the Modify Link button, and move the link endpoint to be more precise. Do the same on the raster image side of the link. Repeat for the other three links.

Digitize new shapefiles hint

Digitize four separate parking lot polygon features in or around campus. You can see the parking lot locations because there are cars in the parking lots in the aerial photo. Digitize bus stops at the intersections of Forbes Avenue with Devon, Morewood, Neville, Beeler, and Margaret Morrison streets.

What to turn in

If you are working in a classroom setting with an instructor, you may be required to submit the exercises you created in tutorial 6. Below are the files you are required to turn in. Be sure to use a compression program such as PKZIP or WinZip to include all three files as one .zip document for review and grading. Include your name and assignment number in the .zip document (YourNameAssn6.zip). *Do not* turn in interim files that are not in your final map.

ArcMap documents

C:\Gistutorial\Answers\Assignment6\Assignment6-1.mxd
C:\Gistutorial\Answers\Assignment6\\Assignment6-2.mxd

Exported maps

C:\Gistutorial\Answers\Assignment6\Assignment6-1.doc
C:\Gistutorial\Answers\Assignment6\Assignment6-2.doc

Shapefiles

C:\Gistutorial\Answers\Assignment6\Beat1.shp
C:\Gistutorial\Answers\Assignment6\Beat2.shp

C:\Gistutorial\Answers\Assignment6\CMUParking.shp
C:\Gistutorial\Answers\Assignment6\CMUBusStops.shp
C:\Gistutorial\Answers\Assignment6\AcademicBldgs.shp

OBJECTIVES

Geocode data by ZIP Code
Geocode to streets
Prepare data and street maps
Interactively locate addresses
Perform batch geocoding
Correct street layer addresses
Use alias files

GIS Tutorial 7

Geocoding

The process used to plot address data as points on a map is referred to as address geocoding. You can geocode addresses to different levels such as ZIP Codes or streets, depending on the type of addresses you want to map and the type of reference data to which you are matching your address records. In this tutorial, you will learn to perform address geocoding using tables of address data, TIGER street centerlines, and ZIP Code polygons obtained from the U.S. Census. You will also learn how to find and fix errors in the address data you used for geocoding.

Geocode data by ZIP Code

Geocoding to ZIP Codes is a common practice for many organizations. For example, many stores ask for your ZIP Code as you check out. This is a useful marketing method to learn where customers reside.

In this chapter, you will match attendees for an art event sponsored by an arts organization in Pittsburgh, Pennsylvania, called FLUX. The event planners of FLUX would like to know from where those attending their functions come. First you will geocode a table of FLUX attendees by ZIP Code, and later in the chapter you will geocode the attendee data using their complete street address.

Map of Pennsylvania ZIP Codes.

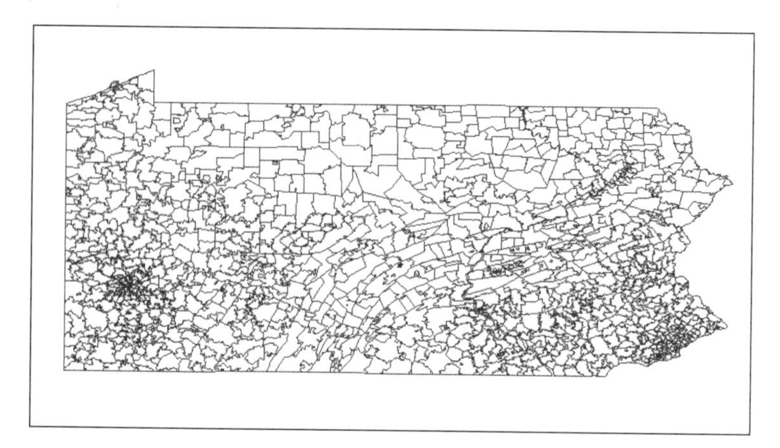

Database table of FLUX attendees.

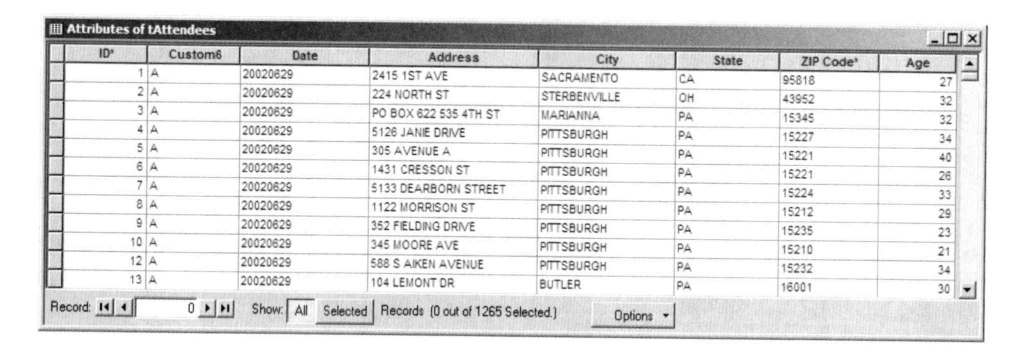

Attributes of tAttendees

ID*	Custom6	Date	Address	City	State	ZIP Code*	Age
1	A	20020629	2415 1ST AVE	SACRAMENTO	CA	95818	27
2	A	20020629	224 NORTH ST	STERBENVILLE	OH	43952	32
3	A	20020629	PO BOX 622 535 4TH ST	MARIANNA	PA	15345	32
4	A	20020629	5126 JANIE DRIVE	PITTSBURGH	PA	15227	34
5	A	20020629	305 AVENUE A	PITTSBURGH	PA	15221	40
6	A	20020629	1431 CRESSON ST	PITTSBURGH	PA	15221	26
7	A	20020629	5133 DEARBORN STREET	PITTSBURGH	PA	15224	33
8	A	20020629	1122 MORRISON ST	PITTSBURGH	PA	15212	29
9	A	20020629	352 FIELDING DRIVE	PITTSBURGH	PA	15235	23
10	A	20020629	345 MOORE AVE	PITTSBURGH	PA	15210	21
12	A	20020629	588 S AIKEN AVENUE	PITTSBURGH	PA	15232	34
13	A	20020629	104 LEMONT DR	BUTLER	PA	16001	30

Record: 0 Show: All Selected Records (0 out of 1265 Selected.) Options

Open the Pennsylvania ZIP Code map

1 **Start ArcMap and open \Gistutorial\Tutorial7-1.mxd.**

Add the FLUX Attendee data file

1 **Click the Add Data button.**

2 **Navigate to \Gistutorial\Flux, double-click FluxEvent.mdb, and click tAttendees.**

3 **Click Add.**

4 **In the table of contents, right-click the tAttendees table, and click Open.**

The attributes of tAttendees contains the addresses and ages of all the attendees of a recent FLUX event.

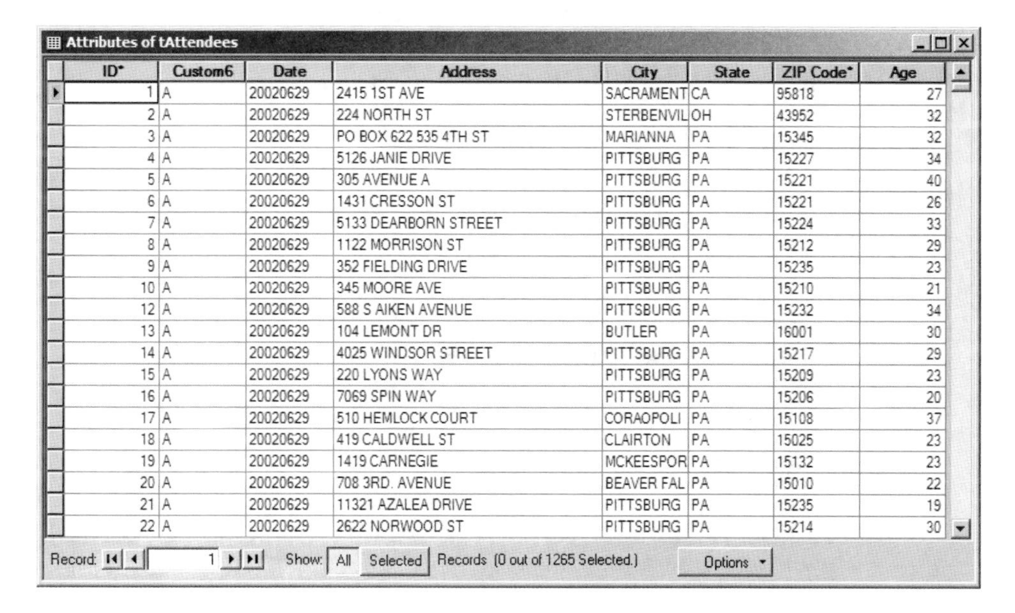

Attributes of tAttendees

ID*	Custom6	Date	Address	City	State	ZIP Code*	Age
1	A	20020629	2415 1ST AVE	SACRAMENT	CA	95818	27
2	A	20020629	224 NORTH ST	STERBENVIL	OH	43952	32
3	A	20020629	PO BOX 622 535 4TH ST	MARIANNA	PA	15345	32
4	A	20020629	5126 JANIE DRIVE	PITTSBURG	PA	15227	34
5	A	20020629	305 AVENUE A	PITTSBURG	PA	15221	40
6	A	20020629	1431 CRESSON ST	PITTSBURG	PA	15221	26
7	A	20020629	5133 DEARBORN STREET	PITTSBURG	PA	15224	33
8	A	20020629	1122 MORRISON ST	PITTSBURG	PA	15212	29
9	A	20020629	352 FIELDING DRIVE	PITTSBURG	PA	15235	23
10	A	20020629	345 MOORE AVE	PITTSBURG	PA	15210	21
12	A	20020629	588 S AIKEN AVENUE	PITTSBURG	PA	15232	34
13	A	20020629	104 LEMONT DR	BUTLER	PA	16001	30
14	A	20020629	4025 WINDSOR STREET	PITTSBURG	PA	15217	29
15	A	20020629	220 LYONS WAY	PITTSBURG	PA	15209	23
16	A	20020629	7069 SPIN WAY	PITTSBURG	PA	15206	20
17	A	20020629	510 HEMLOCK COURT	CORAOPOLI	PA	15108	37
18	A	20020629	419 CALDWELL ST	CLAIRTON	PA	15025	23
19	A	20020629	1419 CARNEGIE	MCKEESPOR	PA	15132	23
20	A	20020629	708 3RD. AVENUE	BEAVER FAL	PA	15010	22
21	A	20020629	11321 AZALEA DRIVE	PITTSBURG	PA	15235	19
22	A	20020629	2622 NORWOOD ST	PITTSBURG	PA	15214	30

Record: 1 Show: All Selected Records (0 out of 1265 Selected.) Options

5 **Close the tAttendees table.**

Build address locators for ZIP Codes

1 **From the ArcMap Standard toolbar, click the ArcCatalog button.** 

2 **In the ArcCatalog catalog tree, expand the Gistutorial folder by clicking its plus sign (+) and click the Flux folder to select it.**

3 **Click File, New, Address Locator.**

4 **In the Create New Address Locator dialog box, scroll down the list of address locators, then locate and click ZIP 5Digit.**

The reference data you will use to geocode the attendee data is the PAZip (Pennsylvania ZIP Codes) layer that's currently in your map. You selected this address locator style because the only address data contained in the PAZip layer is ZIP Codes.

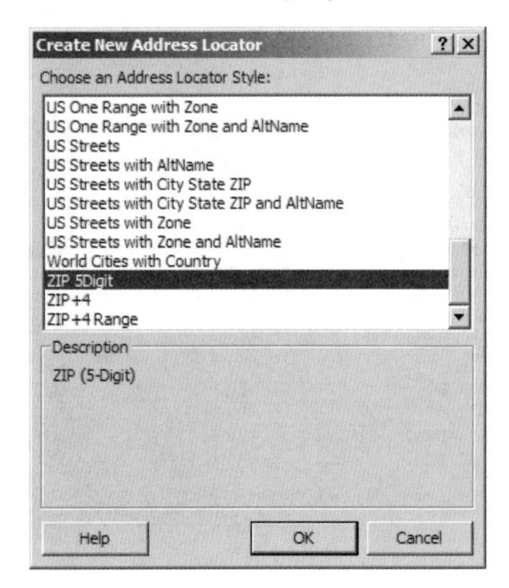

5 **Click OK.**

Specifying a reference table (ZIP Codes)

The next step in the geocoding setup process is to specify the reference data to which the address records will be matched. The reference data, along with the addressing style and its properties that you choose, is referred to as the Address Locator. Once you have set up an Address Locator, it's stored and can be applied to other address-matching projects.

1 **In the upper left corner of the New ZIP 5Digit Address Locator dialog box, change the name from New Address Locator to PAZipCodes.**

2 **In the Primary table tab, click the Browse button.**

3 **Browse to \Gistutorial\UnitedStates\Pennsylvania, click PAZIP.shp, Add.**

4 **Check the settings on your screen to be sure they match the settings below.**

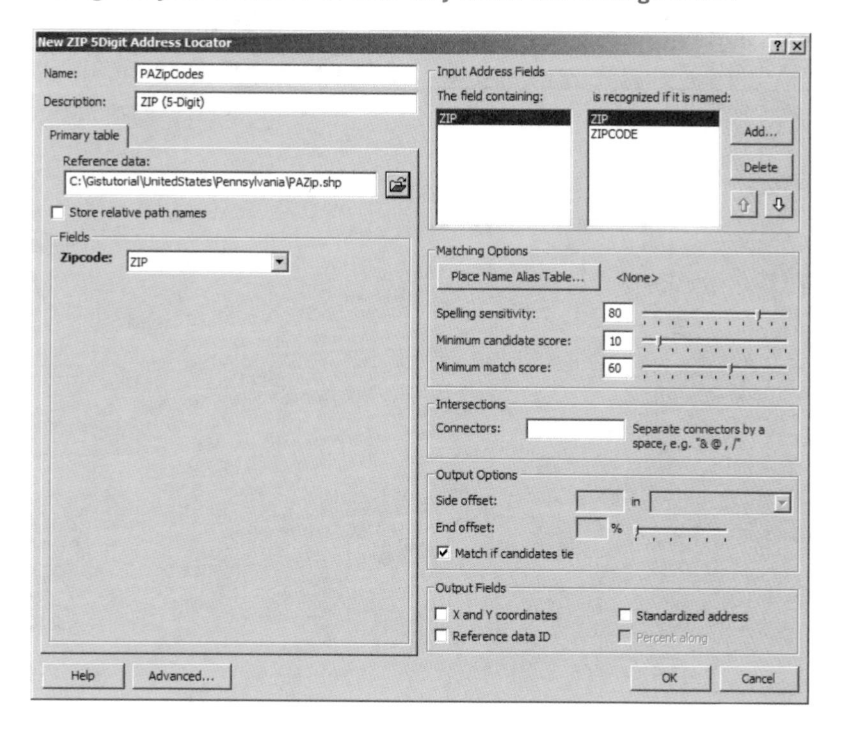

5 **Click OK.**

ArcCatalog creates the new Address Locator. You can now use this address locator to address match any tables containing Pennsylvania ZIP Codes in their records.

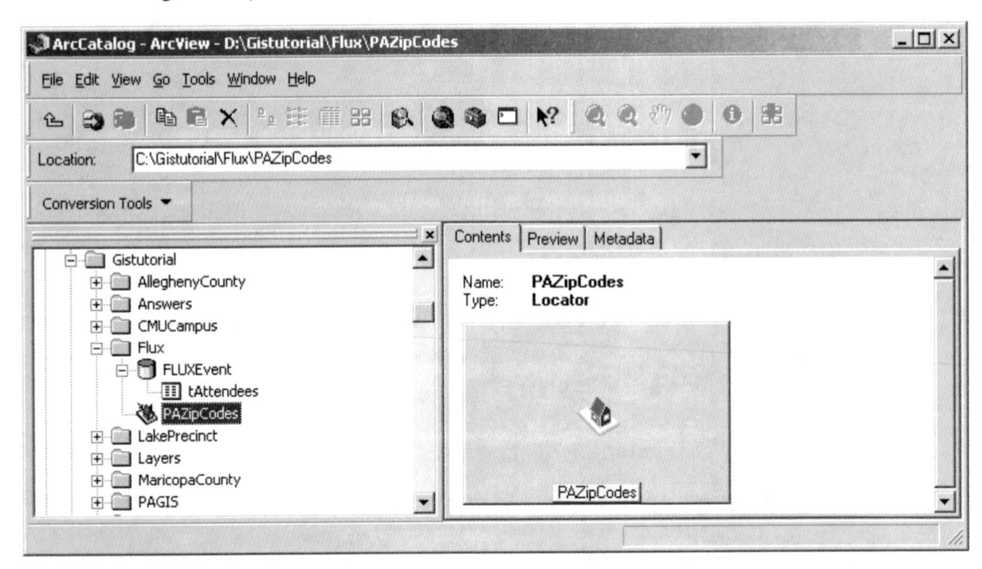

6 Close ArcCatalog.

Add address locator to map

1 In ArcMap, click Tools, Geocoding, Address Locator Manager.

2 In the Address Locator Manager, click Add.

3 Browse to the Flux folder, then click the PAZIPCodes address locator.

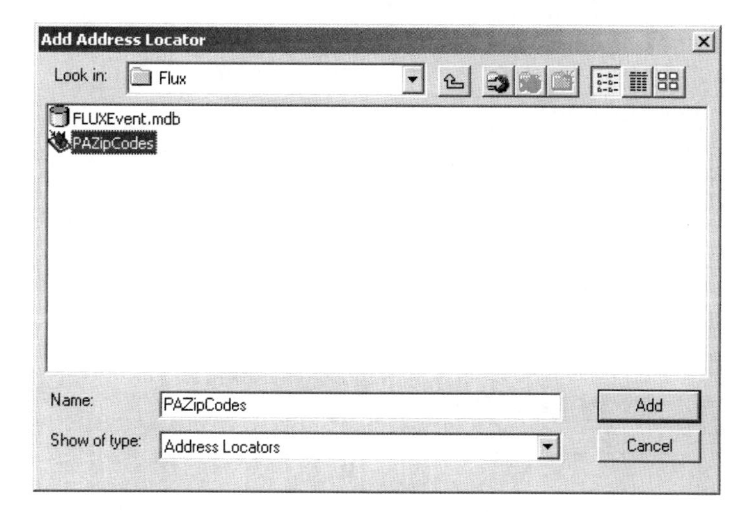

4 Click Add.

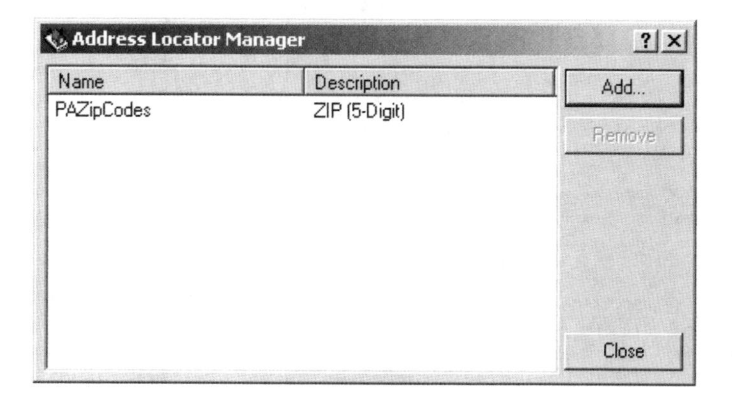

5 Click Close.

Batch match

The tAttendees table can now be address matched to the Pennsylvania ZIP Codes.

Prepare table for geocoding

1 **In the ArcMap table of contents, right-click the tAttendees table.**

2 **Click Geocode Addresses.**

3 **Click PAZipCodes, OK.**

4 **In the Address Input Fields frame, click the ZIP drop-down list and choose ZIP_Code.**

5 **Name the output shapefile FluxAttendeeZIP.shp and save it in the \Gistutorial\Flux folder.**

6 **Make sure your settings match those in the graphic below, then click OK.**

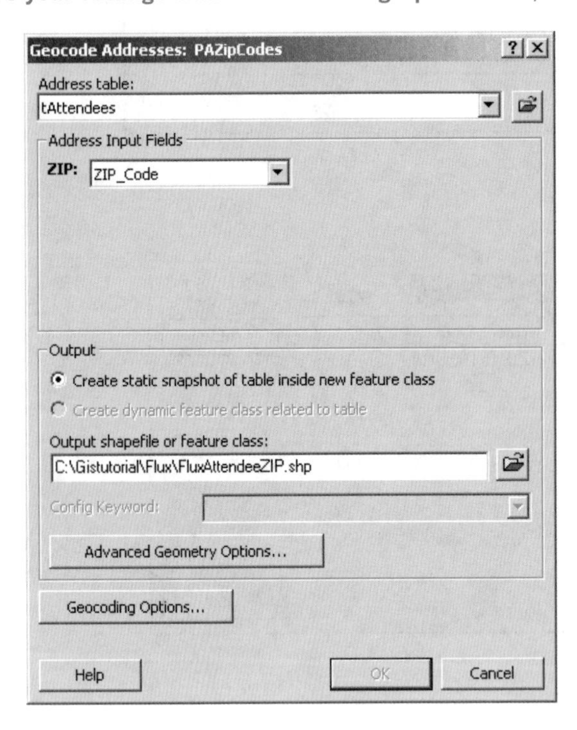

ArcView matches 1,089 attendees (86 percent) of the input records from the tAttendees table to the PAZip layer. ArcView was not able to match 176 (14 percent) of the records. You will look more at these results later.

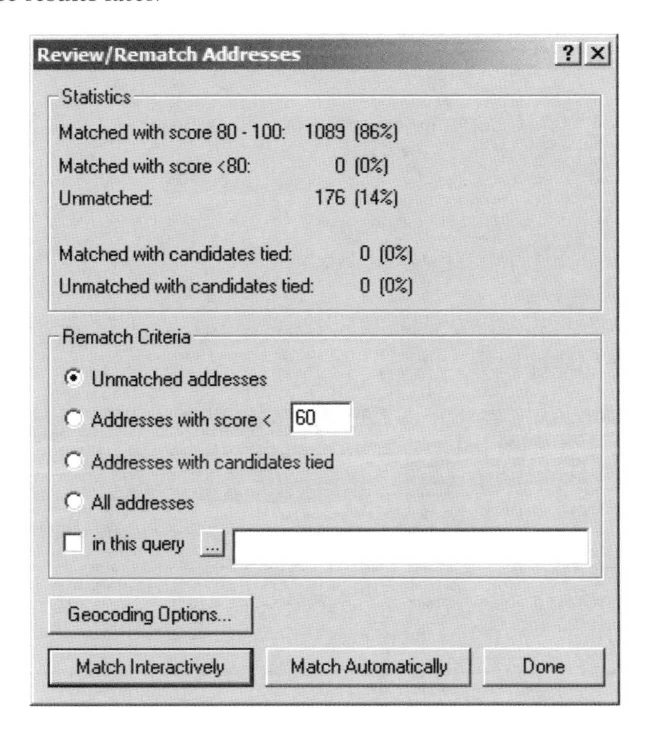

8 Click Done.

The map will display the new point shapefile of the addresses that successfully matched. Change the color and symbol to your liking.

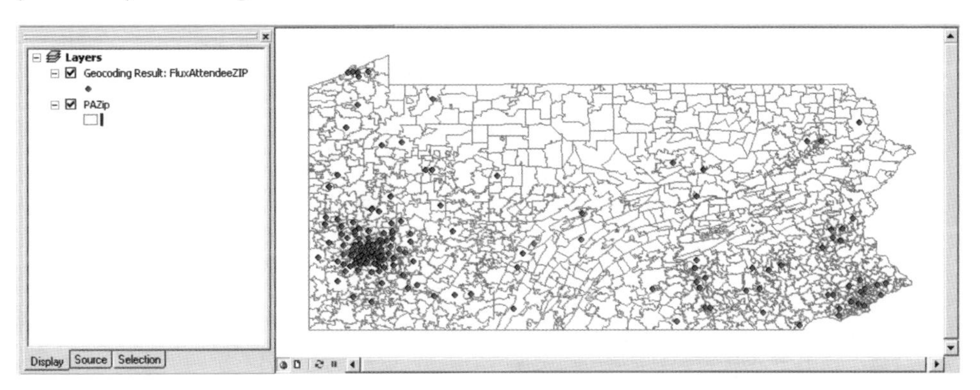

Review unmatched ZIP Codes

There were 176 attendee records that were not matched with a Pennsylvania ZIP Code. At this point, you will review these records to determine why they were not matched.

1. **In the ArcMap table of contents, right-click Geocoding Result: FluxAttendeeZIP.**
2. **Click Open Attribute Table.**
3. **Right-click the Status field name and click Sort Descending.**

Records with a Status value of U are unmatched. Most of these unmatched records are outside of Pennsylvania or have ZIP Codes that are missing.

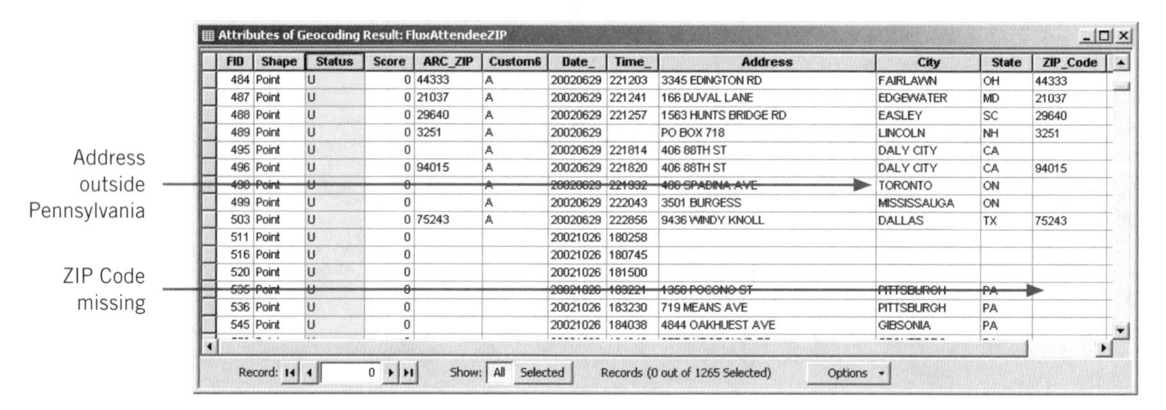

Attributes of Geocoding Result: FluxAttendeeZIP

FID	Shape	Status	Score	ARC_ZIP	Custom6	Date_	Time_	Address	City	State	ZIP_Code
484	Point	U	0	44333	A	20020629	221203	3345 EDINGTON RD	FAIRLAWN	OH	44333
487	Point	U	0	21037	A	20020629	221241	166 DUVAL LANE	EDGEWATER	MD	21037
488	Point	U	0	29640	A	20020629	221257	1563 HUNTS BRIDGE RD	EASLEY	SC	29640
489	Point	U	0	3251	A	20020629		PO BOX 718	LINCOLN	NH	3251
495	Point	U	0		A	20020629	221814	406 88TH ST	DALY CITY	CA	
496	Point	U	0	94015	A	20020629	221820	406 88TH ST	DALY CITY	CA	94015
498	Point	U	0		A	20020629	221932	406 SPADINA AVE	TORONTO	ON	
499	Point	U	0		A	20020629	222043	3501 BURGESS	MISSISSAUGA	ON	
503	Point	U	0	75243	A	20020629	222856	9436 WINDY KNOLL	DALLAS	TX	75243
511	Point	U	0			20021026	180258				
516	Point	U	0			20021026	180745				
520	Point	U	0			20021026	181500				
535	Point	U	0			20021026	183221	1358 POCONO ST	PITTSBURGH	PA	
536	Point	U	0			20021026	183230	719 MEANS AVE	PITTSBURGH	PA	
545	Point	U	0			20021026	184038	4844 OAKHUEST AVE	GIBSONIA	PA	

Record: 0 Show: All Selected Records (0 out of 1265 Selected) Options

Fix and rematch ZIP Codes

1 Click Tools, Editor Toolbar.

2 Click Editor, Start Editing.

3 In the Source column of the Start Editing dialog box, click the C:\Gistutorial\Flux folder and click OK.

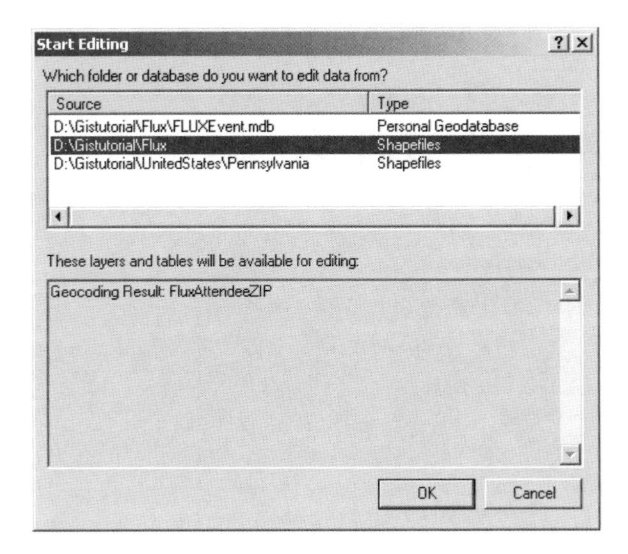

4 In the open attribute table, scroll to and select the record with FID value 49 whose Address value is 414 SOUTH CRAIG STREET.

This record's ZIP_Code value is blank and should be changed to 15213.

5 Change the ZIP Code to 15213.

The missing ZIP Code value has been entered and you can now rematch it using the geocoding tools.

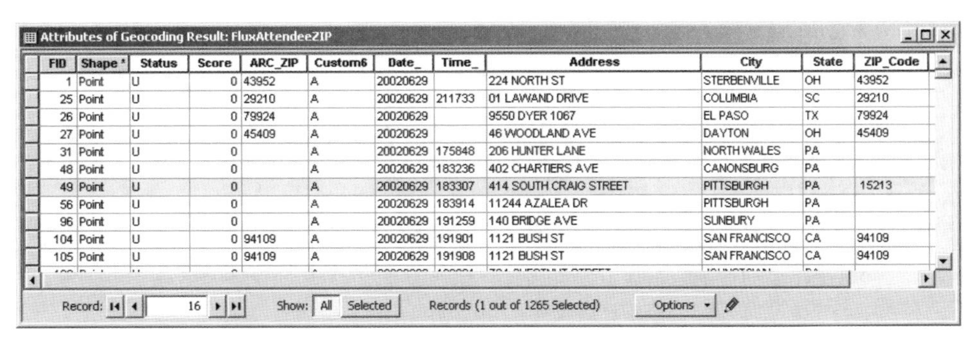
Attributes of Geocoding Result: FluxAttendeeZIP

FID	Shape *	Status	Score	ARC_ZIP	Custom6	Date_	Time_	Address	City	State	ZIP_Code
1	Point	U	0	43952	A	20020629		224 NORTH ST	STERBENVILLE	OH	43952
25	Point	U	0	29210	A	20020629	211733	01 LAWAND DRIVE	COLUMBIA	SC	29210
26	Point	U	0	79924	A	20020629		9550 DYER 1067	EL PASO	TX	79924
27	Point	U	0	45409	A	20020629		46 WOODLAND AVE	DAYTON	OH	45409
31	Point	U	0		A	20020629	175848	206 HUNTER LANE	NORTH WALES	PA	
48	Point	U	0		A	20020629	183236	402 CHARTIERS AVE	CANONSBURG	PA	
49	Point	U	0		A	20020629	183307	414 SOUTH CRAIG STREET	PITTSBURGH	PA	15213
56	Point	U	0		A	20020629	183914	11244 AZALEA DR	PITTSBURGH	PA	
96	Point	U	0		A	20020629	191259	140 BRIDGE AVE	SUNBURY	PA	
104	Point	U	0	94109	A	20020629	191901	1121 BUSH ST	SAN FRANCISCO	CA	94109
105	Point	U	0	94109	A	20020629	191908	1121 BUSH ST	SAN FRANCISCO	CA	94109

Record: 16 Show: All Selected Records (1 out of 1265 Selected) Options

6 On the Editor toolbar, click Editor, Save Edits, Stop Editing. Close the Editor toolbar.

Rematch ZIP Codes

1 **Click Tools, Geocoding, Review/Rematch Addresses, Geocoding Result: FluxAttendeeZIP.**

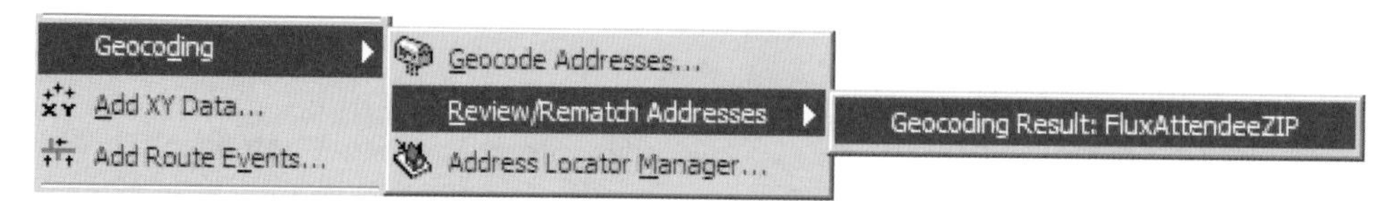

2 **Click Match Automatically.**

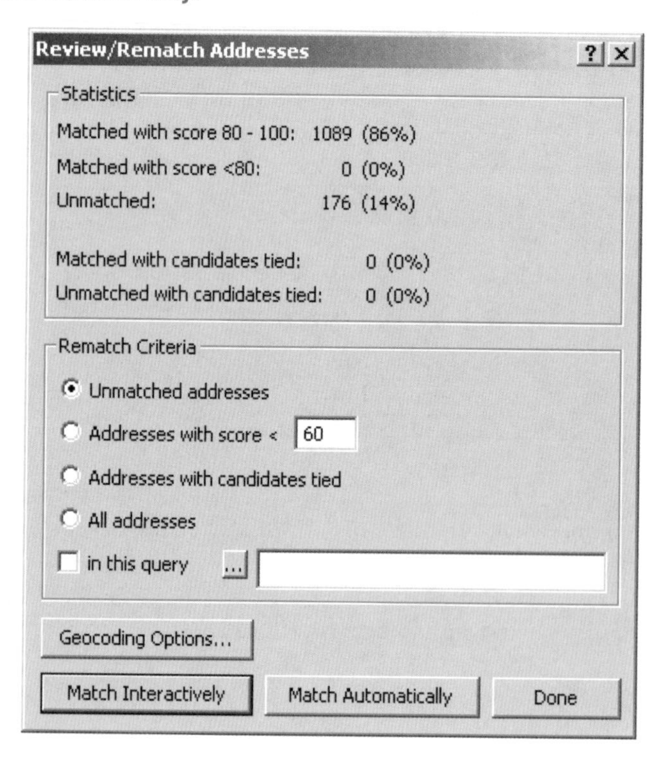

The matched score increases by 1 to 1089.

3 **Click Done.**

YOUR TURN

Use the U.S. Postal Service's ZIP Code Lookup at *http://zip4.usps.com/zip4/welcome.jsp* to get the missing ZIP Code for the record with FID value 56, 11244 Azalea Dr, Pittsburgh, PA. Enter the ZIP Code that you find and rematch.

Geocode to streets

In this exercise, you will again geocode the FLUX attendee records, but this time you will geocode them at the street level for Pittsburgh, Pennsylvania. Geocoding to the street level requires, at the least, that the records you wish to map contain attributes for the street name and house number. In this case, you will also incorporate the ZIP Code value into the address locator, because some addresses may have the same house number and street name but exist in different ZIP Codes. This happens frequently in study areas that have two or more municipalities, such as a county.

When preparing to geocode addresses, it is important to obtain good and reliable datasets for both the input and the reference datasets. Before setting up the Address Locator and running the geocoding command, it's common practice to review and clean the involved tables. For example, you could delete records outside of your study area or check fields to be sure the addresses were entered properly.

Street centerline map (reference data).

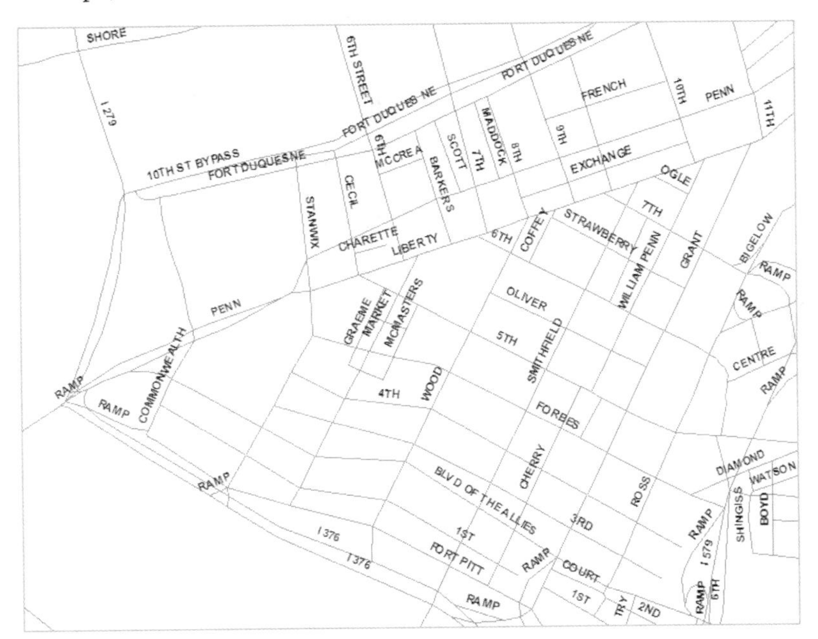

Database table of FLUX attendees (input data).

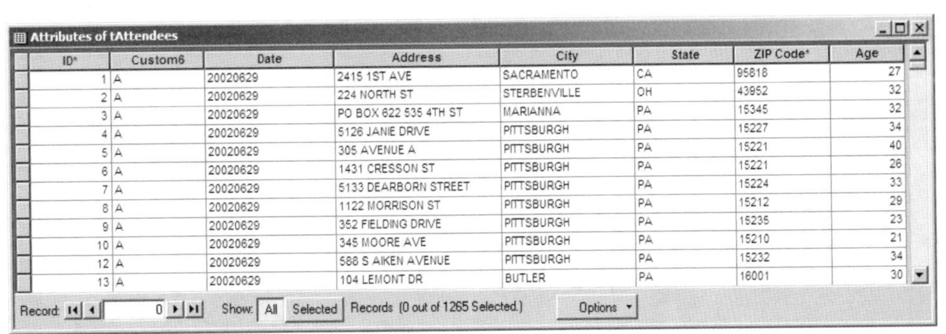

Attributes of tAttendees

ID*	Custom6	Date	Address	City	State	ZIP Code*	Age
1	A	20020629	2415 1ST AVE	SACRAMENTO	CA	95818	27
2	A	20020629	224 NORTH ST	STERBENVILLE	OH	43952	32
3	A	20020629	PO BOX 622 535 4TH ST	MARIANNA	PA	15345	32
4	A	20020629	5126 JANIE DRIVE	PITTSBURGH	PA	15227	34
5	A	20020629	305 AVENUE A	PITTSBURGH	PA	15221	40
6	A	20020629	1431 CRESSON ST	PITTSBURGH	PA	15221	26
7	A	20020629	5133 DEARBORN STREET	PITTSBURGH	PA	15224	33
8	A	20020629	1122 MORRISON ST	PITTSBURGH	PA	15212	29
9	A	20020629	352 FIELDING DRIVE	PITTSBURGH	PA	15235	23
10	A	20020629	345 MOORE AVE	PITTSBURGH	PA	15210	21
12	A	20020629	588 S AIKEN AVENUE	PITTSBURGH	PA	15232	34
13	A	20020629	104 LEMONT DR	BUTLER	PA	16001	30

Record: 0 Show: All Selected Records (0 out of 1265 Selected.) Options

Prepare data and street maps

As you did in the previous exercise, you must define an address locator before you can geocode the data.

1 **Open Tutorial7-2.mxd from the \Gistutorial folder.**

Tutorial7-2 contains a streets and neighborhoods layer for the City of Pittsburgh and the FLUX Attendees table. Click the Source tab below the table of contents if you do not see the tAttendees table.

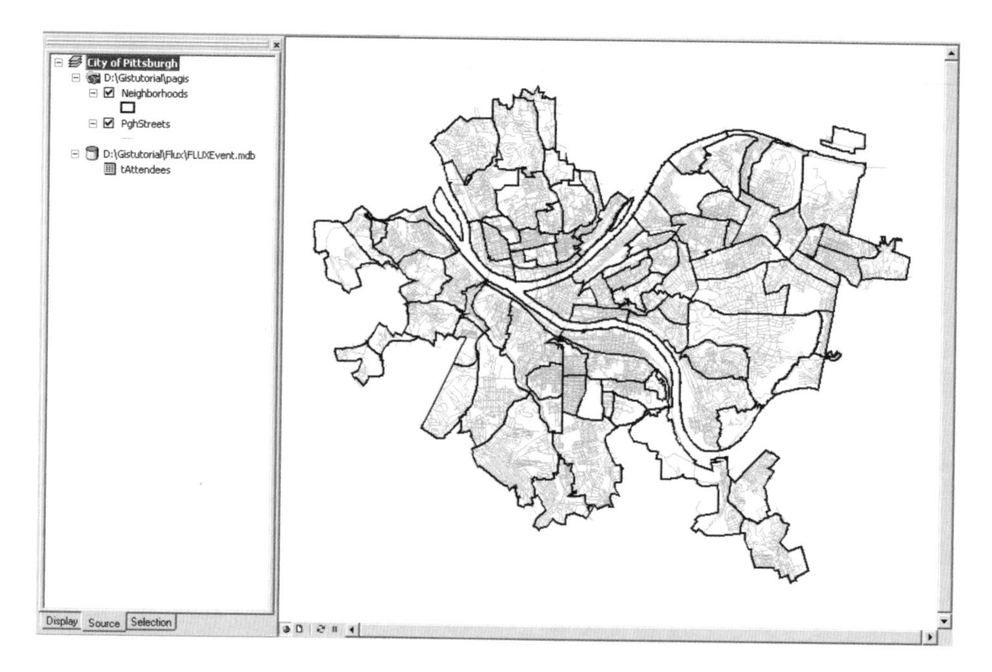

2 **Open the attribute tables for Pgh Streets and tAttendees and review their contents, especially addresses. Close the tables when you are finished reviewing them.**

Create an address locator for streets

1 Click the ArcCatalog button to open ArcCatalog.

2 In the catalog tree, right-click the PAGIS folder and click New, Address Locator.

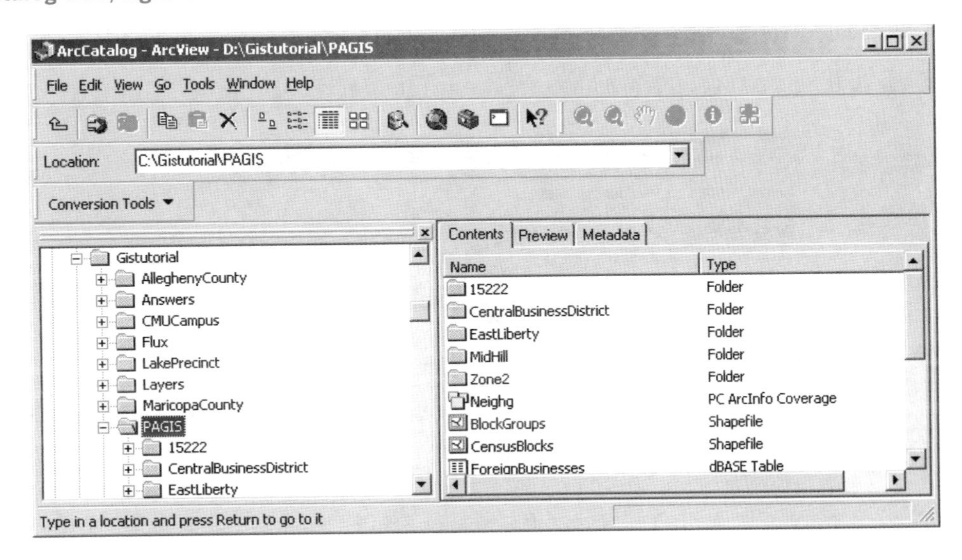

3 Scroll down the list of Address Locators, then locate and click US Streets with Zone.

This Address Locator will geocode records containing street addresses and ZIP Code (Zones).

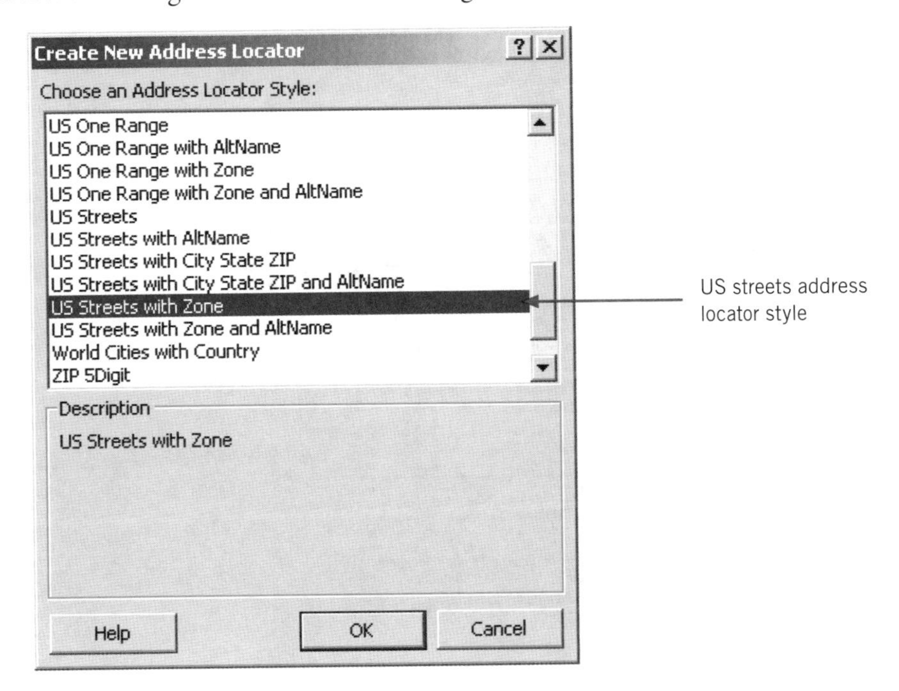

4 Click OK.

Specify a reference table (streets)

At this point, you need to define the properties associated with the US Streets with Zone Address Locator. These properties include the name of the address locator, the reference data, and the fields from the reference data that store the address ranges.

1 **Change the name of the new locator to PittsburghStreets.**

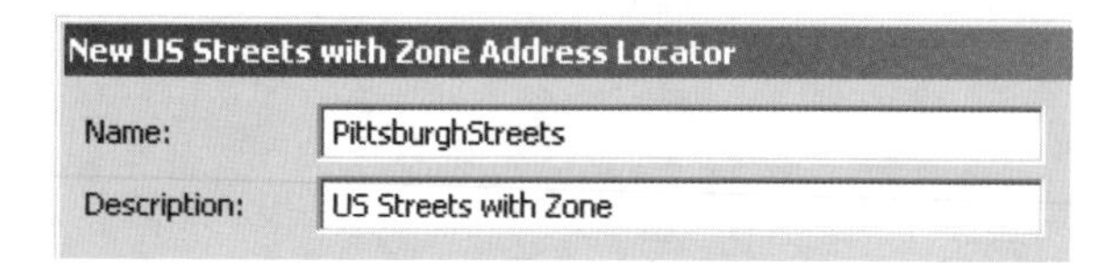

2 **In the Primary tab, click the Browse button next to the Reference data box.**

3 **In the Choose Reference Data dialog box, browse to the \Gistutorial\PAGIS folder, click PghStreets.shp, then click Add.**

4 **Use the graphic below to verify that the correct fields are chosen from their respective drop-down lists in the Fields frame.**

The Fields list automatically recognizes the correct address fields because the field names in the PghStreets shapefile conform to industry standards for address data.

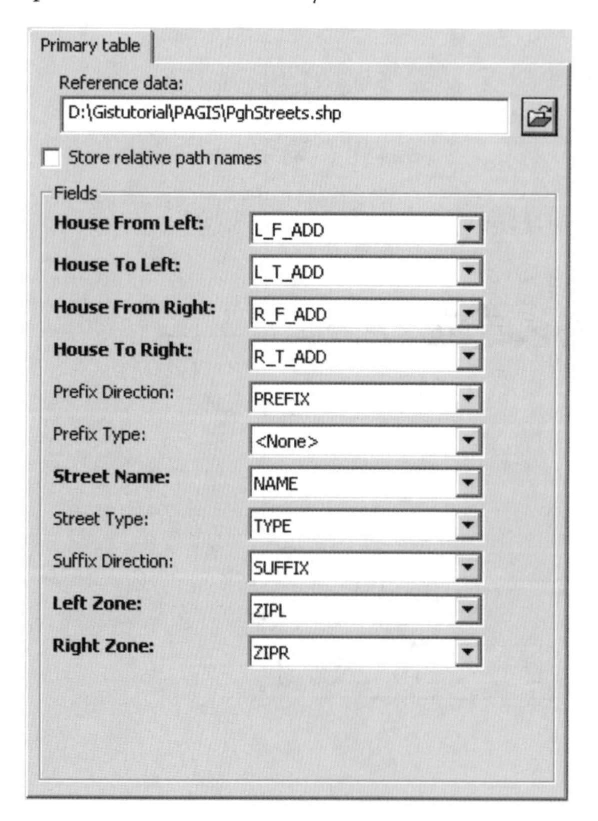

Create address locator

1 Verify that the settings in your dialog box match those in the graphic below, then click OK.

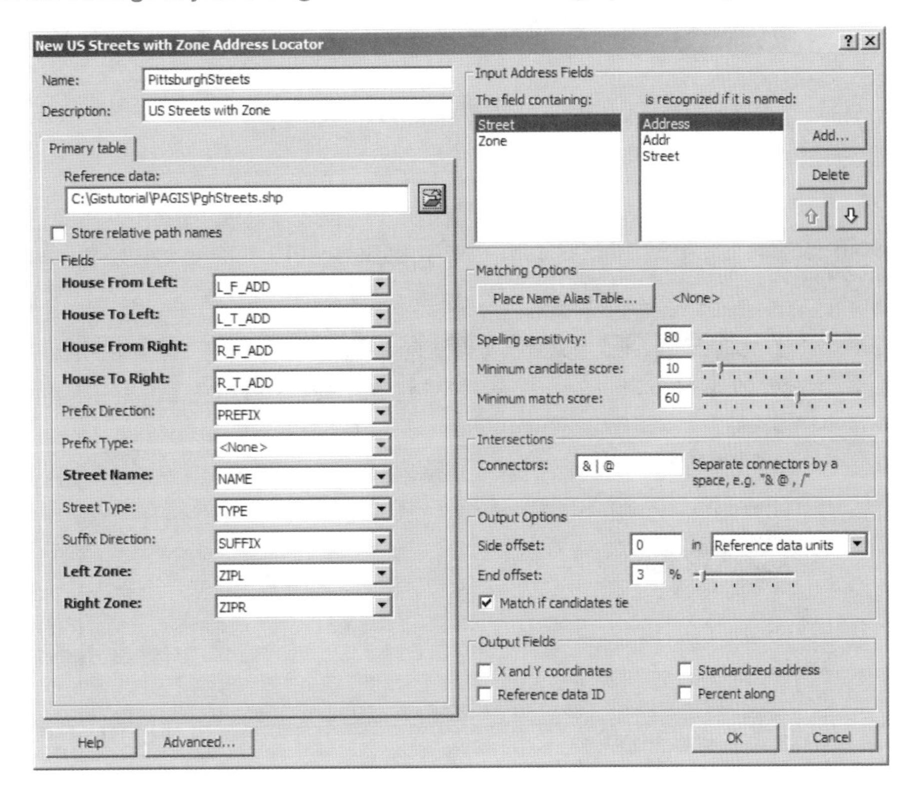

The Pittsburgh Streets Address Locator is now available for geocoding Pittsburgh addresses to Pittsburgh streets.

2 Close ArcCatalog.

Add address locator to a map

Now that an Address Locator has been created for the streets, it can be used to geocode a batch of addresses from a table or to interactively locate addresses using the Find tool in ArcMap.

1 **In ArcMap click Tools, Geocoding, Address Locator Manager.**

2 **Click Add.**

3 **From the Look in drop-down list, choose C:\Gistutorial\PAGIS, then click the PittsburghStreets address locator.**

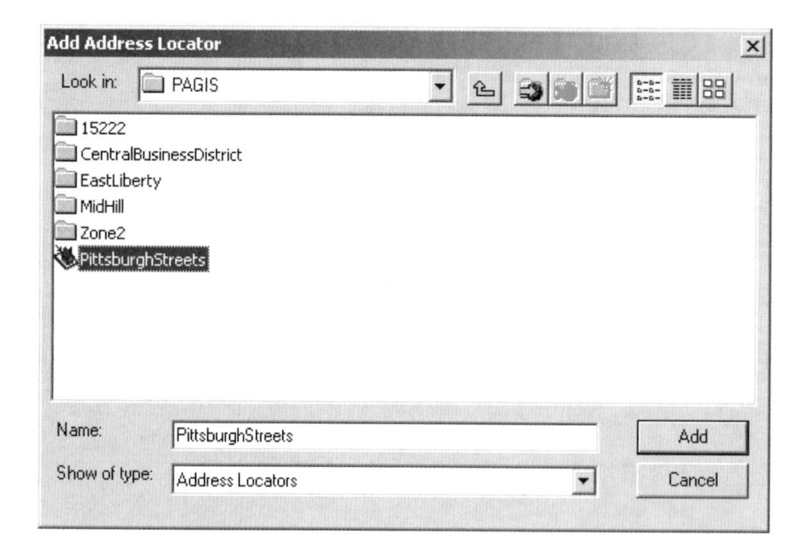

4 **Click Add.**

5 **In the Address Locator Manager, click Close.**

Interactively locate addresses

Before address matching the entire tAttendee database to the streets layer, use the Find tool in ArcMap to locate individual addresses.

1 **In ArcMap, click the Find button on the Tools toolbar.**

2 **In the Find dialog box, click the Addresses tab.**

3 **Type 3609 Penn Ave in the Street or Intersection box.**

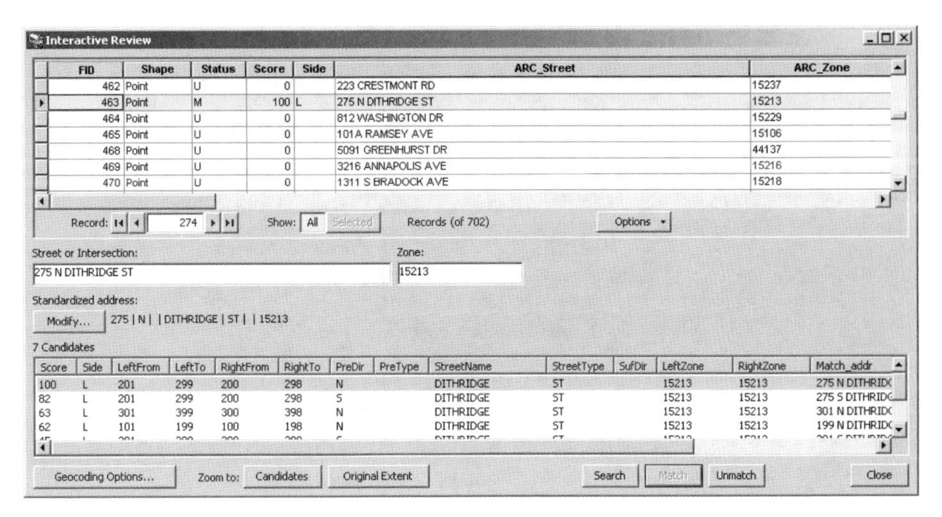

4 **Click Find.**

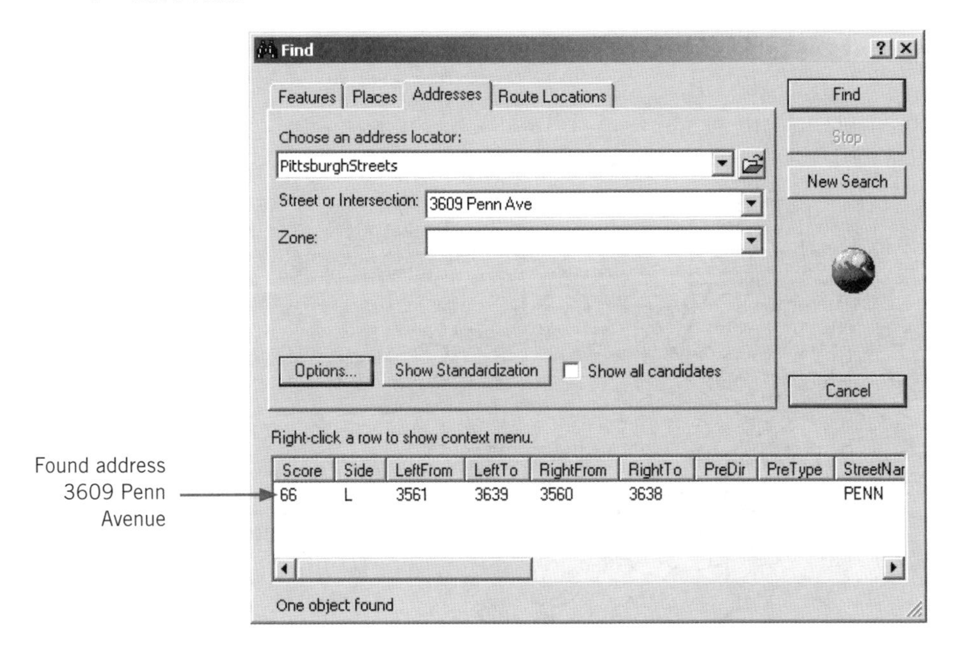

Found address 3609 Penn Avenue

Show address on map

1 **Right-click the potential address shown near the bottom of the Find dialog box.**

2 **Click Add Label Point.**

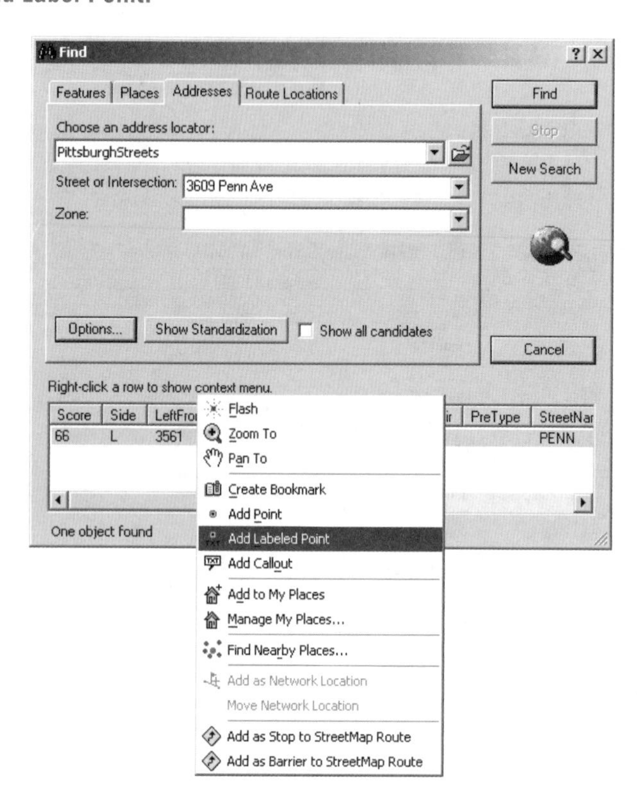

3 **Close the Find window.**

A graphic point marker and label appears on the map at the location of the address.

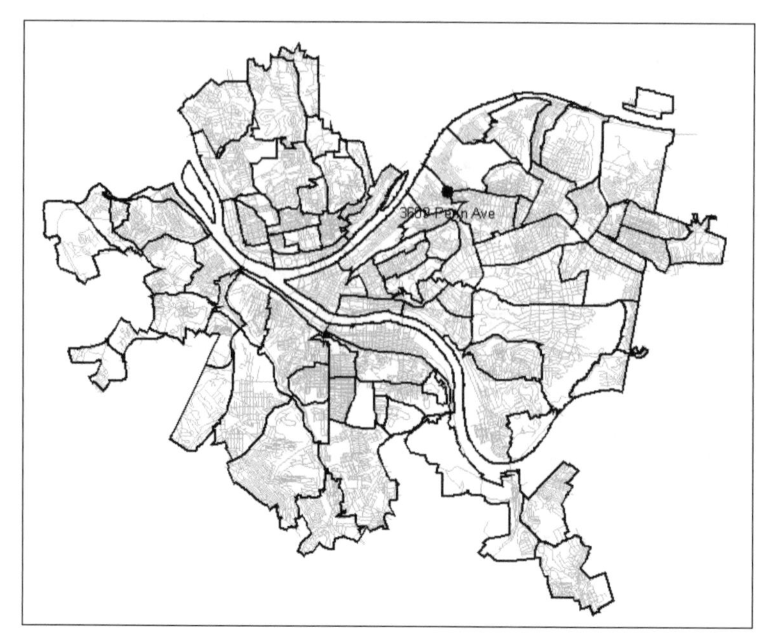

YOUR TURN

Using the Find tool, locate the following addresses:

- 1920 S 18th ST
- 255 Atwood ST
- 3527 Beechwood BLVD

Select the graphics and delete them before continuing.

Perform batch geocoding

The entire tAttendees table can now be geocoded to the PittsburghStreets layer.

Prepare table for geocoding

1 In the ArcMap table of contents, right-click the tAttendees table.

2 Click Geocode Addresses.

3 Click PittsburghStreets address locator, then click OK.

4 In the Address Input Fields frame, click the Zone drop-down list and choose ZIP_CODE.

5 Name the Output **FluxAttendeeStreets.shp** and save it in the **\Gistutorial\Flux** folder.

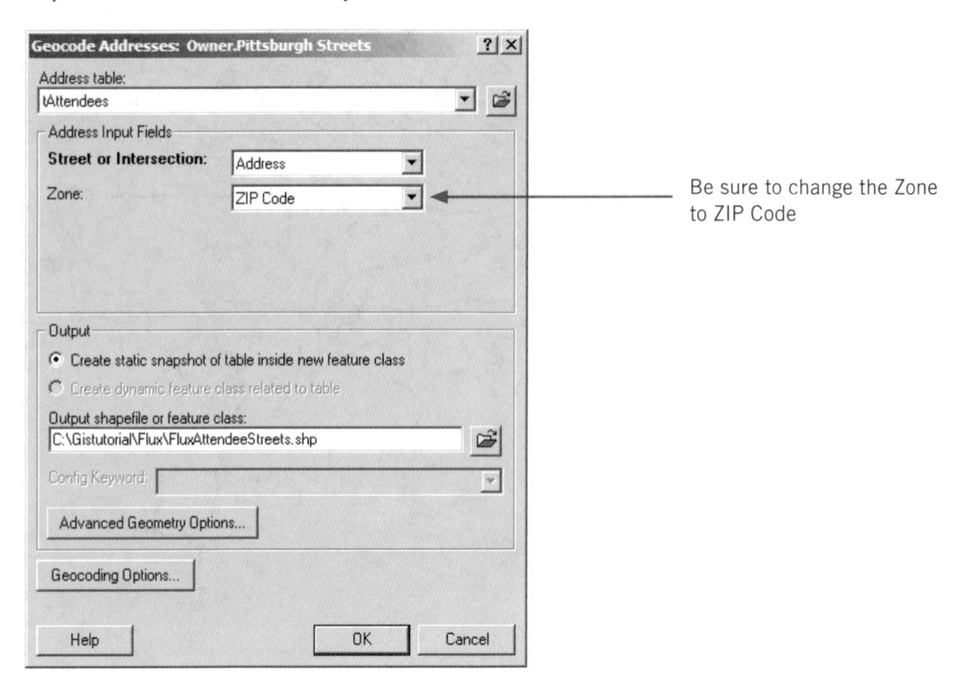

Be sure to change the Zone to ZIP Code

6 Click Geocoding Options and change the Spelling sensitivity to 75.

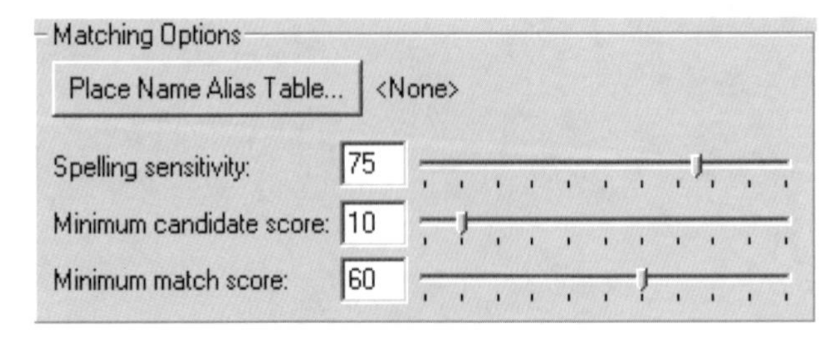

7 Click OK twice to geocode the addresses.

A total of 542 addresses (43 percent) in the tAttendees table was matched to the PghStreets layer. A total of 704 addresses (56 percent) was not matched. Although it appears that many did not match, most of the unmatched records are outside of Pittsburgh. You will look more at the results later.

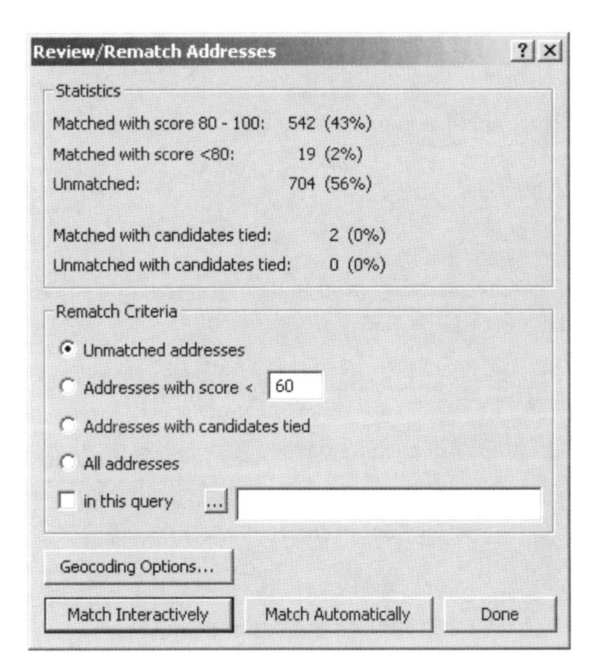

8 Click Done.

The map now displays the new point shapefile containing the addresses that were successfully matched. Change the point marker symbol to your liking.

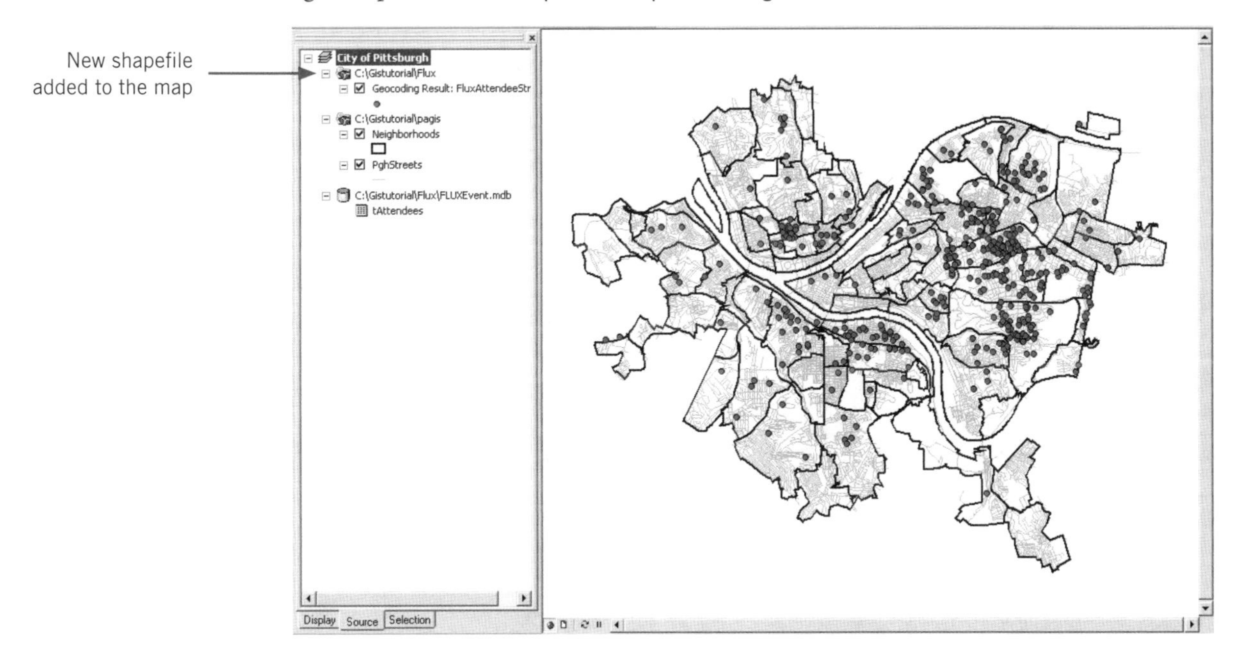

Correct addresses using interactive rematch

Many of the addresses in the tAttendees table were not matched to the PghStreets layer because they are outside of Pittsburgh. Several others records did not match due to spelling errors or data omissions in either the input table (tAttendees) or the reference data (PghStreets). Resolving these types of errors requires investigation to identify the nature of the problem and make the necessary corrections. Making corrections to the address and street data mostly depends on the user's knowledge of the streets and access to the correct data. In this exercise, you will use an interactive review process to correct then match the unmatched records.

Match interactively

1 **In ArcMap, click Tools, Geocoding, Review/Rematch Addresses, Geocoding Result: FluxAttendeeStreets.**

2 **In the Review/Rematch Addresses dialog box, click Match Interactively.**

This option shows each unmatched record individually in the Interactive Review dialog box, and allows you to manually edit the address values. Horizontally scroll the fields so you can see the Address, City, State, and Zip_Code fields. Notice that many of the unmatched records are outside of Pittsburgh.

3 **In the Interactive Review dialog box, scroll down the list of unmatched records, then locate and click the record with the FID value 45 and address 1915 PENN AVE.**

This record and the next have the incorrect ZIP Code value 15221. It should be 15222.

4 **Type 15222 in the Zone box, press Tab, and click the resulting candidate record with Score 100.**

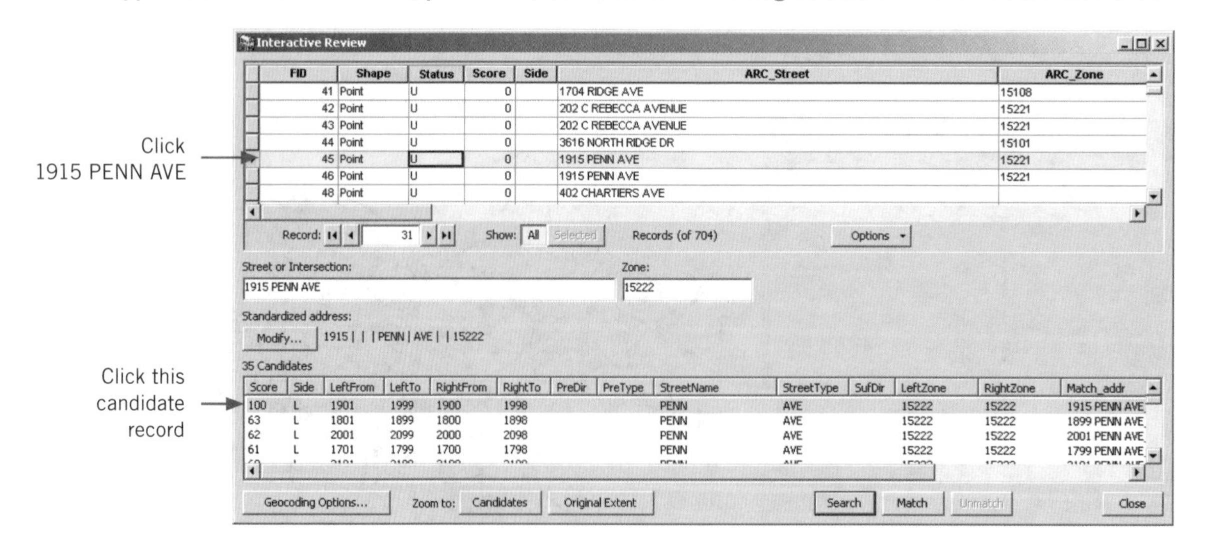

5 Click Match and repeat step 4 and this step for the next record with FID value 46.

6 In the list of unmatched records, locate and then click the record with FID value 463 and Address 275 N.DITHRIDGE ST.

This record did not match because of the period after the "N" in N.DITHRIDGE ST. Removing this period will allow the record to find its match in the street data.

7 In the Street or Intersection box, replace the period in 275 N.DITHRIDGE ST with a blank space, then press the Tab key.

8 In the Candidate list, click the first candidate with a Score of 100, then click Match.

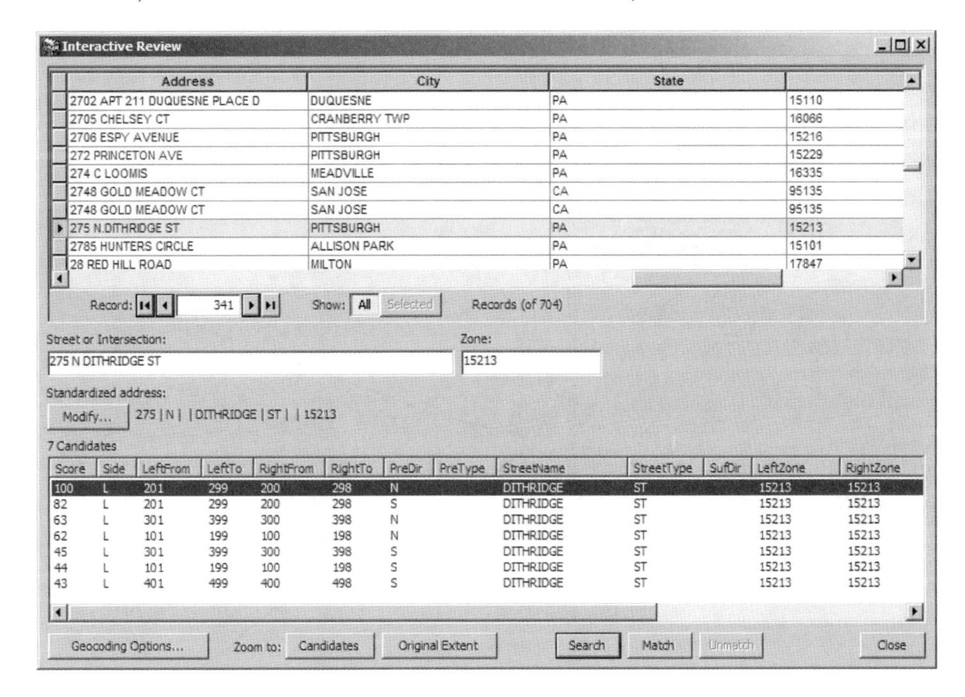

9 In the Interactive Review dialog box, click Close.

10 In the Review/Rematch Addresses dialog box, click Done.

New point features were created in the FluxAttendeeStreets shapefile at the two address locations that you rematched.

11 Save your map document.

Correct street layer addresses

In this exercise, you will learn how to find and fix an incorrect address in a street layer used for geocoding. To do this, you will examine unmatched user addresses, identify candidate streets, then examine the attributes of the streets to look for misspellings or data omissions.

Open a new map

1 **If necessary, start ArcMap.**

2 **Open Tutorial7-3.mxd from the \Gistutorial folder.**

3 **If necessary, click the Source tab at the bottom of the table of contents.**

Tutorial7-3 contains a table of clients and a layer containing the streets in Pittsburgh's Central Business District.

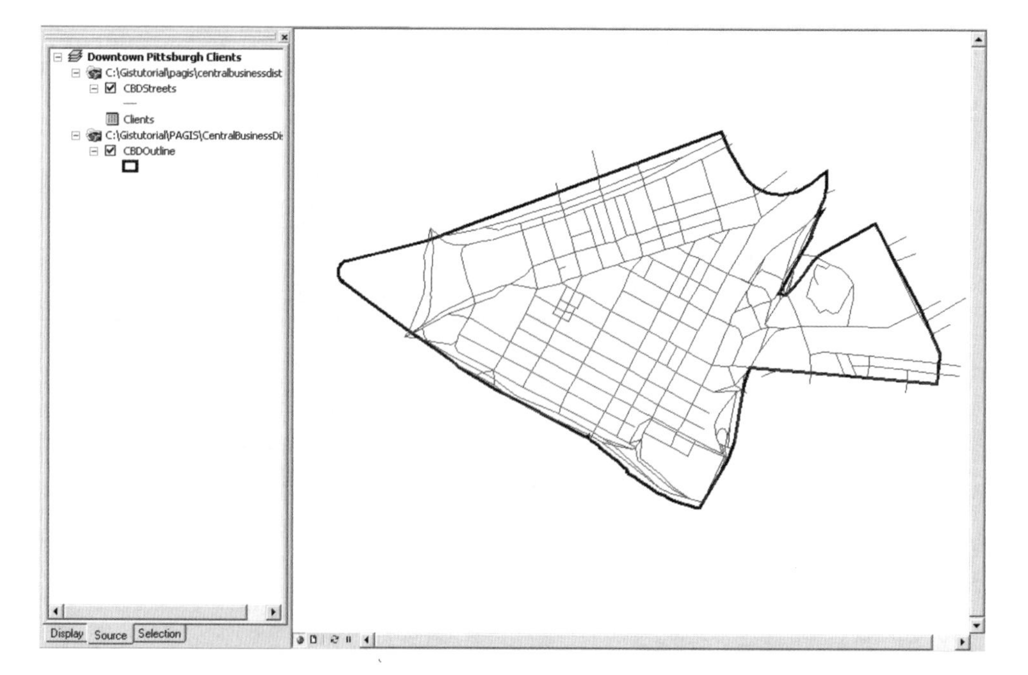

Create a new address locator

1 Start ArcCatalog.

2 Create a new address locator in the **Gistutorial\PAGIS\CentralBusinessDistrict** folder that uses the US Streets style.

3 Name the address locator **Central Business District Streets**.

4 Set the address style's reference data to **Gistutorial\PAGIS\CentralBusinessDistrict\CBDStreets.shp**.

5 Verify that your settings match those in the graphic below.

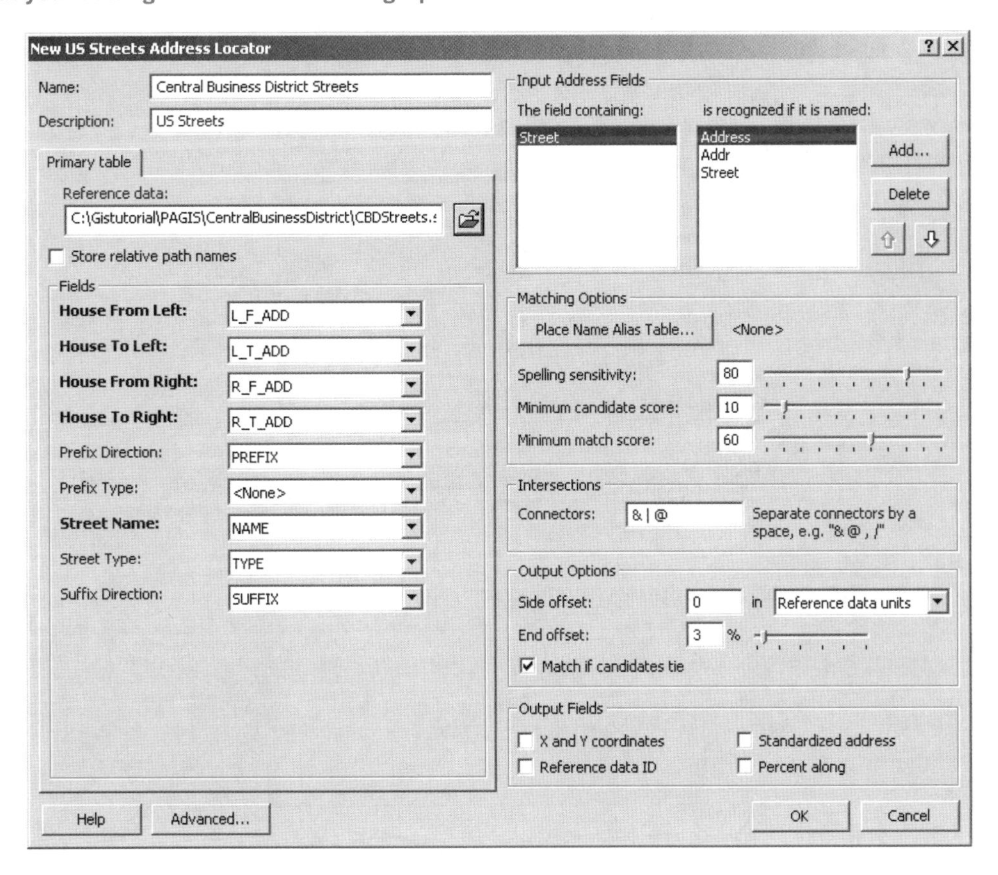

6 Click OK.

7 Close ArcCatalog.

Geocode clients' addresses to CBD streets

1 **In ArcMap, click Tools, Geocoding, Address Locator Manager, and add the Central Business District Streets address locator to your map.**

2 **Geocode the records in the Clients table using selections in the following graphic.**

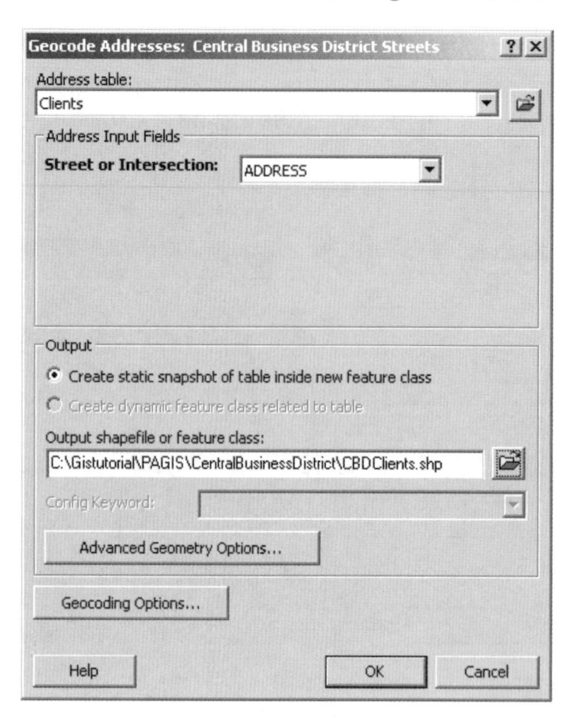

3 **When your geocode settings are complete, click OK, Done.**

4 **Change the point symbology of the CBDClients layer to suite your preferences.**

ArcView initially matched 15 (56 percent) of the 27 records, partially matched 1 (4 percent), and could not match 11 (41 percent).

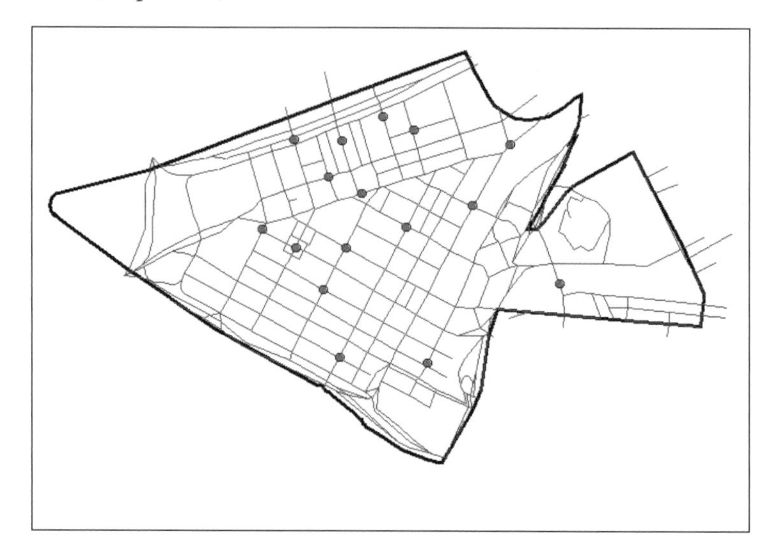

Identify and isolate streets for unmatched address

1 **Click the Display tab at the bottom of the table of contents, right-click the Geocoding Results: CBDClients layer, and click Open Attribute table.**

2 **In the attribute table, right-click the Status field name and click Sort Descending.**

You will investigate why 490 Penn Ave was unmatched by taking a look the street records with Penn as their Name.

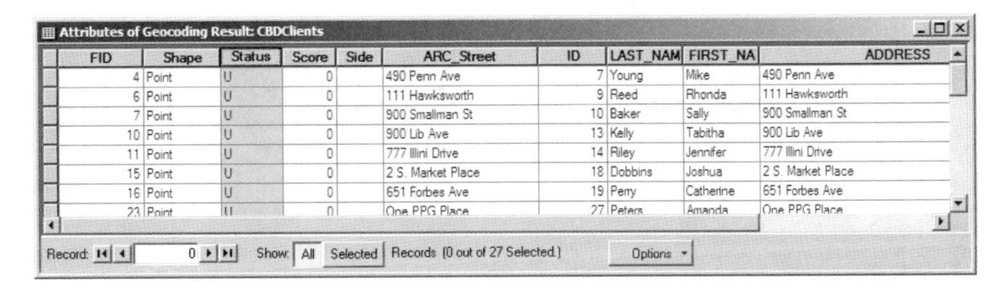

Attributes of Geocoding Result: CBDClients

FID	Shape	Status	Score	Side	ARC_Street	ID	LAST_NAM	FIRST_NA	ADDRESS
4	Point	U	0		490 Penn Ave	7	Young	Mike	490 Penn Ave
6	Point	U	0		111 Hawksworth	9	Reed	Rhonda	111 Hawksworth
7	Point	U	0		900 Smallman St	10	Baker	Sally	900 Smallman St
10	Point	U	0		900 Lib Ave	13	Kelly	Tabitha	900 Lib Ave
11	Point	U	0		777 Illini Drive	14	Riley	Jennifer	777 Illini Drive
15	Point	U	0		2 S. Market Place	18	Dobbins	Joshua	2 S. Market Place
16	Point	U	0		651 Forbes Ave	19	Perry	Catherine	651 Forbes Ave
23	Point	U	0		One PPG Place	27	Peters	Amanda	One PPG Place

Record: 0 Show: All Selected Records (0 out of 27 Selected.) Options

3 **In the table of contents, right-click the CBDStreets layer and click Open Attribute Table.**

4 **Click Options, Select By Attributes.**

5 **In the panel with field names, scroll down, double-click "Name", click the = button, click Get Unique Values, scroll down in the Unique Values box, and double-click 'Penn'.**

6 **Verify that your settings match those in the following graphic.**

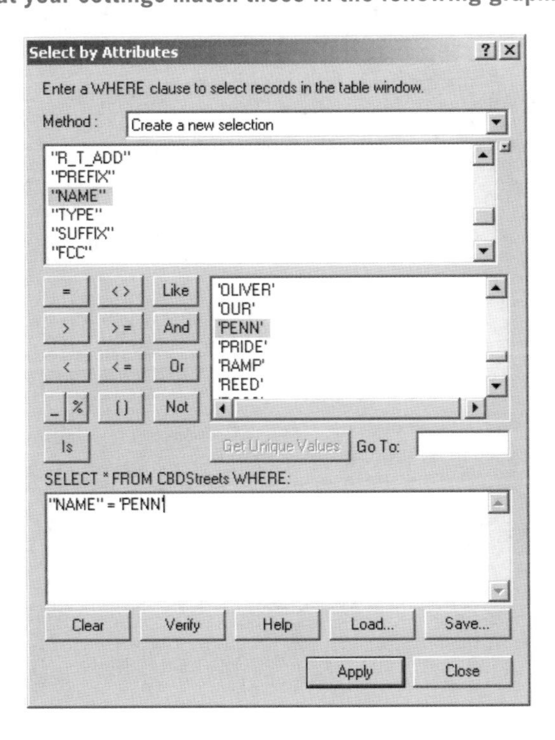

7 **Click Apply and Close.**

Identify and isolate unmatched streets

1 **In the Attributes of CBDStreets table, click Selected.**

2 **Scroll to the right in the Attributes of CBDStreets table until the L_F_ADD, L_T_ADD, R_F_ADD, and R_T_ADD are all visible at the same time.**

3 **Right-click the L_F_ADD column heading and click Sort Ascending.**

4 **Determine if there is a record with an address range that contains 490.**

You will find that this record does not exist in the table. There are, however, records with no address range values.

5 **Click the gray row selector square directly to the left of the row with the blank range values.**

The row you selected is highlighted in yellow, as is the corresponding feature in the map. Suppose that you obtained information from another map suggesting that this is the sought street segment.

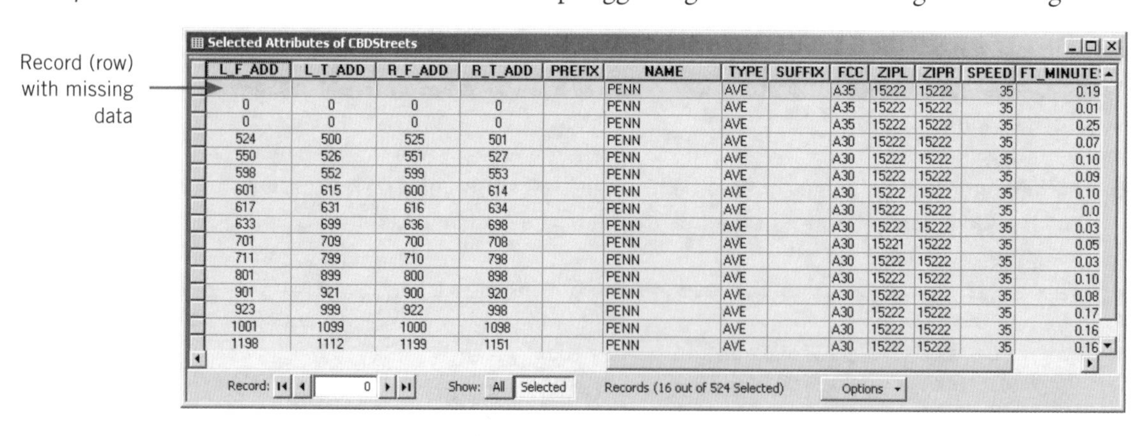

L_F_ADD	L_T_ADD	R_F_ADD	R_T_ADD	PREFIX	NAME	TYPE	SUFFIX	FCC	ZIPL	ZIPR	SPEED	FT_MINUTE:
					PENN	AVE		A35	15222	15222	35	0.19
0	0	0	0		PENN	AVE		A35	15222	15222	35	0.01
0	0	0	0		PENN	AVE		A35	15222	15222	35	0.25
524	500	525	501		PENN	AVE		A30	15222	15222	35	0.07
550	526	551	527		PENN	AVE		A30	15222	15222	35	0.10
598	552	599	553		PENN	AVE		A30	15222	15222	35	0.09
601	615	600	614		PENN	AVE		A30	15222	15222	35	0.10
617	631	616	634		PENN	AVE		A30	15222	15222	35	0.0
633	699	636	698		PENN	AVE		A30	15222	15222	35	0.03
701	709	700	708		PENN	AVE		A30	15221	15222	35	0.05
711	799	710	798		PENN	AVE		A30	15222	15222	35	0.03
801	899	800	898		PENN	AVE		A30	15222	15222	35	0.10
901	921	900	920		PENN	AVE		A30	15222	15222	35	0.08
923	999	922	998		PENN	AVE		A30	15222	15222	35	0.17
1001	1099	1000	1098		PENN	AVE		A30	15222	15222	35	0.16
1198	1112	1199	1151		PENN	AVE		A30	15222	15222	35	0.16

Modify the attributes of CBDStreets

1 **Move or minimize the Attributes tables so you can see the map.**

2 **Click the Identify button, then click the street segment to the right of the selected segment.**

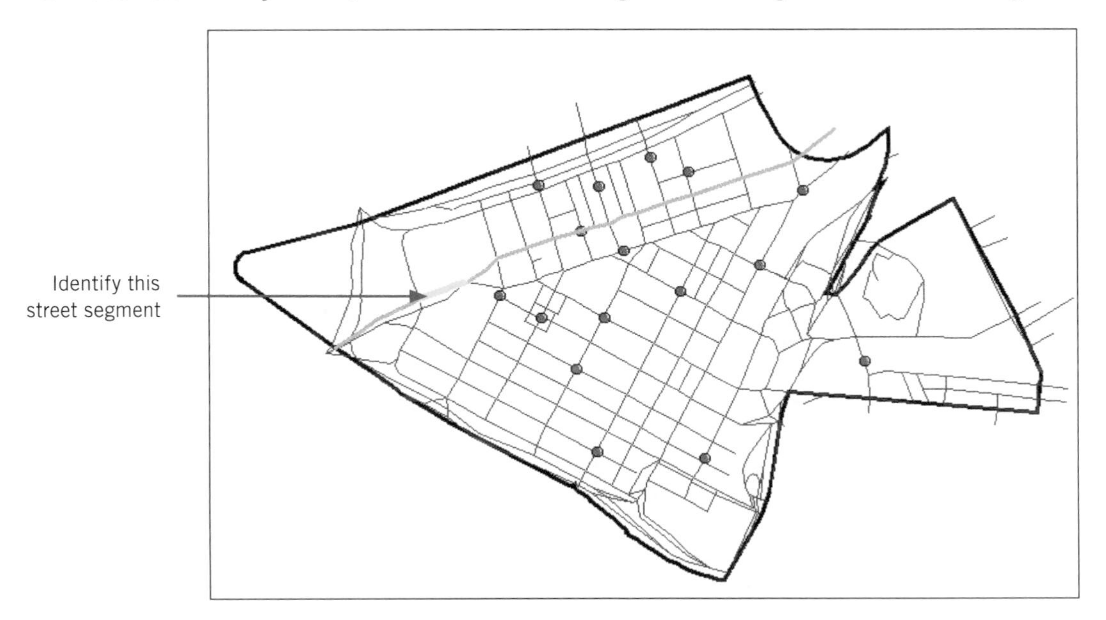

Note the values for L_F_ADD, L_T_ADD, R_F_ADD, and R_T_ADD. The segment to the east (right) contains address values ranging from 500 to 525.

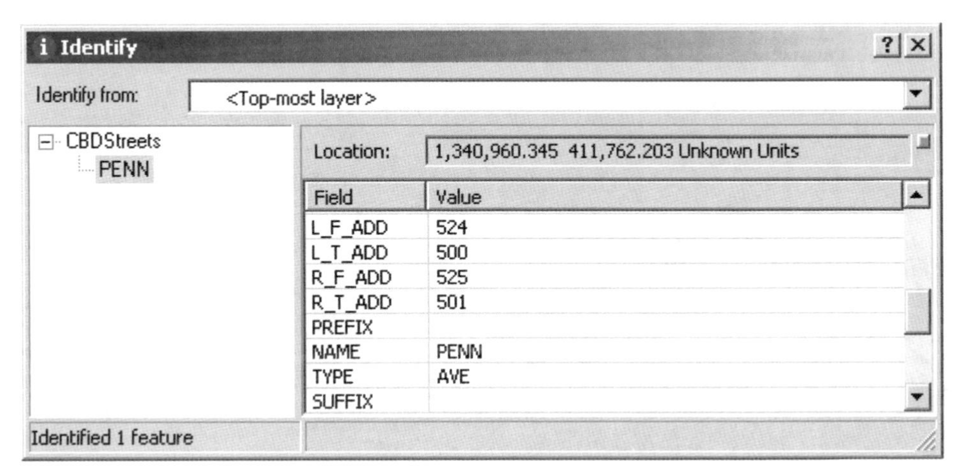

Suppose further that you have obtained valid numbers for the street segment's missing attributes: 498 to 474 on the left side and 499 to 475 on the right side.

3 **Close the Identify dialog box.**

4 **Click Tools, Editor Toolbar to open the Editor toolbar.**

5 **Click Editor, Start Editing, then in the Start Editing dialog box click the source folder containing the CBDStreets shapefile and click OK.**

6 **In the Attributes of CBDStreets table, add the following values to the record highlighted in yellow: L_F_ADD=498, L_T_ADD=474, R_F_ADD=499, AND R_T_ADD=475.**

7 **Click Editor, Stop Editing, Yes, and close the Editor toolbar.**

8 **Click Selection, Clear Selected Features, then click the All button.**

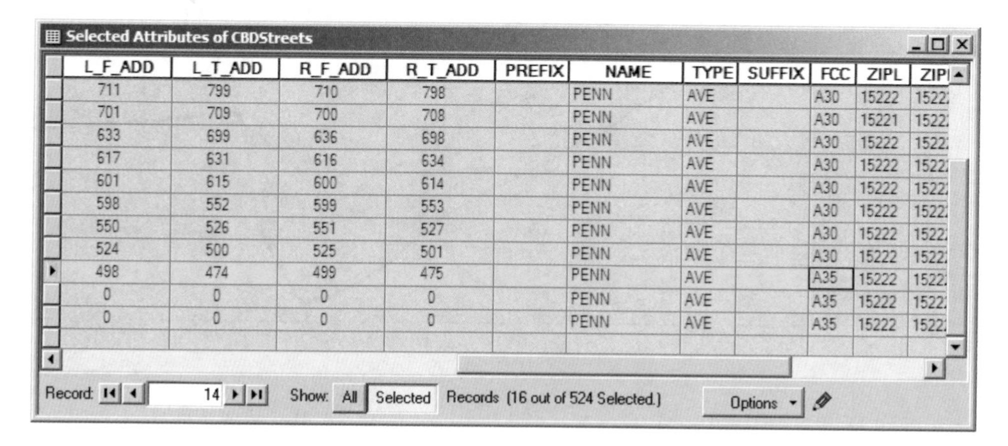

Selected Attributes of CBDStreets

L_F_ADD	L_T_ADD	R_F_ADD	R_T_ADD	PREFIX	NAME	TYPE	SUFFIX	FCC	ZIPL	ZIPI
711	799	710	798		PENN	AVE		A30	15222	1522
701	709	700	708		PENN	AVE		A30	15221	1522
633	699	636	698		PENN	AVE		A30	15222	1522
617	631	616	634		PENN	AVE		A30	15222	1522
601	615	600	614		PENN	AVE		A30	15222	1522
598	552	599	553		PENN	AVE		A30	15222	1522
550	526	551	527		PENN	AVE		A30	15222	1522
524	500	525	501		PENN	AVE		A30	15222	1522
498	474	499	475		PENN	AVE		A35	15222	1522
0	0	0	0		PENN	AVE		A35	15222	1522
0	0	0	0		PENN	AVE		A35	15222	1522

Record: 14 Show: All Selected Records (16 out of 524 Selected.) Options

9 **Close the Attributes of CBDStreets table.**

Now, 490 Penn Ave, one of the unmatched client addresses, will geocode the next time you attempt to rematch the addresses.

YOUR TURN

Geocode CBDClients to add the client at 490 Penn Ave to the map. To accomplish this, you will have to build a new locator, Central Business District Streets 2, and geocode the client table from scratch to create CBDClient2.shp. The new locator will index and include the street segment you edited.

Use alias tables

Some locations are more commonly located by their landmark name instead of their street address. For example, the White House may be listed in a table as "White House" instead of 1600 Pennsylvania Avenue NW Washington, D.C., 20500. In the following exercise, you will use an alias table to geocode records that are identified by their landmark name rather than their street address.

Add an alias table and rematch addresses

1 **Click the Add Data button and add the BldgNameAlias.dbf table from the \Gistutorial\PAGIS\CentralBusinessDistrict folder.**

2 **In the table of contents, right-click BldgNameAlias and click Open.**

Browse the data in the table. The table contains the alias name and street address for each record.

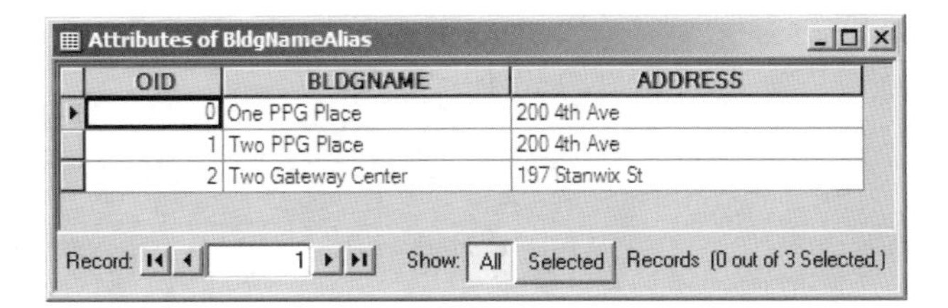

3 **Close the alias table.**

4 **Click Tools, Geocoding, Review/Rematch Addresses, Geocoding Result:CBDClients.**

5 **Click Geocoding Options.**

6 **Click Place Name Alias Table.**

7 **From the Alias Table drop-down list, click BldgNameAlias.**

8 **From the Alias field drop-down list, click BLDGNAME.**

9 **Verify that your settings match those in the following graphic, and click OK.**

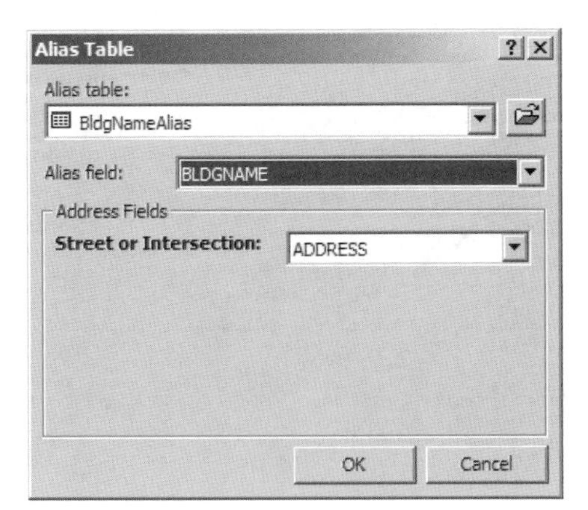

10 Click OK to close the Geocoding Options dialog box.

11 Click Match Automatically.

ArcMap will match the records listed as PPG Place and Gateway Center (building names in downtown Pittsburgh). The remaining unmatched addresses are outside of Pittsburgh's central business district.

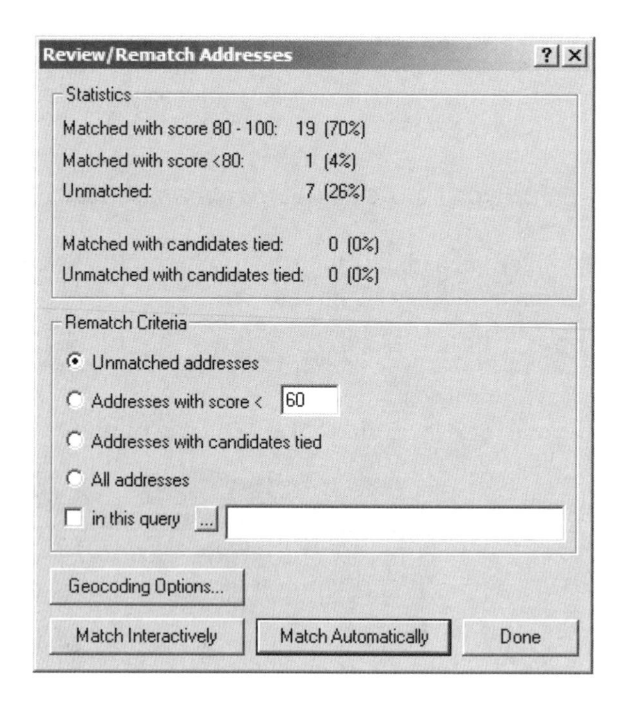

12 Click Done.

You will see newly added points on the map as a result of rematching.

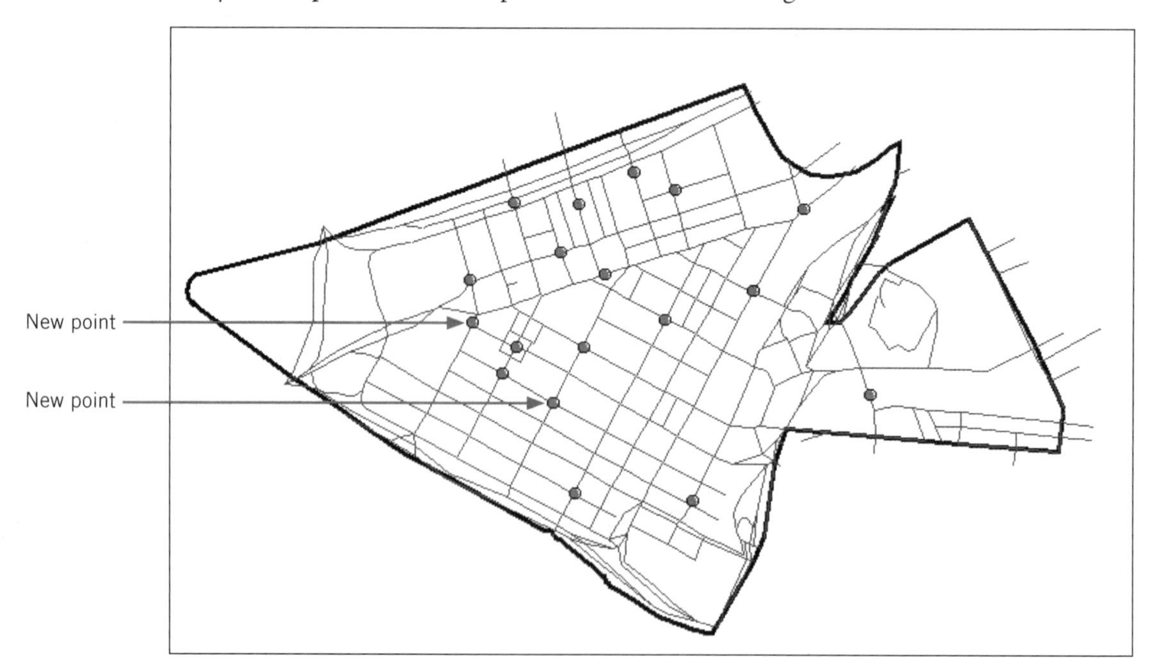

13 Save your map as \Gistutorial\Tutorial7-3.mxd.

Assignment 7-1

Geocode household hazardous waste participants to ZIP Codes

Many county, city, and local environmental organizations receive inquiries from residents asking how they can dispose of household hazardous waste (HHW) materials that cannot be placed in regular trash or recycling collections. Homeowners continually search for environmentally responsible methods for disposing of common household products such as paint, solvents, automotive fluids, pesticides, insecticides, and cleaning chemicals.

The Pennsylvania Resources Council (PRC) *(www.prc.org)* is a nonprofit organization dedicated to protecting the environment. The PRC facilitates meetings, organizes collection events, spearheads fundraising and volunteer efforts, and develops education and outreach materials in response to the HHW problem.

At each event, the PRC collects residence data from participants. In this exercise, you will geocode participants by ZIP Code for a recent Allegheny County event.

Start with the following:

- **C:\Gistutorial\UnitedStates\Pennsylvania\ HHWZIPCodes.dbf**—table of ZIP Codes for a HHW Allegheny County event collected by the PRC. *Note: All attributes except ZIP Code have been suppressed to protect confidentiality.*
- **C:\Gistutorial\UnitedStates\Pennsylvania\PAZIP.shp**—polygon layer of Pennsylvania ZIP Codes used for address matching.
- **C:\Gistutorial\UnitedStates\Pennsylvania\PACounties.shp**—polygon layer of Pennsylvania counties.

Create a choropleth map of HHW participants by ZIP Code

In ArcCatalog, create an address locator to use when geocoding HHW participants to ZIP Codes. Store the locator as C:\Gistutorial\Answers\Assignment7\HHWZIP.loc.

In ArcMap, create a new map document called C:\Gistutorial\Answers\Assignment7\Assignment7-1.mxd that uses the UTM Zone 17N projection and includes a choropleth map in a layout showing the number of Household Hazardous Waste participants by ZIP Code in Pennsylvania. Add the PA County shapefile as a thick, dark outline. Label counties with county names.

Export a layout with title, map, and legend to a file called C:\Gistutorial\Answers\Assignment7\Assignment7-1.jpg. Include the layout image in a Word document, C:\Gistutorial\Answers\Assignment7\Assignment7-1.doc, in which you describe residence patterns of HHW event attendees.

Geocoding and spatial joining hints

- Geocode the ZIP Codes in HHWZIPCodes.dbf to create a point layer of participants by ZIP Code, called C:\Gistutorial\Answers\Assignment7\GeocodedParticipants.shp. You should match 1,464 (99 percent) and not match 14 (1 percent).
- Spatially join the new geocoded points to PAZip.shp to create a polygon layer called C:\Gistutorial\Answers\Assignment7\HHWParticipants.shp that includes a count of the number of participants in each ZIP Code. *Hint: Start by right-clicking PAZip.shp in the table of contents when joining.*

Assignment 7-2

Geocode ethnic businesses to Pittsburgh streets

As the 2000 Census attests to, immigrants are largely becoming one of the most salient indicators of growth and wealth in a region. By looking at the immigrants who live in a city and analyzing where they decide to set up their businesses, city planners can investigate why certain neighborhoods are more immigrant-friendly than others, and in turn, focus on the qualities that make a neighborhood open and diverse.

According to the 2000 Census, Pittsburgh, Pennsylvania, ranked twenty-fifth of all metropolitan areas in the number of immigrants who live there. GIS can geocode as points where these immigrants create businesses and then aggregate this to neighborhoods. The data used in this exercise is a sample of businesses generated by CMU students that focused on foreign-run high-tech firms, restaurants, and grocery stores. A more complete listing of businesses would need to be collected from a variety of sources. For more details on foreign businesses in the region, visit the Pittsburgh Regional Alliance at *www.pittsburghregion.org* and Global Pittsburgh at *www.globalpittsburgh.org*. In addition to mapping the businesses, an expansion of the project could be to download foreign-born population from the U.S. Census SF3 tables, aggregating this to neighborhoods as well.

Start with the following:

- **C:\Gistutorial\PAGIS\ForeignBusinesses.dbf**—sample database of Pittsburgh foreign-owned businesses.
- **C:\Gistutorial\PAGIS\PghStreets.shp**—TIGER line layer of Pittsburgh street centerlines.
- **C:\Gistutorial\PAGIS\Neighborhoods.shp**—polygon layer of Pittsburgh neighborhoods.

Create a pin (point) map of geocoded foreign-owned businesses and choropleth map of businesses by neighborhood

In ArcCatalog, create an address locator with the US Streets with Zone locator style to use when geocoding foreign business locations to streets. Store the locator as C:\Gistutorial\Answers\Assignment7\PghStreets.loc.

In ArcMap, create a new map document called C:\Gistutorial\Answers\Assignment7\Assignment7-2.mxd that includes a layout showing a pin map of foreign business locations in Pittsburgh and a choropleth map showing the count of the number of businesses in each neighborhood.

Export the map as a file called C:\Gistutorial\Answers\Assignment7\Assignment7-2.jpg.

Geocoding hints

- Geocode the foreign businesses to Pittsburgh streets, creating a file called C:\Gistutorial\Answers\Assignment7\GeocodedForeignBusinesses.shp. You should get about 26 percent that do not match. There are a variety of reasons for this: they are outside the City of Pittsburgh, have business locations that are landmark names (e.g., Four Gateway Center), or have addresses or ZIP Codes that are simply entered incorrectly.
- Use PghStreets.shp or Internet sites such as *www.usps.com* or *maps.google.com* to rematch at least five unmatched addresses.
- Keep a log of steps you took to try and rematch addresses and turn this in with your assignment as C:\Gistutorial\Answers\Assignment7\Assignment7-2.doc. For each address investigated, give the original address, the problem information, source, and correction.

Spatial join hint

- Spatially join the geocoded businesses to the neighborhoods in a new shapefile called C:\Gistutorial\Answers\Assignment7\ForeignBusinessses.shp.

What to turn in

If you are working in a classroom setting with an instructor, you may be required to submit the exercises you created in tutorial 7. Below are the files you are required to turn in. Be sure to use a compression program such as PKZIP or WinZip to include all three files as one .zip document for review and grading. Include your name and assignment number in the .zip document (YourNameAssn7.zip).

ArcMap documents

C:\Gistutorial\Answers\Assignment7\Assignment7-1.mxd
C:\Gistutorial\Answers\Assignment7\Assignment7-2.mxd

Exported maps

C:\Gistutorial\Answers\Assignment7\Assignment7-1.jpg
C:\Gistutorial\Answers\Assignment7\Assignment7-2.jpg

Shapefiles

C:\Gistutorial\Answers\Assignment7\GeocodedParticipants.shp
C:\Gistutorial\Answers\Assignment7\HHWParticipants.shp
C:\Gistutorial\Answers\Assignment7\GeocodedForeignBusinesses.shp
C:\Gistutorial\Answers\Assignment7\ForeignBusinesses.shp

Word documents

C:\Gistutorial\Answers\Assignment7\Assignment7-1.doc
C:\Gistutorial\Answers\Assignment7\Assignment7-2.doc

OBJECTIVES

Extract features to create a new shapefile
Clip streets to match a polygon boundary
Dissolve polygons based on their ZIP Code value
Append several polygon layers into one new shapefile
Create a model that uses the Clip and Union tool

GIS Tutorial 8

Spatial Data Processing

Basemaps are available from many sources. Often, however, you will need to modify available maps for use in a specific project. In this tutorial, you will learn how to extract a subset of spatial features from a map using either attribute or spatial queries. You will also learn how to aggregate polygons and how to append two or more layers into a single layer. These functions are referred to as geoprocessing functions and are commonly strung together to perform different types of spatial analysis. One way to build, share, and document your GIS work flows is by creating models, and within this tutorial you will learn how to create and run a simple GIS work-flow model.

Open an existing map

1 From the Windows taskbar, click Start, All Programs, ArcGIS, ArcMap.

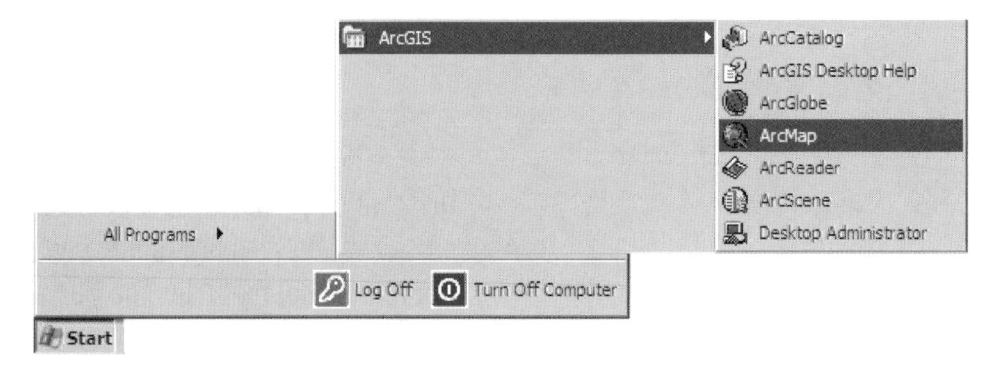

2 Click the An existing map radio button in the ArcMap dialog box.

3 Click OK.

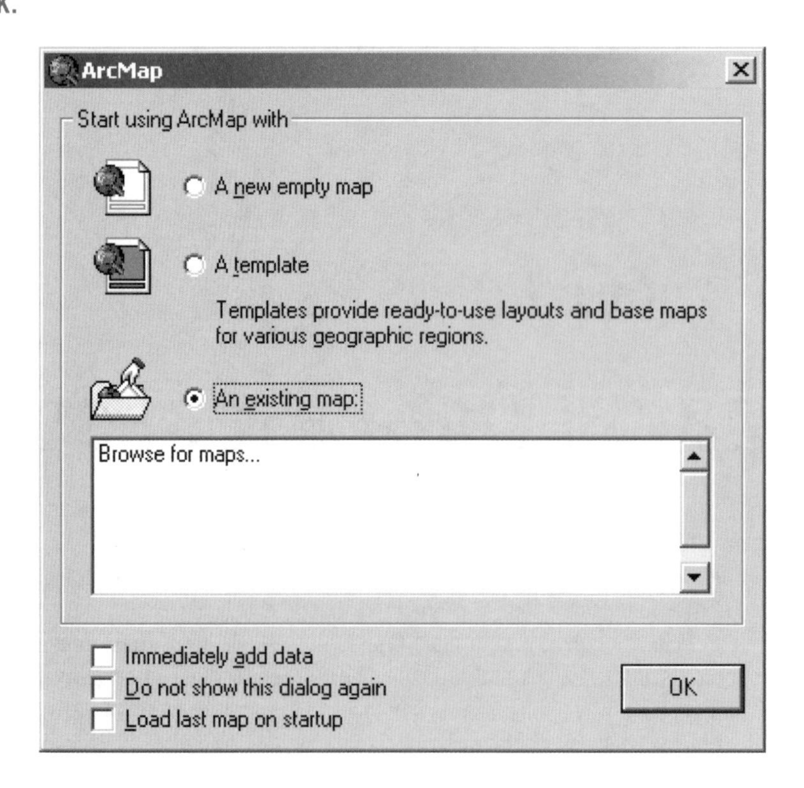

4 **Browse to the drive on which the Gistutorial folder has been installed (e.g., C:\Gistutorial), select the Tutorial8-1.mxd, and click the Open button.**

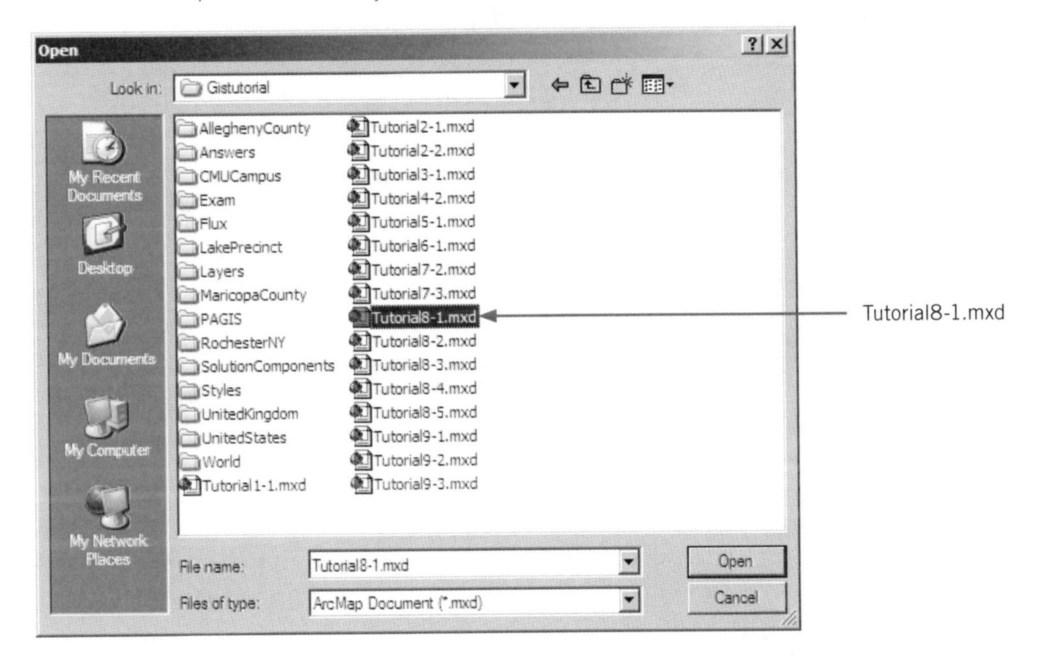

The Tutorial8-1.mxd file opens in ArcMap showing a map of the New York City Metro Area including Manhattan, Brooklyn, the Bronx, Staten Island, and Queens.

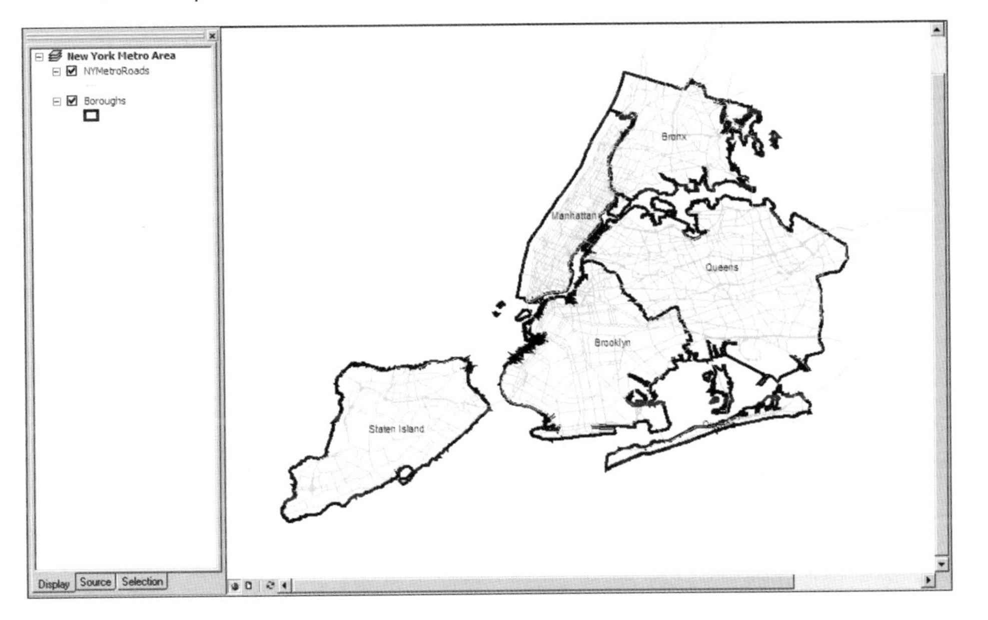

Use data queries to extract features

Use the ArcMap Select By Attributes dialog box

Here you will use ArcMap's Select By Attributes function to create a study area for Manhattan that will be extracted from the NY Boroughs layer.

1 **Click Selection, Select by Attributes.**

2 **From the Layer drop-down list, click Boroughs.**

3 **In the Fields box, double-click "NAME".**

4 **Click the = button.**

5 **Click the Get Unique Values button and then, in the Unique Values box, double-click 'Manhattan'.**

Based on this query expression, the Select by Attributes dialog box will select the Manhattan borough feature.

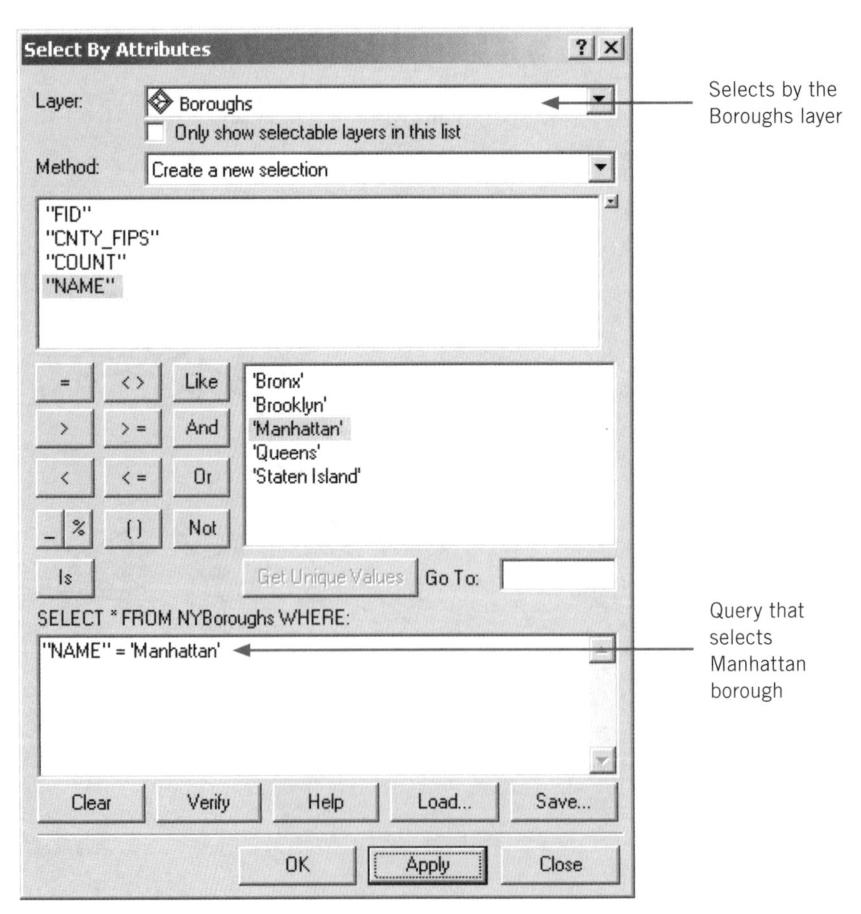

6 **Click Apply and Close.**

Show selected features and convert to shapefile

1 Click View, Zoom Data, Zoom to Selected Features.

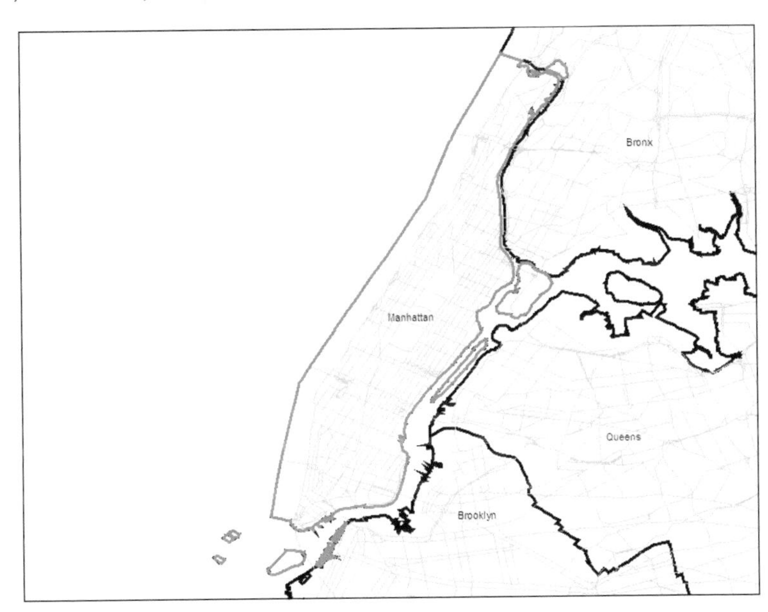

2 In the table of contents, right-click the Boroughs layer, click Data, Export Data.

3 Save the output shapefile as **\Gistutorial\UnitedStates\NewYork\Manhattan.shp**.

4 Click OK, then click Yes to add the layer to the map.

Your map now contains a new shapefile containing only the borough of Manhattan.

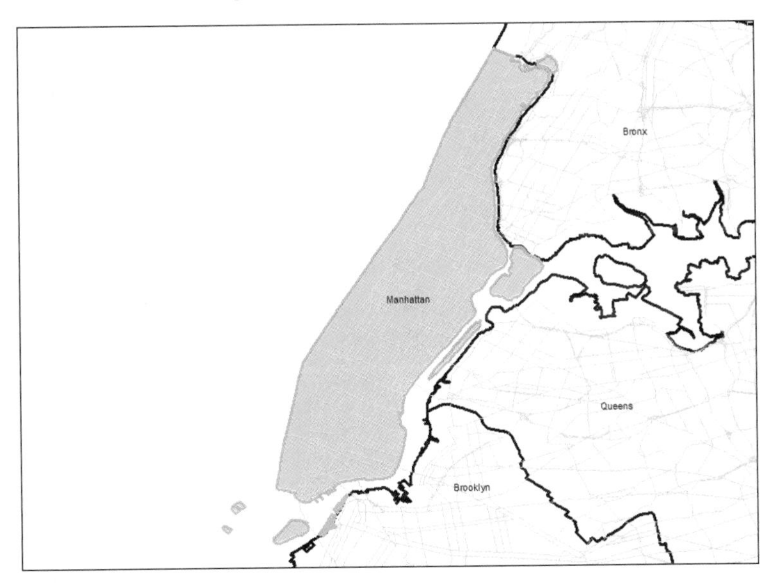

Use the ArcMap Select Features tool

In the previous steps, you used an attribute query to select the feature you wanted to extract. Sometimes, however, it's easier to manually select the feature(s) you want to extract directly from the map display instead of building a query expression in the Select By Attributes dialog box.

1 **Click the Full Extent button,**  **then click the Select Features button.** 

2 **Click once inside the polygon feature for Brooklyn.**

3 **In the table of contents, right-click the Boroughs layer, click Data, Export Data.**

4 **Save the output shapefile as \Gistutorial\UnitedStates\NewYork\Brooklyn.shp.**

5 **Click OK, then click Yes to add the layer to the map.**

Your map now contains another new shapefile, this one containing only the borough of Brooklyn.

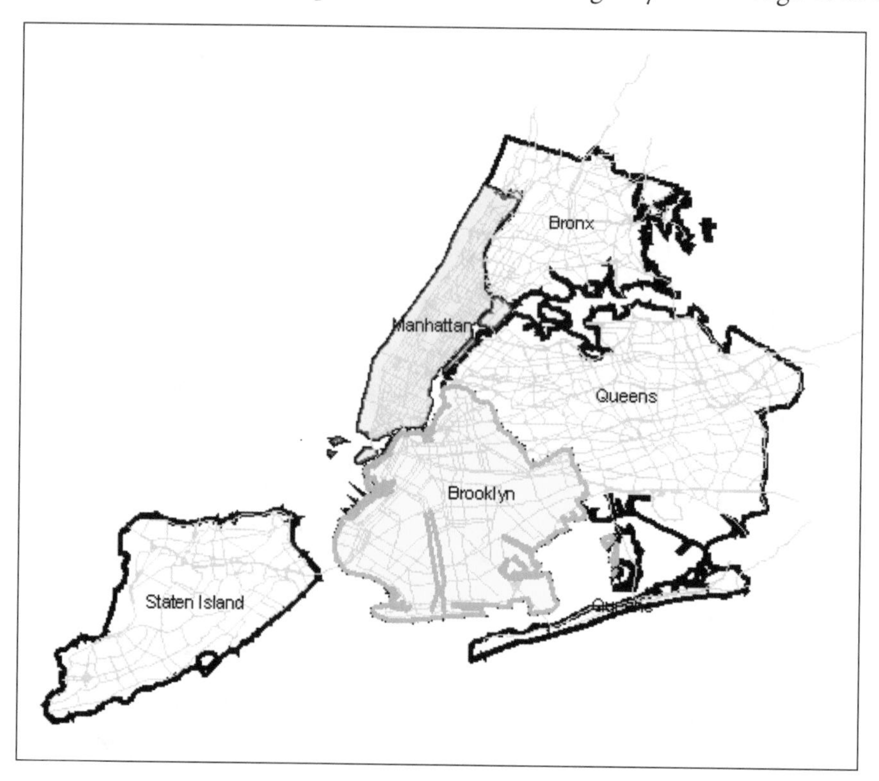

YOUR TURN

Use either the Select By Attributes dialog box or the Select Features tool to create study area shapefiles for Queens, the Bronx, and Staten Island. When finished, clear all selections.

Clip features

Use the ArcMap Select By Location tool

In the following steps, you will use the ArcMap Select By Location dialog box to select the roads in the Manhattan borough. After selecting the roads, you will create a new shapefile from them.

1 **Click Selection, Select by Location.**

2 **If necessary, click the I want to drop-down list and choose select features from.**

3 **Check NYMetroRoads as the layer from which to select features.**

4 **From the selection method (that) drop-down list, choose intersect.**

5 **From the third drop-down list in the dialog box, choose Manhattan.**

Once made, the settings in this dialog box can be read as a sentence. In this case the sentence reads as follows: "I want to select features from the NYMetroRoads layer that intersect the features in the Manhattan layer." The result of these settings will be the selection of all the roads contained by or crossing the boundary of the Manhattan borough.

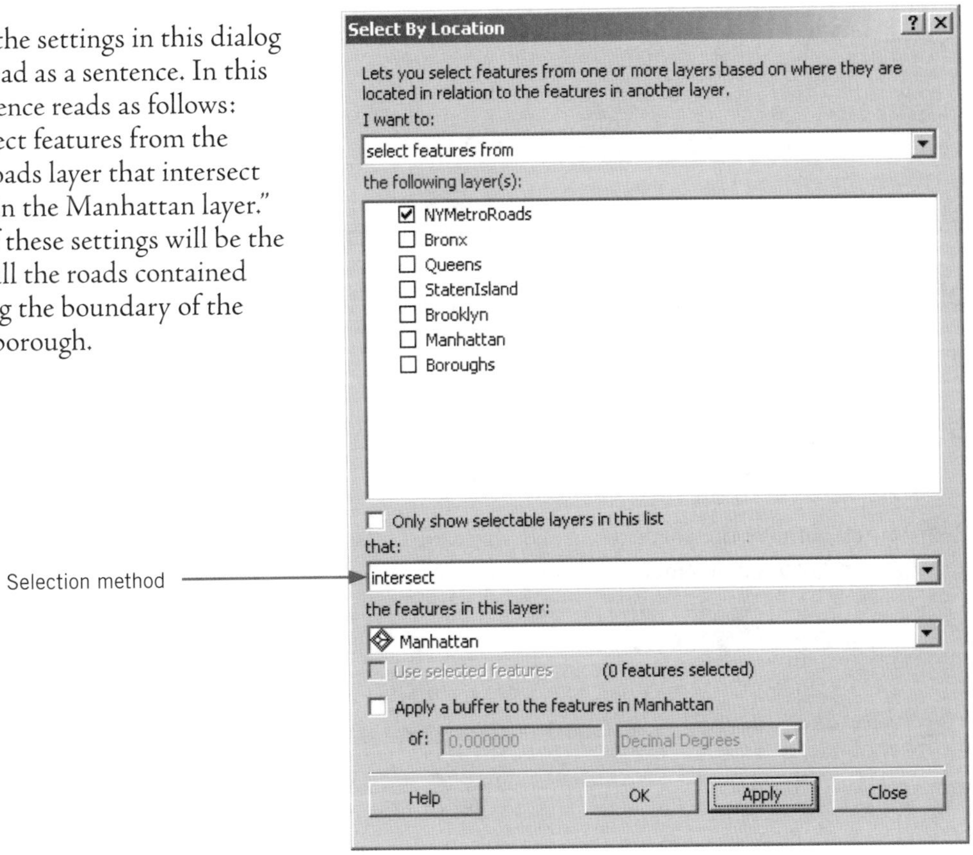

Select By Location

Lets you select features from one or more layers based on where they are located in relation to the features in another layer.

I want to:

select features from

the following layer(s):

- [x] NYMetroRoads
- [ ] Bronx
- [ ] Queens
- [ ] StatenIsland
- [ ] Brooklyn
- [ ] Manhattan
- [ ] Boroughs

Only show selectable layers in this list

that:

intersect

Selection method

the features in this layer:

Manhattan

Use selected features (0 features selected)

Apply a buffer to the features in Manhattan

of: 0.000000 Decimal Degrees

Help OK Apply Close

6 **Verify that your settings match those in the above dialog box, click Apply, then click Close.**

Show selected features and convert to shapefile

1 **Click View, Zoom Data, Zoom to Selected Features.**

The selected roads are only those within or those that intersect the Manhattan borough.

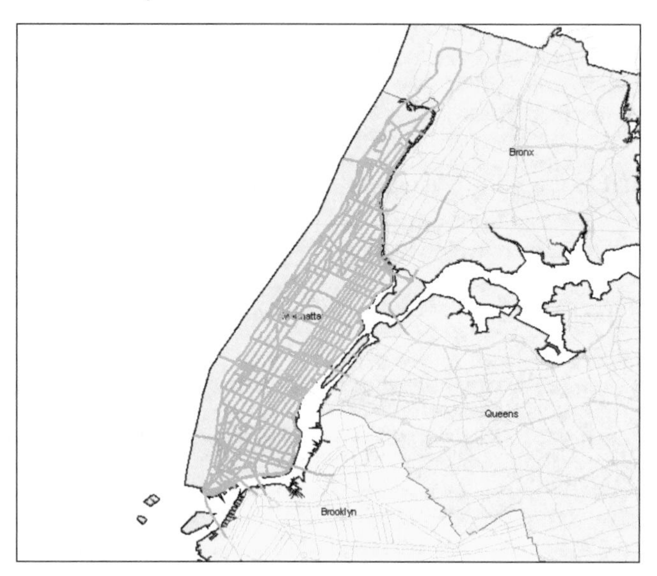

2 **In the table of contents, right-click the NYMetroRoads layer, click Data, Export Data.**

3 **Save the output shapefile as \Gistutorial\UnitedStates\NewYork\ManhattanRoads.shp.**

4 **Click OK, then click Yes to add the layer to the map.**

5 **Turn off the NYMetroRoads layer so the only roads visible in the map are those in the ManhattanRoads layer.**

Notice that some of the roads in the ManhattanRoads layer extend or "dangle" beyond the Manhattan borough outline.

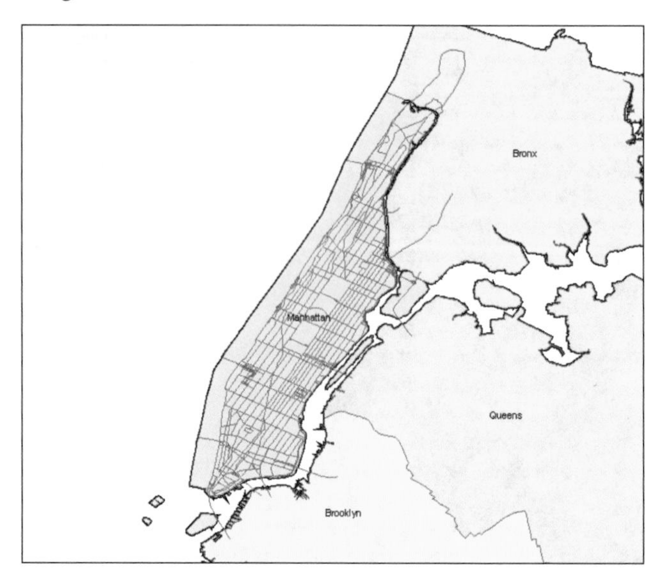

Clip the Manhattan streets

Next, you will open the Clip tool from ArcToolbox and use it to cut off the NYMetroRoads street segments using the Manhattan shapefile. Once this is done, the streets in the ManhattanRoads layer will be clipped exactly to the edge of the Manhattan borough, with no dangling lines, and saved within a new shapefile.

1 **Click the ArcToolbox button.** 

2 **In ArcToolbox, click the plus (+) sign beside the Analysis Tools toolbox, then click the plus (+) sign to expand the Extract toolset.**

3 **In the Extract toolset, double-click the Clip tool to open its dialog box.**

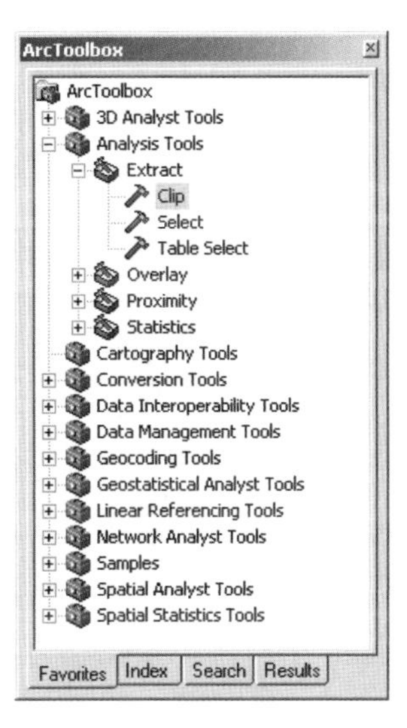

4 **Click the Input Features drop-down list and choose NYMetroRoads.**

5 **Click the Clip Features drop-down list and choose Manhattan.**

6 **Save the Output Feature Class as \Gistutorial\UnitedStates\NewYork\ClippedManhattanRoads.shp.**

7 **Verify that your Clip settings match those in the graphic below, then click OK. Click Close when the process completes.**

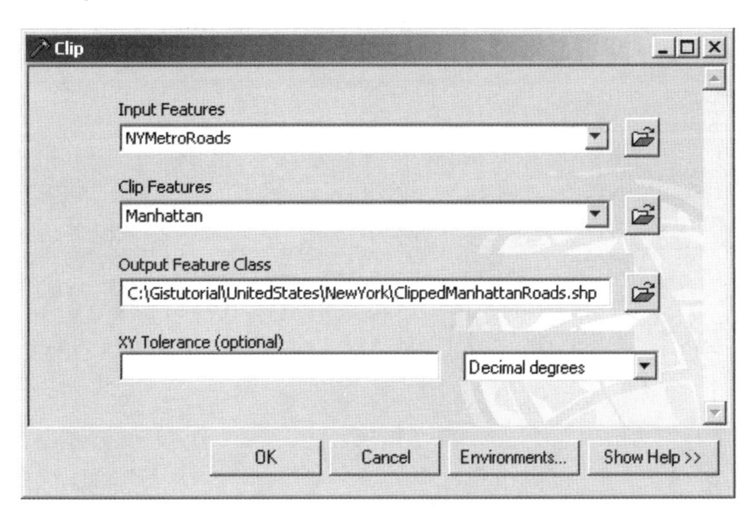

8 Turn on the ClippedManhattanRoads layer and turn off the ManhattanRoads layer.

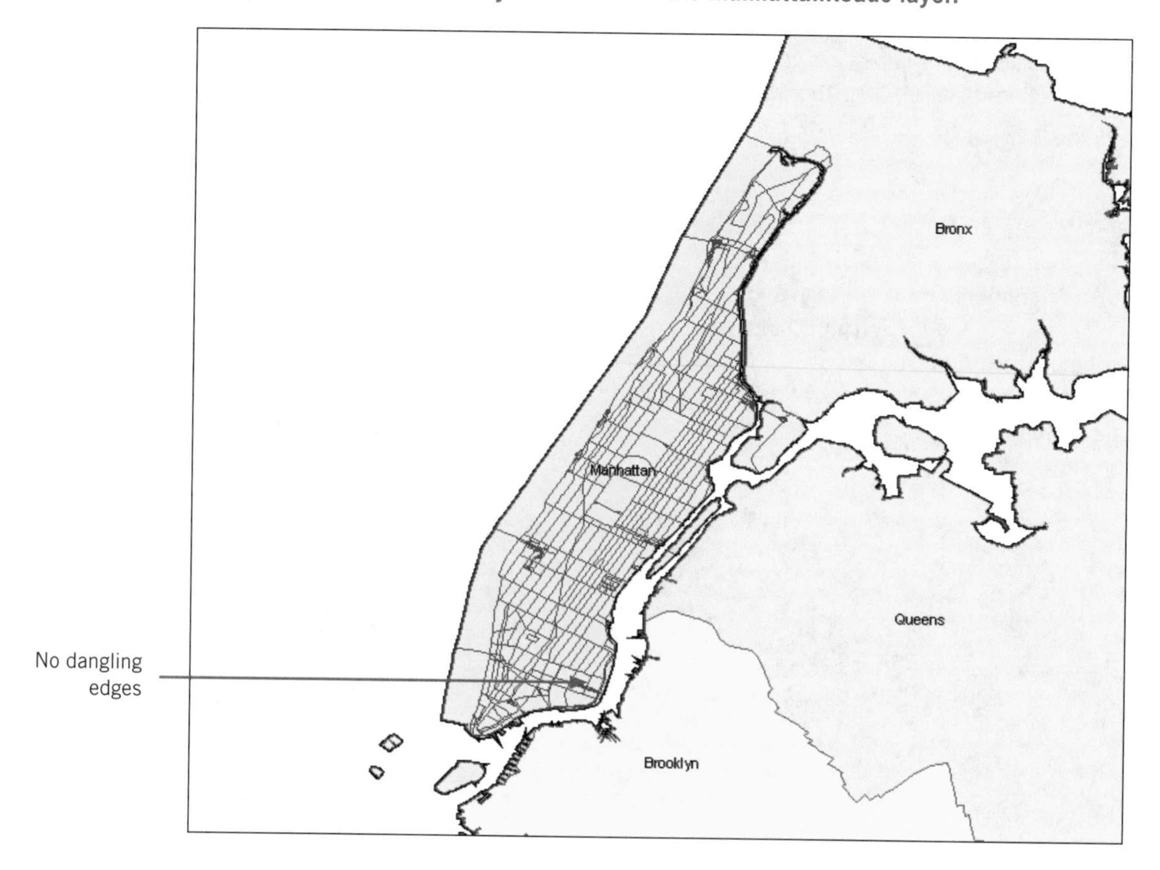

The streets in ClippedManhattanRoads layer do not cross the borough of Manhattan's boundary.

YOUR TURN

Using the Select By Location function in combination with the Clip tool, clip the NYMetroRoads roads to one of the other New York boroughs. When finished, close ArcToolbox and save your map document.

Dissolve features

You can form administrative or other types of boundaries by merging polygons in a feature class that share a common attribute. This type of a merge is called a dissolve, and in this section you will use the Dissolve tool to dissolve ZIP Code polygons based on their borough name.

Open an existing map

1 **In ArcMap open Tutorial8-2.mxd from the \GISTutorial folder.**

Tutorial8-2 contains a map of the New York City Metro Area ZIP Codes, including Manhattan, Brooklyn, Staten Island, the Bronx, and Queens.

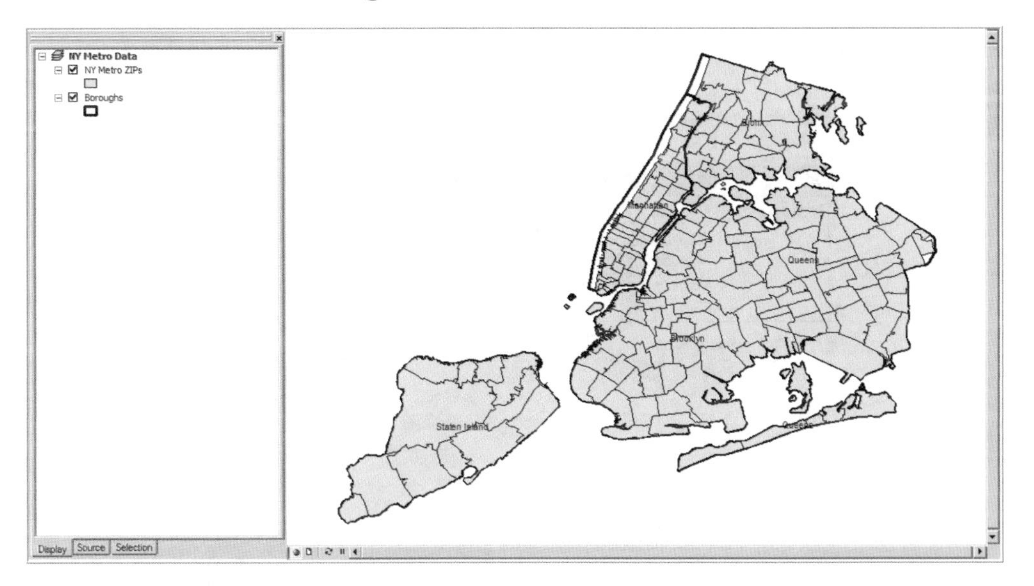

Dissolve ZIP Codes using the ArcMap command line

In ArcToolbox, the Dissolve tool exists inside the Generalization toolset, which is inside the Data Management toolbox. All the geoprocessing tools, however, can be run from the Command Line window in ArcMap or ArcCatalog. Running a tool from the command line allows you to bypass opening the tool from ArcToolbox and interfacing with its dialog box. Instead, you type the name of the tool and its parameters as a string, then execute the tool by pressing Enter. This is a more direct route to a tool and its functionality, especially if you are already familiar with the tool and its parameter values.

1 **Click Window, Command Line.**

2 **In the top half of the Command Line window, type Dissolve.**

3 **Press the space bar, choose NY Metro ZIPS from the pop-up list, then press Enter.**

This is your input feature class.

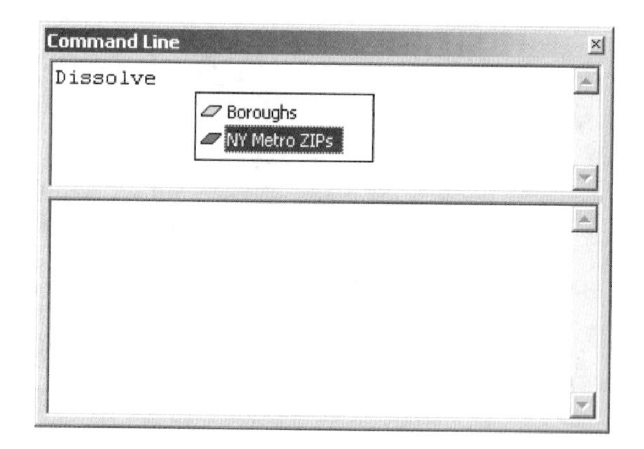

4 **Press the space bar and type C:\Gistutorial\UnitedStates\NewYork\DissolvedNYBoroughs.shp.**

This is your output feature class.

5 **Press the space bar and choose PO_NAME from the pop-up list. Press Enter.**

This is the field upon which the dissolve will be based.

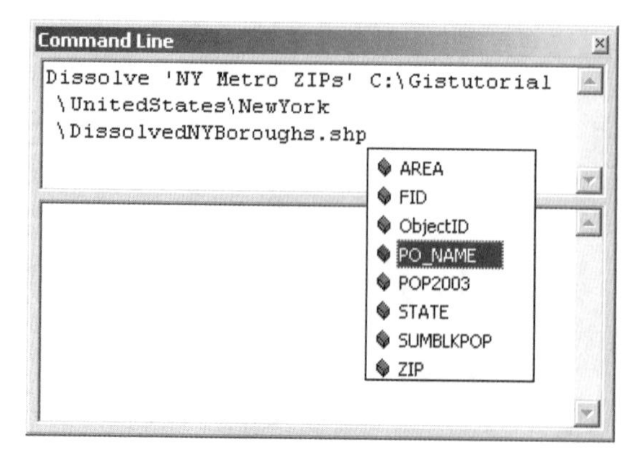

6 Press the space bar and choose POP2003.

This is an optional setting. When the dissolve runs, the values in the POP2003 field will be summarized for each group of polygons with the same PO_NAME value. In other words, it will summarize the population for each new polygon feature.

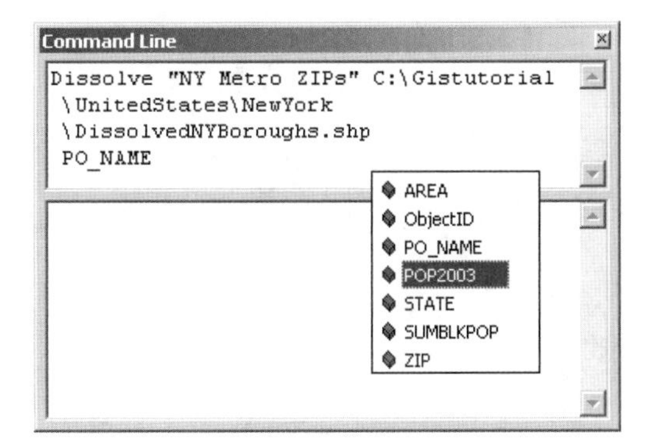

7 Type SUM, then place quotation marks around the last two parameter values so they read "POP2003 SUM".

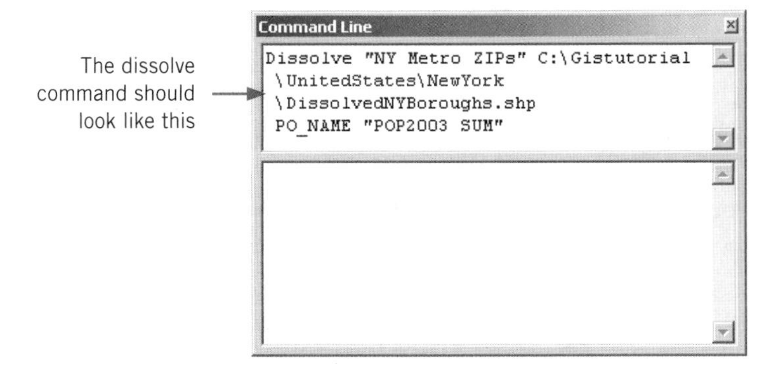

8 Press Enter.

The lower half of the Command Line window reports the status of the dissolve process and whether or not it successfully executes.

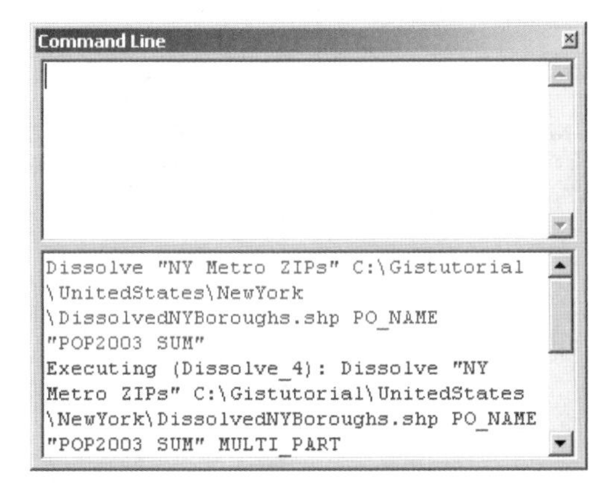

Once the dissolve process completes, the DissolvedNYBoroughs shapefile is automatically added to the map. This new shapefile contains the boundaries created from the dissolved ZIP Codes polygons. The ZIP Codes for the Queens borough were not dissolved, because the ZIP Codes in that borough had unique values in the PO_NAME field.

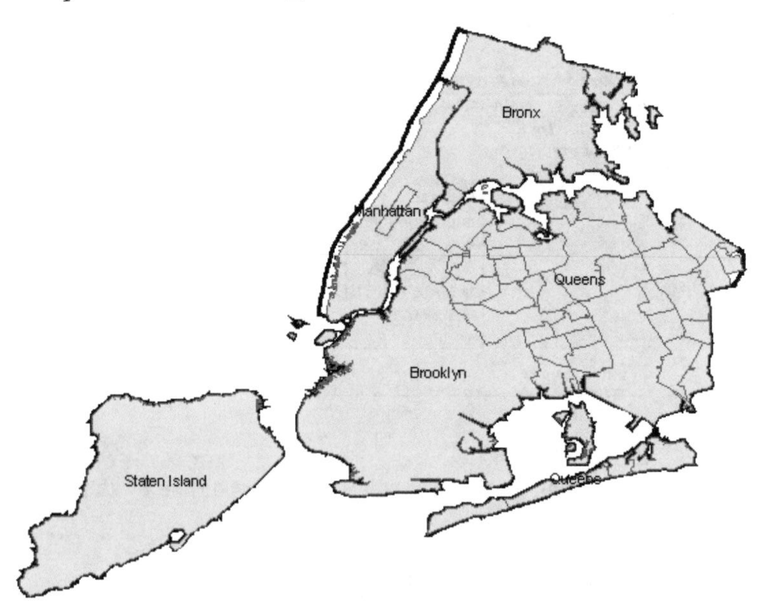

9 Use the Identify button to view the attribute information for each dissolved borough.

In addition to the name of each borough, you will see a population value in the SUM_POP200 field, which was derived from the POP2003 values stored with the polygons that were dissolved.

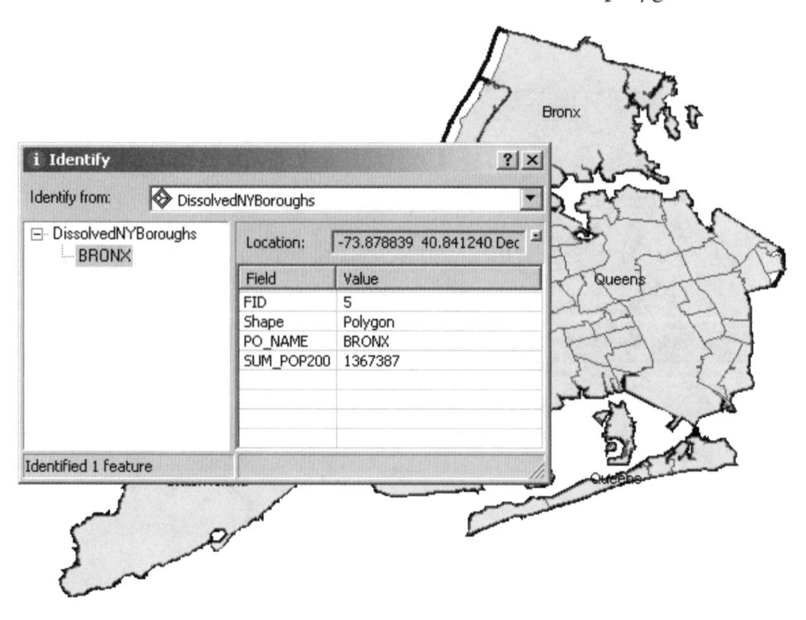

10 Save your map document.

Append layers

Sometimes it is necessary to append (or merge) two or more separate layers into a single layer. For example, you may want to build a single soils layer for an environmental study that includes two adjacent counties, and you already have a soils layer for each county. Using the Append tool, you could merge these two soil layers into a single layer, then use the complete soil mosaic for further analysis. When running the Append command, input feature classes (layers) must have the same geometry type (point, line, or polygon). You can append feature classes that do not have identical attribute fields, but the differing fields will be dropped from the output.

Open an existing map

1 In ArcMap, open **Tutorial8-3.mxd** from your **\Gistutorial** folder.

Tutorial8-3.mxd contains a map of the New York City Metro Area's boroughs. Each borough in the map exists in a separate shapefile.

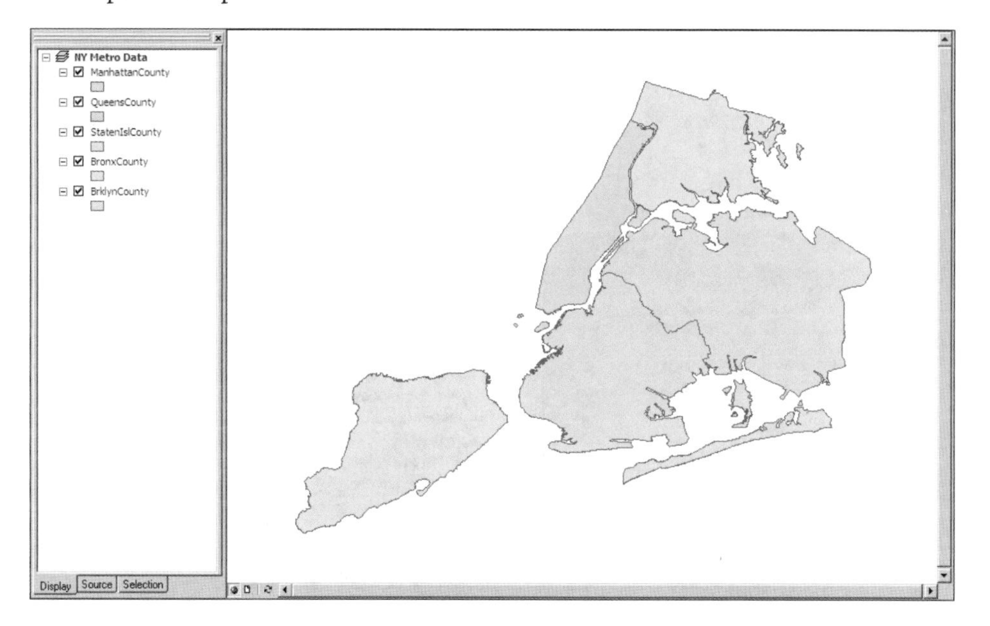

Create an empty polygon layer for appending

1 On the Standard toolbar, click the ArcCatalog button. 

2 In ArcCatalog, create a new polygon shapefile named **AppendedNYBoroughs.shp** in the **\Gistutorial\UnitedStates\NewYork** folder.

3 Close ArcCatalog.

4 Add **AppendedNYBoroughs.shp** to ArcMap.

Append several shapefiles into one shapefile

1 If necessary, open ArcToolbox in ArcMap by clicking the Show/Hide ArcToolbox button on the Standard toolbar. 

2 In ArcToolbox, expand the Data Management Tools toolbox, then expand the General toolset.

3 Double-click Append.

4 Click the Input Features drop-down list and, one at a time, add all five New York Borough shapefiles to the Input Features list, one at a time.

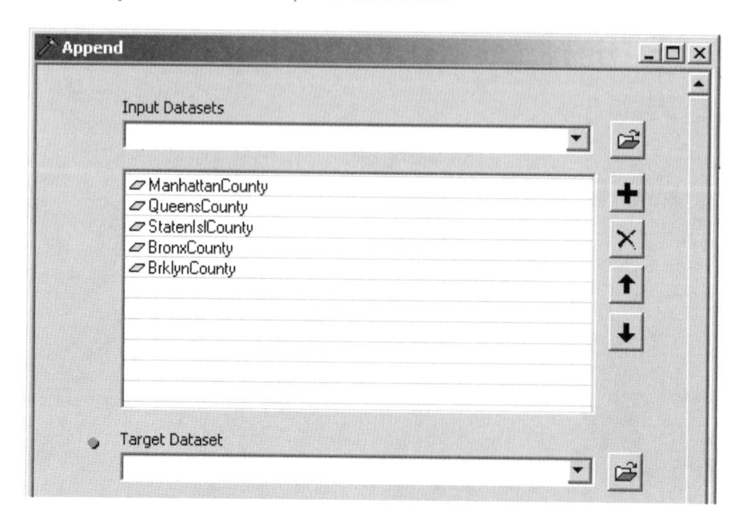

5 **Click the Target Dataset drop-down list and choose AppendedNYBoroughs.**

6 **Click the Schema Type drop-down list and choose NO_TEST.**

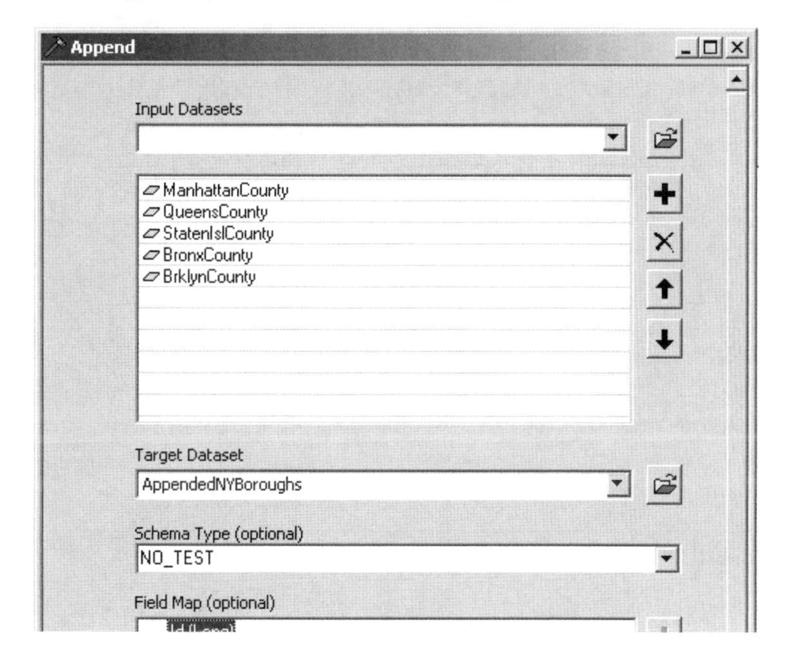

7 **Click OK.**

The AppendedNYBoroughs layer now contains all five boroughs of New York.

YOUR TURN

Practice using the Append tool by using it to append two New York boroughs. When finished, save your map document.

Union layers

The Union tool performs a type of overlay in which the geometry and attributes of two input polygon layers are combined to generate a new output polygon layer. The output from a union contains polygon features derived from the geometric intersection of the input polygons; the output also contains the attributes from both inputs. For example, you could union a vegetation and soils layer, then query the output to find polygons with a specific vegetation and soil type.

In this example, you will union a layer containing the ZIP Codes in the borough of Queens with a layer containing the boundary Queens. The output of the union will contain all the ZIP Codes in Queens, and each ZIP Code will have its borough name (Queens) assigned to it.

Open an existing map

1 **In ArcMap, open Tutorial8-4.mxd from your \Gistutorial folder.**

Tutorial8-4.mxd contains the borough of Queens and its ZIP Codes.

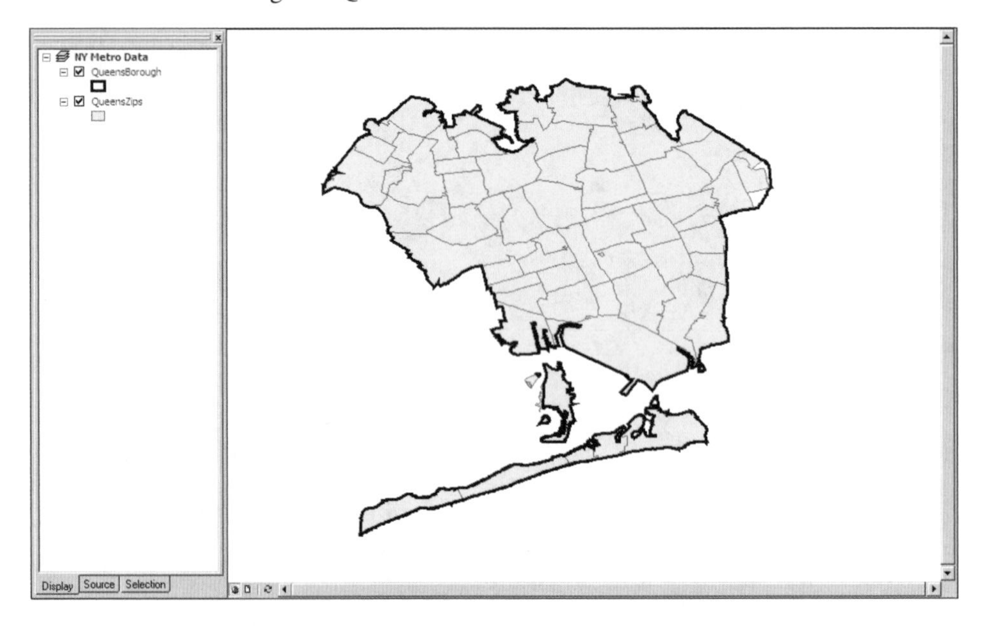

Union shapefiles

1 If necessary, open ArcToolbox.

2 In ArcToolbox, expand the Analysis Tools toolbox, then expand the Overlay toolset.

3 Double-click the Union tool.

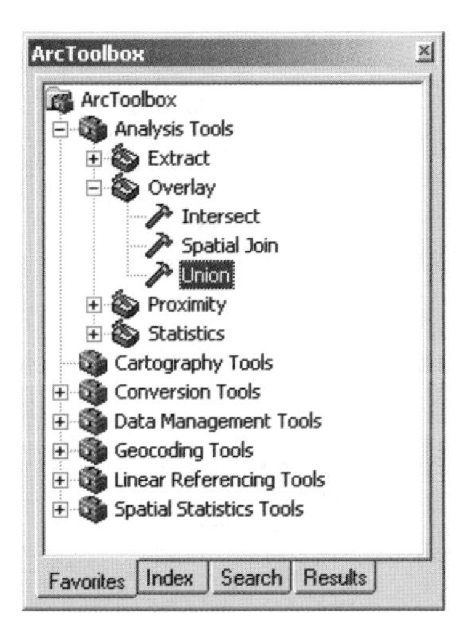

4 From the Input Features drop-down list choose QueensBorough and QueensZips, one at a time.

5 Save the Output Feature Class as **\Gistutorial\UnitedStates\NewYork\QueensBoroughZips_Union.shp**.

6 From the Join Attributes drop-down list, choose ALL.

7 Verify that your settings match those in the graphic below.

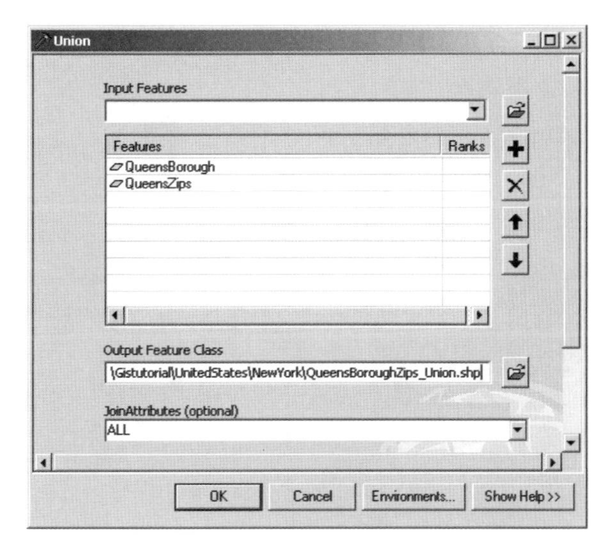

8 Click OK, Close.

The output added to your map contains the ZIP Code polygons with the borough name (Queens) now included in its attributes table.

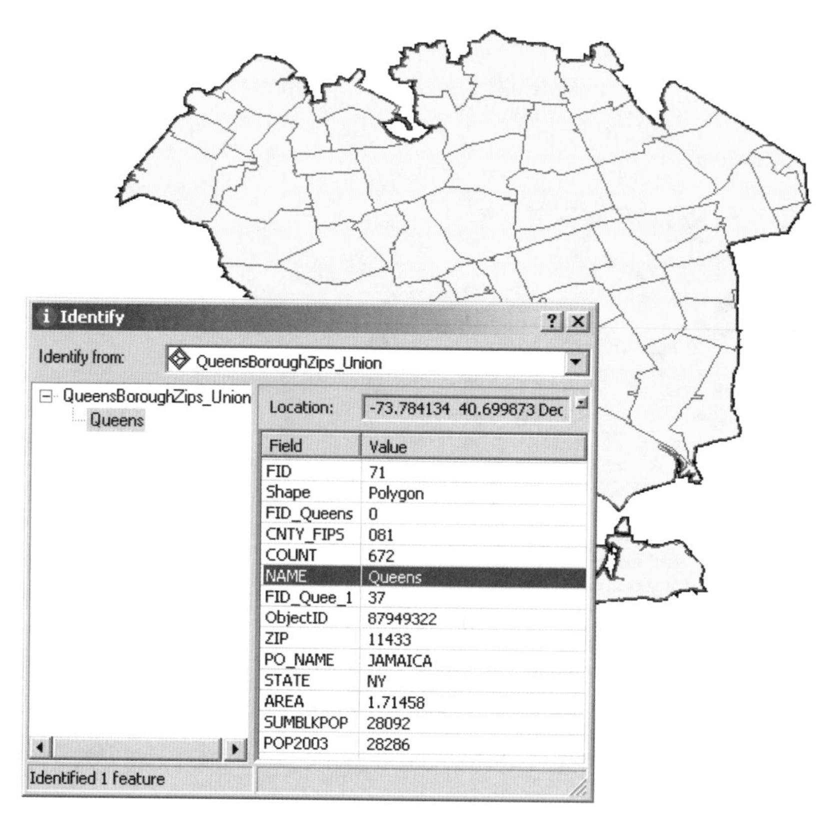

9 **Save your map document.**

ModelBuilder

The geoprocessing tools in ArcGIS are often used together in sequence to perform spatial analysis. Sometimes there are many steps involved, making it difficult to keep track of the tools used, the datasets involved, and the parameters defined within the overall work flow.

ModelBuilder is an application within ArcView that you can use to document and automate your geoprocessing work flows. Within ModelBuilder, you construct model diagrams from the data and geoprocessing tools needed for your analysis or work flow. Once the model is built, you can run it once or save it and run it again using different input data parameters.

In this exercise, you will build a model that will clip then union a set of census tracts to a neighborhood layer.

Open an existing map

1 In ArcMap, open **Tutorial8-5.mxd** from your **\Gistutorial** folder.

Tutorial8-5 contains the Pennsylvania census tracts and Pittsburgh neighborhoods. Although census tracts aggregate to neighborhoods, you will notice that some borders do not match exactly.

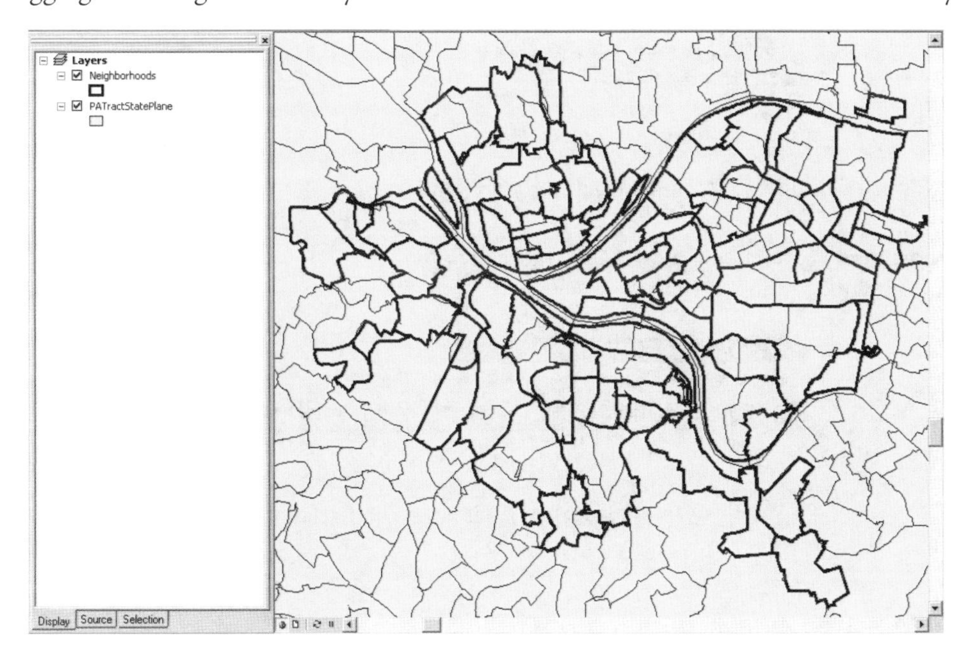

Create a new model

1 **Open ArcToolbox.**

2 **Right-click any white space inside ArcToolbox, and choose New Toolbox.**

3 **Name the new toolbox ModelBuilder.**

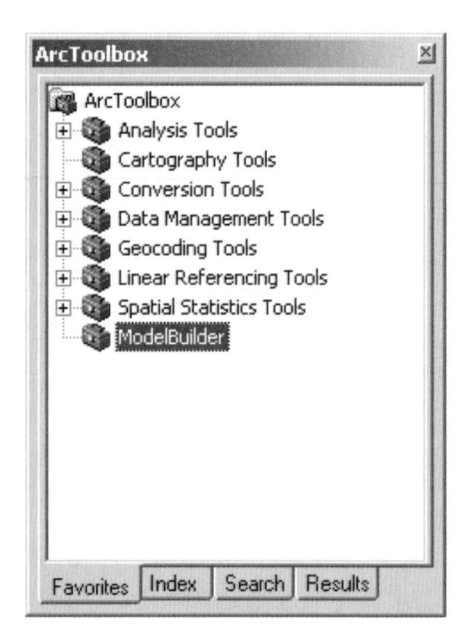

4 **Right-click the ModelBuilder toolbox and click New, Model.**

Choosing to create a new model automatically opens the Model window. You will use this window to create your model.

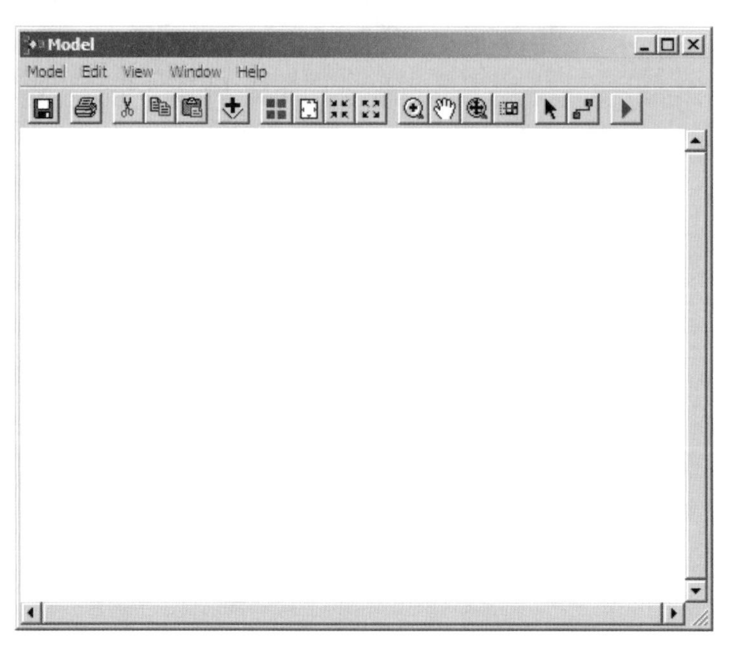

Add a Clip operation to the model

1 If necessary, position the Model window where you can see it simultaneously with the ArcMap table of contents and ArcToolbox. In the ArcMap table of contents, click the PATractStatePlane layer, drag it into the Model window, and drop it.

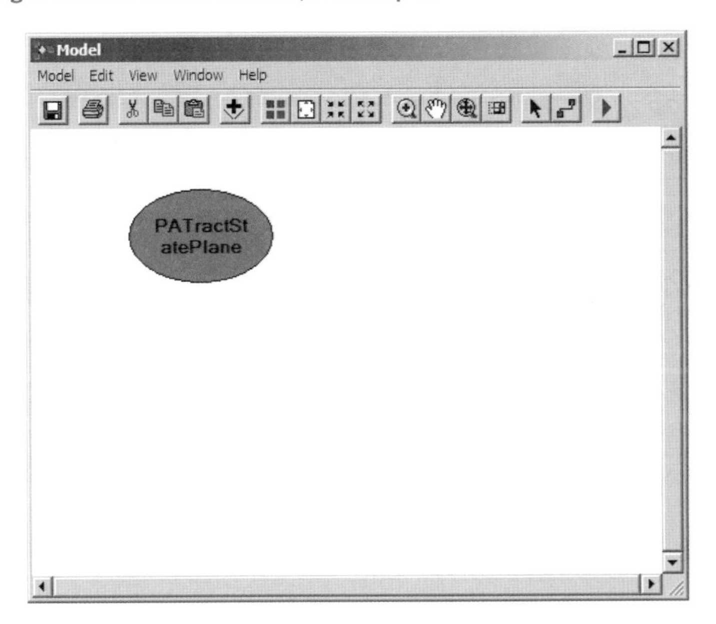

2 In ArcToolbox, expand the Analysis Tools toolbox, then expand the Extract toolset.

3 In the Extract toolset, click the Clip tool, drag it into the Model window, and drop it.

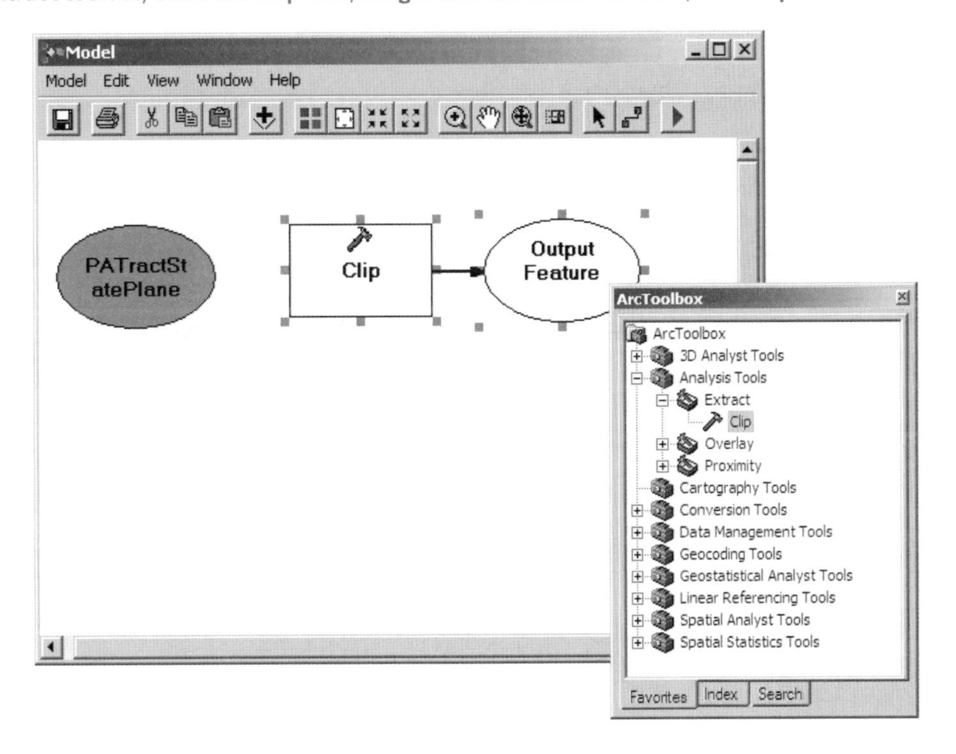

Connect PATractStatePlane layer to Clip tool

1 **From the Model toolbar, click the Add Connection button.** 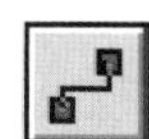

2 **In the Model window, click the PATractStatePlane layer and drag a line to the Clip tool.**

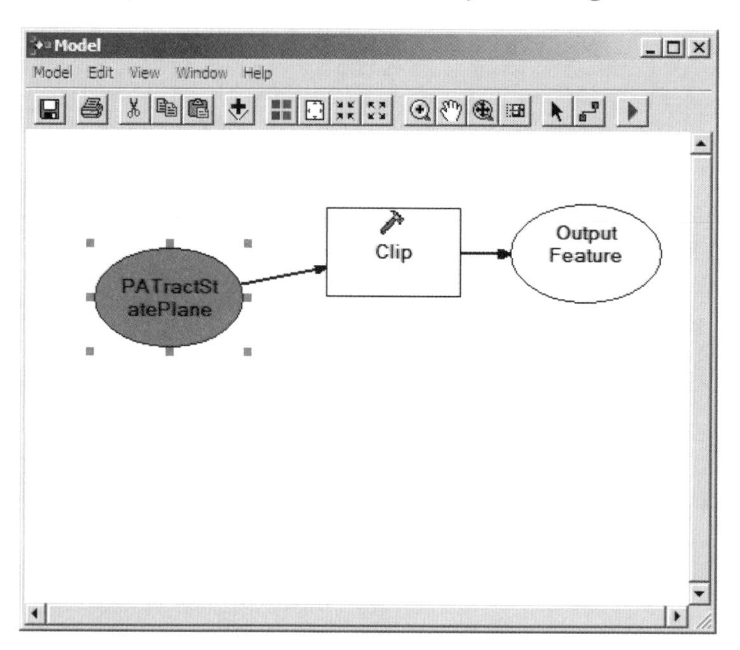

3 **From the Model toolbar, click the Select button.** 

4 **In the Model window, double-click the Clip tool.**

The Clip tool's dialog box appears with the Input Features property already defined (PATractStatePlane).

5 **From the Clip features drop-down list, select the Neighborhoods layer.**

6 **Save the Output Feature Class as \Gistutorial\UnitedStates\Pennsylvania\PATractStatePlane_Clip.shp.**

7 **Verify that your settings match those in the graphic shown at right, then click OK.**

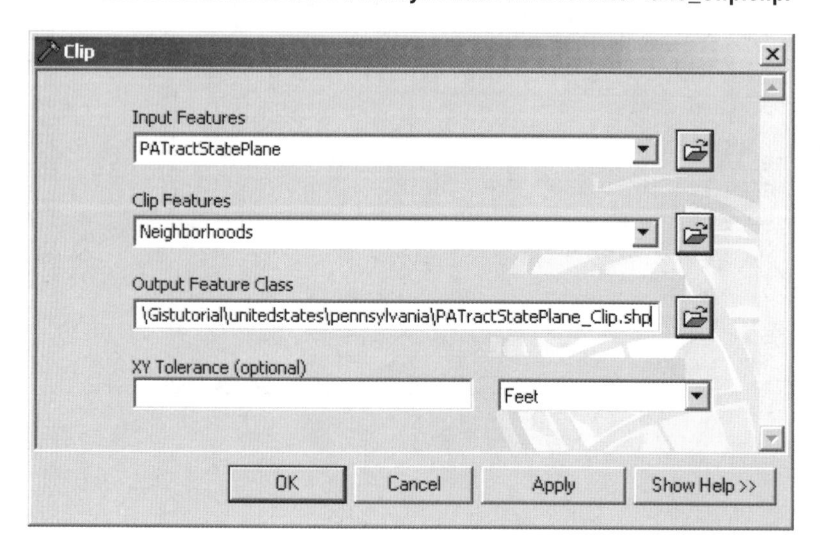

Connect Union function to layers

1 In ArcToolbox, expand the Overlay toolset.

2 In the Overlay toolset, click the Union tool, then drag it into the Model window.

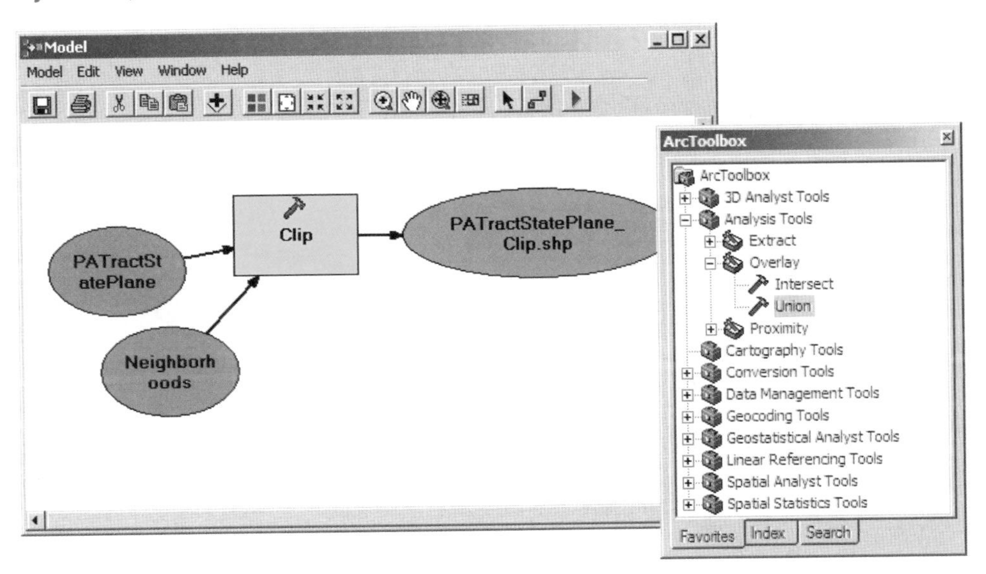

3 From the Model toolbar, click the Add Connection button.

4 In the Model window, click PATractStatePlane_Clip.shp and drag a line to the Union tool.

5 In the Model window, click Neighborhoods and drag a line to the Union tool.

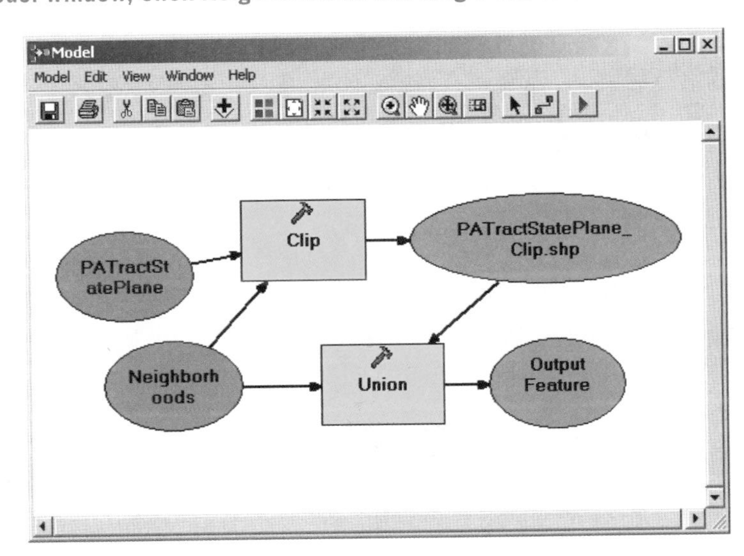

6 From the Model toolbar, click the Select button, then double-click the Union tool in the Model window.

7 Name the Output Features Class **TractNeigh_Clip_Union.shp** and save it in the **\Gistutorial\UnitedStates\Pennsylvania** folder.

8 Verify that your settings match those in the graphic below, then click OK.

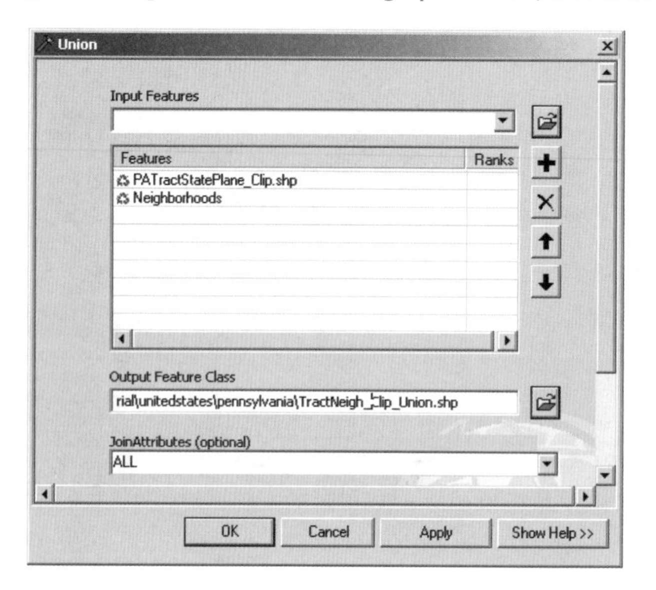

9 From the Model toolbar, click the **Auto Layout** button,  then click the **Full Extent** button. 

Clicking the Auto Layout button followed by the Full Extent button is a good way to quickly organize and get a full view of your model.

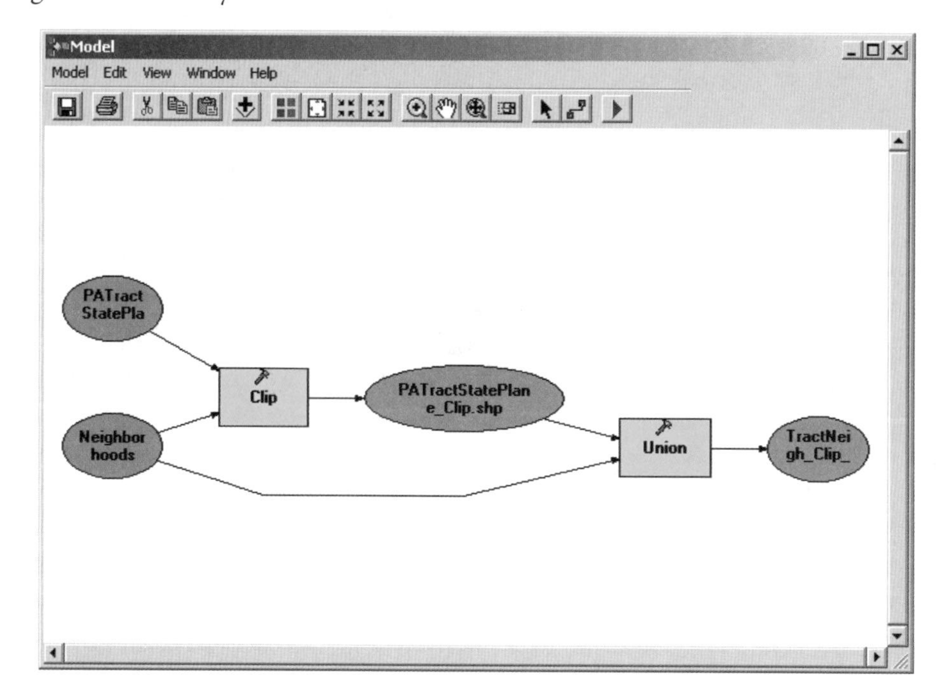

Run the model

1 From the Model menu, click Model, Run.

The functions will process and the following messages will appear, indicating that the model was successful.

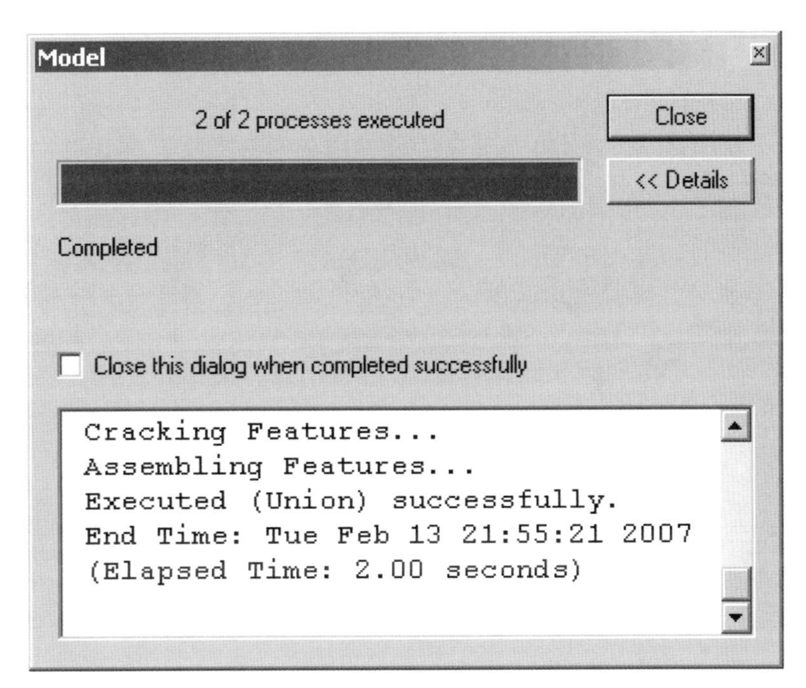

2 Click Close when the process is complete.

Display the new shapefile

1 In the Model window, right-click **TractNeigh_Clip_Union.shp** and choose Add to Display.

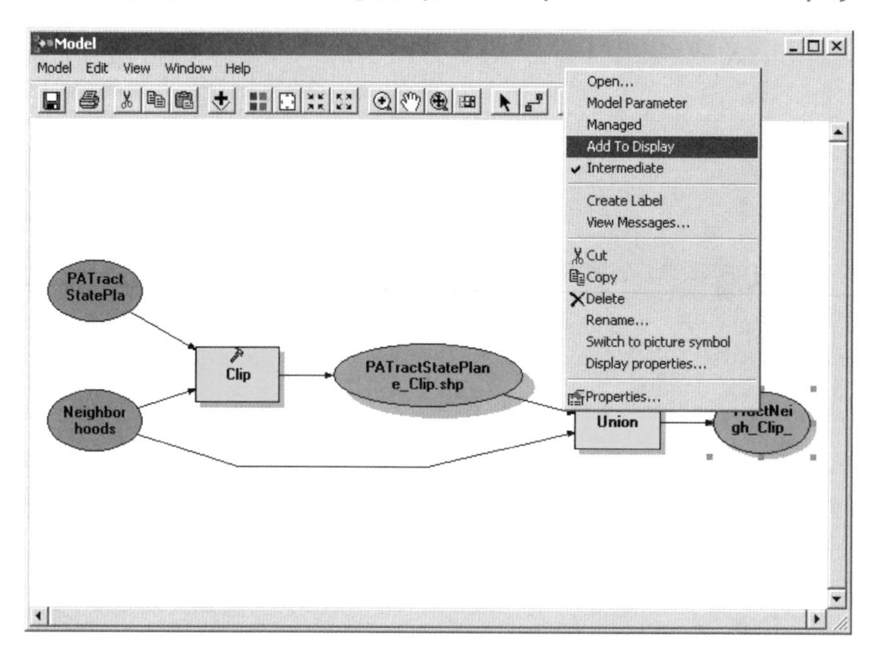

2 Close the Model window and save your changes to the model.

3 In the ArcMap table of contents, uncheck the Neighborhoods and PATractStatePlane layers.

The TractNeigh_Clip_Union layer contains the census tracts that exist within the boundaries defined by the neighborhoods layer. This new layer also contains all the attributes from the Neighborhoods and PATractStatePlane layers.

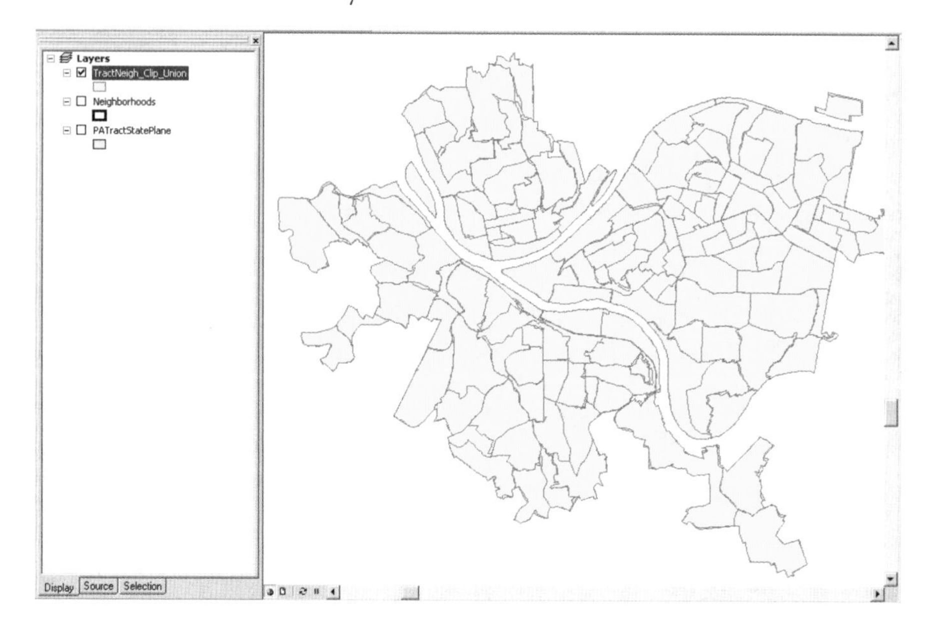

4 **Use the Identify tool**  **to view the neighborhood name and information for each census tract.**

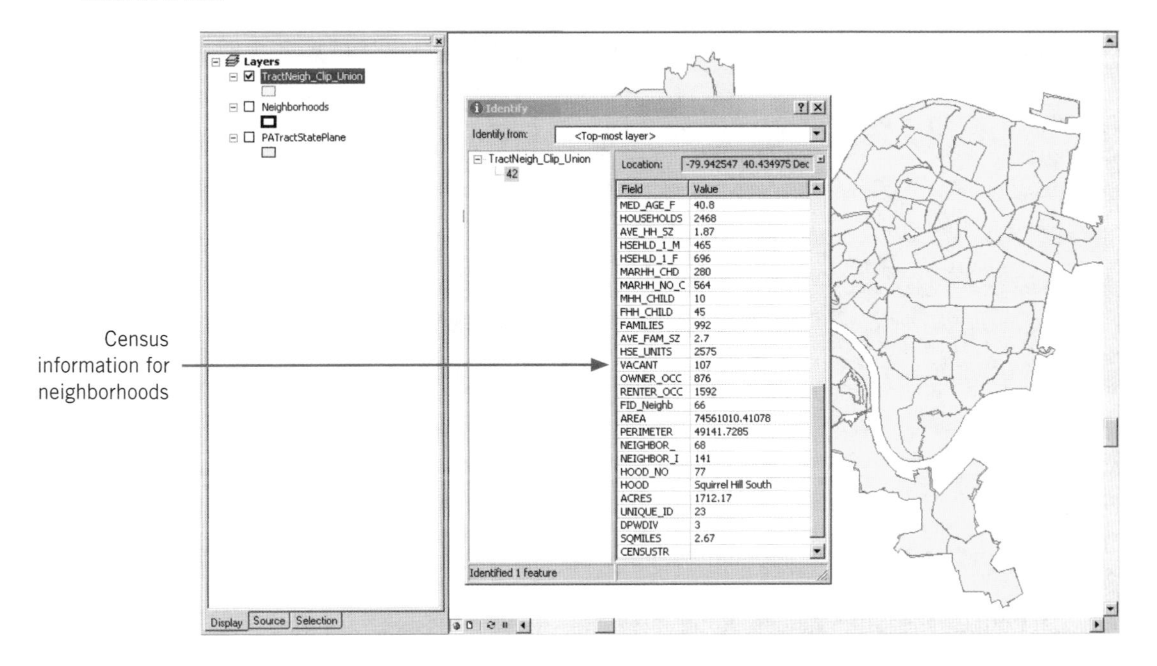

YOUR TURN

Open the Model window (in ArcToolbox, right-click your new model and choose Edit), add the Dissolve tool to the model, and dissolve the tracts based on neighborhood names. Also, when setting up the dissolve options, choose to summarize the population within each neighborhood.

Assignment 8-1

Build a study region for Colorado counties

GIS layers are typically available at a county or state level, but study areas sometimes need to be created for smaller areas such as neighborhoods or regions. A study area may also span across more than one county. GIS specialists can use the ArcGIS geoprocessing tools such as Clip, Union, and Dissolve to extract or build study area layers.

In this assignment, you will create a study area for two rapidly growing counties in Colorado, Denver and Jefferson counties. You will create new shapefiles for an urban area study using polygon shapefiles downloaded from the U.S. Census Web site. Because we want to study two counties, some of the shapefiles need to be joined together, and some need to be clipped from the state level to the smaller study area.

Start with the following:

- **C:\Gistutorial\UnitedStates\Colorado\Counties.shp**—polygon shapefile of Colorado counties.
- **C:\Gistutorial\UnitedStates\Colorado\Streets.shp**—TIGER line shapefile of Jefferson County streets.
- **C:\Gistutorial\UnitedStates\Colorado\Streets2.shp**—TIGER line shapefile of Denver County streets.
- **C:\Gistutorial\UnitedStates\Cities_dtl.shp**—point shapefile of detailed cities.
- **C:\Gistutorial\UnitedStates\Colorado\Urban.shp**—urban area polygons for Jefferson County, Colorado.
- **C:\Gistutorial\UnitedStates\Colorado\Urban2.shp**—urban area polygons for Denver County, Colorado.

Create study area map layers

Use ArcCatalog to copy C:\Gistutorial\UnitedStates\Colorado\Streets.shp and C:\Gistutorial\UnitedStates\Colorado\Urban.shp to C:\Gistutorial\Answers\Assignment8\. Then append C:\Gistutorial\UnitedStates\Colorado\Streets2.shp and C:\Gistutorial\UnitedStates\Colorado\Urban2.shp to the respective copies of Streets.shp and Urban.shp. For example, make C:\Gistutorial\UnitedStates\Colorado\Streets2.shp the input dataset and C:\Gistutorial\Answers\Assignment8\Streets.shp the target dataset in the Append user dialog.

Extract Jefferson and Denver county polygons from C:\Gistutorial\UnitedStates\Colorado\Counties.shp to yield C:\Gistutorial\Answers\Assignment8\StudyCounties.shp. Likewise extract cities from C:\Gistutorial\UnitedStates\Cities_dtl.shp for Jefferson and Denver counties to yield C:\Gistutorial\Answers\Assignment8\StudyCities.shp.

In ArcCatalog, create a new personal geodatabase called C:\Gistutorial\Answers\Assignment8\UrbanAreaStudy.mdb and import C:\Gistutorial\Answers\Assignment8\Streets.shp, C:\Gistutorial\Answers\Assignment8\Urban.shp, C:\Gistutorial\Answers\Assignment8\StudyCities.shp, and C:\Gistutorial\Answers\Assignment8\StudyCounties.shp. Use the shapefile names as layer names.

Create a map document

Create C:\Gistutorial\Answers\Assignment8\Assignment8-1.mxd and set map document properties to store relative paths. Add C:\Gistutorial\UnitedStates\Colorado\Counties.shp and all of the layers in your geodatabase.

Use good cartographic practices to symbolize layers. Label counties and cities. Set the visible scale range for Cities, Streets, and Urban Areas to display only when zoomed in to the study area. Hint: Add 1 to the visible scale so that if you zoom to the study area, the detailed layers display.

Create an 8.5 × 11-inch landscape layout with map zoomed in to the study area, legend, and title. Export the layout as C:\Gistutorial\Assignment\Assignment8\Assignment8-1.jpg. Create a Word document, saved as C:\Gistutorial\Assignment5\Assignment8-1.doc, that has a title, your name, and your map layout image.

Assignment 8-2

Dissolve property parcels to create a zoning map

In this assignment, you will dissolve a parcel map to create a zoning map that highlights a proposed commercial development in what is now a residential area. A commercial company wants to apply for a zoning variance so that it can use the land in residential parcels with FID values 628, 647, 656, 661, 666, and 675 for a commercial purpose. Change the zoning code of these properties to X and highlight them on your map with a red color fill. The Zoning Department wants the map for a public hearing on the proposal and will use it in a PowerPoint presentation.

Start with the following:

- **C:\Gistutorial\PAGIS\EastLiberty\EastLib**—ArcINFO coverage for the East Liberty neighborhood boundary.
- **C:\Gistutorial\PAGIS\EastLiberty\Parcels**—ArcINFO coverage for land parcels in the East Liberty neighborhood of Pittsburgh.
- **ZON_CODE**—an attribute with zoning code values.
- **TAX_AREA, TAX_BLDG, and TAX_LAND_A**—have annual property tax components.
- **C:\Gistutorial\PAGIS\EastLiberty\Curbs**—ArcINFO coverage that has street curbs and annotation with street names.

Prepare map layers

Convert the input coverage for parcels to a shapefile called C:\Gistutorial\Answers\Assignment8\Parcels.shp.

Notice in the attributes of the parcels shapefile that groups of the zoning codes start with the same letter:

A – development
C – commercial
M – industrial
R – residential
S – special

The digit or characters following the first letter further classify land uses. For example, R4 is a residential dwelling with four units. You will create an aggregate-level zoning code by adding a new field to the parcels attribute table that has just the first character of the full zoning code.

Call the new field Zone with text data type and length 1. Use the Field calculator on the new field. Click the String radio button, click the Left ( ) function, and type inside the parentheses of the function in the Zone = panel to yield Left ([ZON_CODE],1), and click OK. The left function extracts the number of characters entered, 1 in this case, starting on the left of the input field, ZON_CODE. Edit the new field to change the Zone values of the parcels in the proposal to X.

Dissolve the parcels shapefile using the Dissolve tool under Data Management, Generalization in ArcToolbox. Use your new field, Zone, as the dissolve field and add SUM statistics for the three tax fields in parcels. Click in the Statistics Type cells to select SUM. Make the output feature class be C:\Gistutorial\Answers\Assignment8\Zoning.shp. Create a new personal geodatabase called C:\Gistutorial\Answers\Assignment8\ZoningMap.mdb and import Zoning.shp into it.

Map document

Create a new map document saved as C:\Gistutorial\Answers\Assignment8\Assignment8-2.mxd. Add the new ZoningMap.mdb to your map document, as well as the curbs arcs and annotation, and East Liberty outline. Use categories, unique values for symbolizing the Zone field. Use pastel colors for the various Zone values, with green for residential but bright red for the proposal. Output Assignment8-2.jpg from a nice 8.5 × 11-inch landscape layout, zoomed in to the upper left quarter of the neighborhood and including a legend.

What to turn in

If you are working in a classroom setting with an instructor, you may be required to submit the exercises you created in tutorial 8. Below are the files you are required to turn in. Be sure to use a compression program such as PKZIP or WinZip to include all three files as one .zip document for review and grading. Include your name and assignment number in the .zip document (YourNameAssn8.zip). *Do not* turn in interim files that are not in your final map (e.g., HHW participants geocoded point shapefile).

Note: *Do not* submit any of the original files that queries were performed from or .zip files that were downloaded for assignment 8-1. Only submit final shapefiles, projects, and Word documents.

ArcMap documents

C:\Gistutorial\Answers\Assignment8\Assignment8-1.mxd
C:\Gistutorial\Answers\Assignment8\Assignment8-2.mxd

Exported maps

C:\Gistutorial\Answers\Assignment8\Assignment8-1.doc
C:\Gistutorial\Answers\Assignment8\Assignment8-2.jpg

Personal geodatabases

Important note: It is especially important to compact the personal geodatabases in this exercise. Compact them in ArcCatalog by right-clicking and choosing “Compact Database.”
C:\Gistutorial\Answers\Assignment8\UrbanAreaStudy.mdb
C:\Gistutorial\Answers\Assignment8\ZoningMap.mdb

OBJECTIVES

Create buffer points for proximity analysis
Conduct a site suitability analysis
Apportion data for noncoterminous polygons

GIS Tutorial 9

Spatial Analysis

The coordinates associated with spatial data permits the use of special algorithms designed especially for GIS applications. For instance, it is possible to place buffers around features and retrieve nearby features for proximity analysis (for example, find crime locations near properties with crime-prone land uses such as bars). It is also possible to use spatial joins to carry out complex spatial processing, such as apportioning census tract data to police administrative areas. In this case, census tracts may be subdivided among two or more police administrative areas, so an approximation is needed to split up or apportion tract data to the police administrative areas. GIS is also useful for site selection analysis, especially when it involves several selection criteria, such as being in a business area, on a major street, and centrally located.

Buffer points for proximity analysis

Some land uses, such as bars, attract crime. It is a good idea for police to keep track of crimes in the vicinity of bars, and it is possible to do this with GIS using circular buffers. Because police want to display buffers on maps, you will take the approach of creating a new buffer map layer.

Open map

1 **From the Windows taskbar, click Start, All Programs, ArcGIS, ArcMap.**

2 **Click the An existing map radio button in the ArcMap dialog box and click OK.**

3 **Browse to the drive on which the Gistutorial folder has been installed (e.g., C:\Gistutorial), select Tutorial9-1.mxd, and click Open.**

Tutorial9-1 contains a map of the Lake Precinct of the Rochester, New York, Police Department. Shown are assault crime offense points, bars, police car beats (with one patrol car assigned to each beat), and street centerlines.

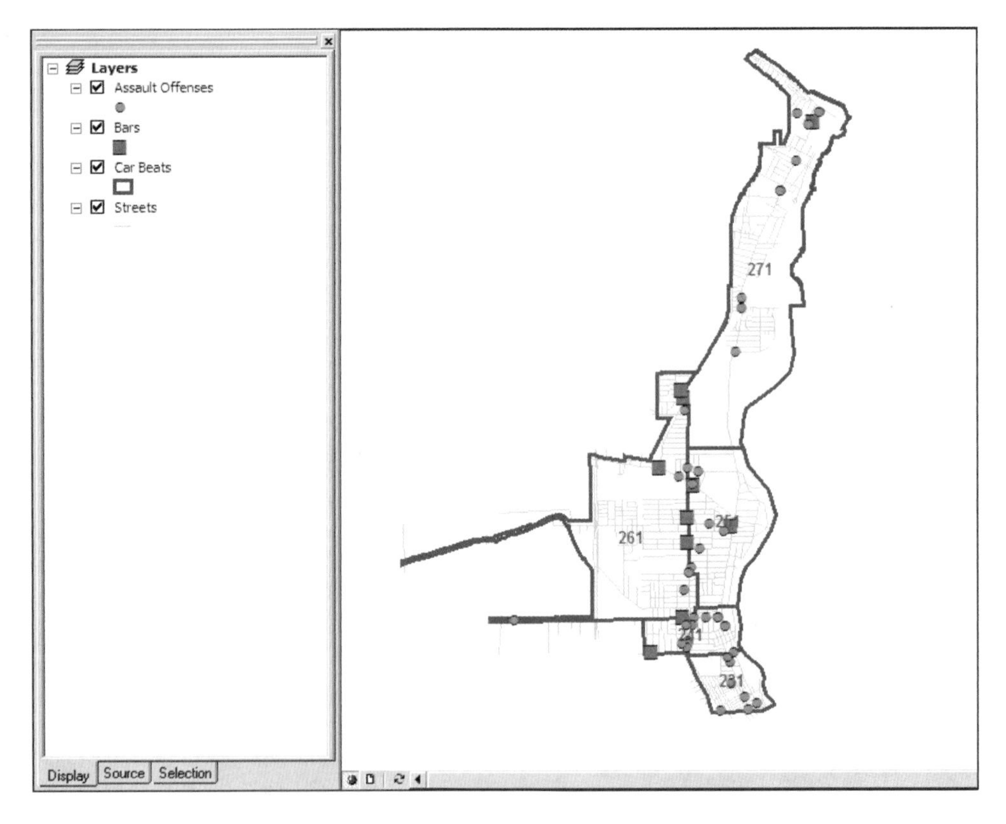

Buffer bars

1 Click the Show/Hide ArcToolbox Window button  to open ArcToolbox.

2 In ArcToolbox, click the plus (+) sign beside the Analysis Tools toolbox to expand it, then expand the Proximity toolset.

3 Double-click the Buffer tool.

4 From the Input Features drop-down list, choose Bars.

5 Save the Output Feature Class as **\Gistutorial\LakePrecinct\Lake.mdb\Buffer_of_Bars**.

6 Click the Linear Unit radio button and type **0.25** in the field below it, then select Miles from the Units drop-down list to the right.

7 From the Dissolve Type drop-down list, choose ALL.

8 Click OK, then click Close when the process is complete.

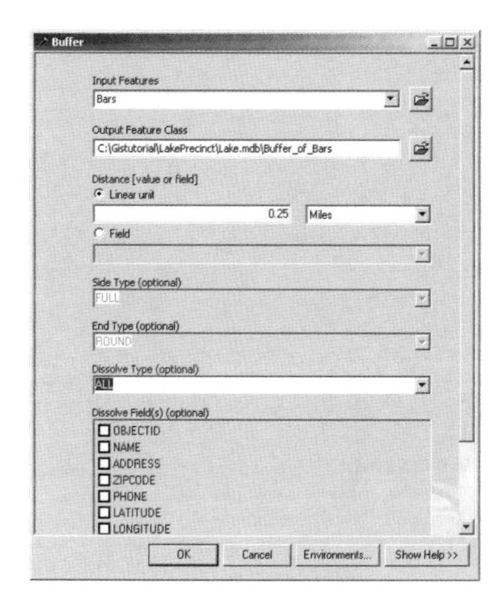

9 Change the symbology of the Buffer_of_Bars layer to a hollow fill.

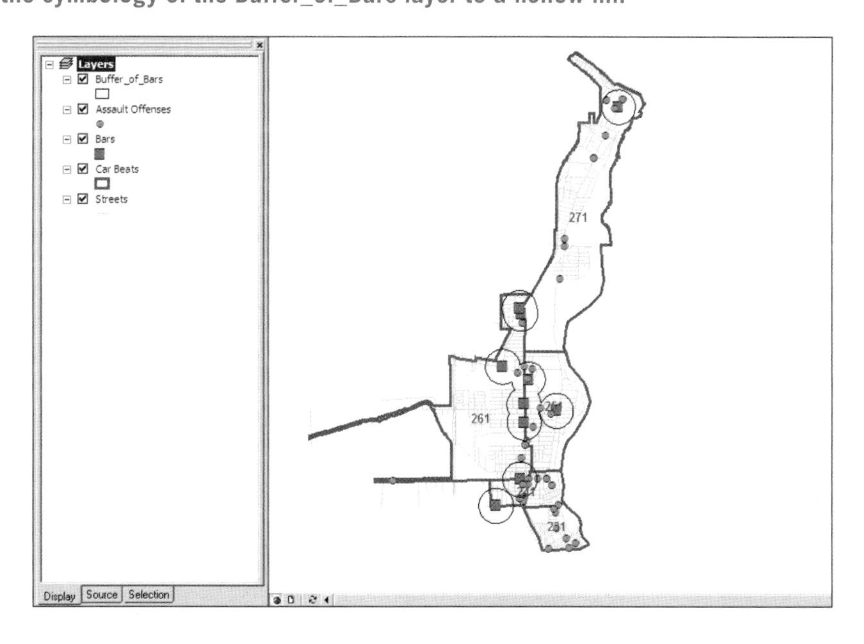

Extract assault offenses in bar buffers

1. **In ArcMap, click Selection, Select by Location.**

2. **In the Select By Location dialog box, set the options to select features from the Assault Offenses layer that intersect the features in the Buffer_of_Bars layer.**

3. **Click Apply, Close.**

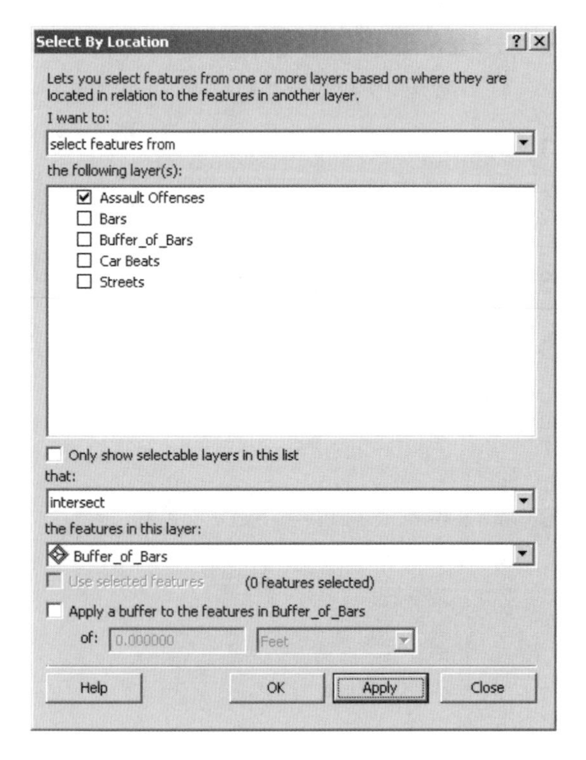

After running the Select By Location command, all the assault points within bar buffer zones are selected.

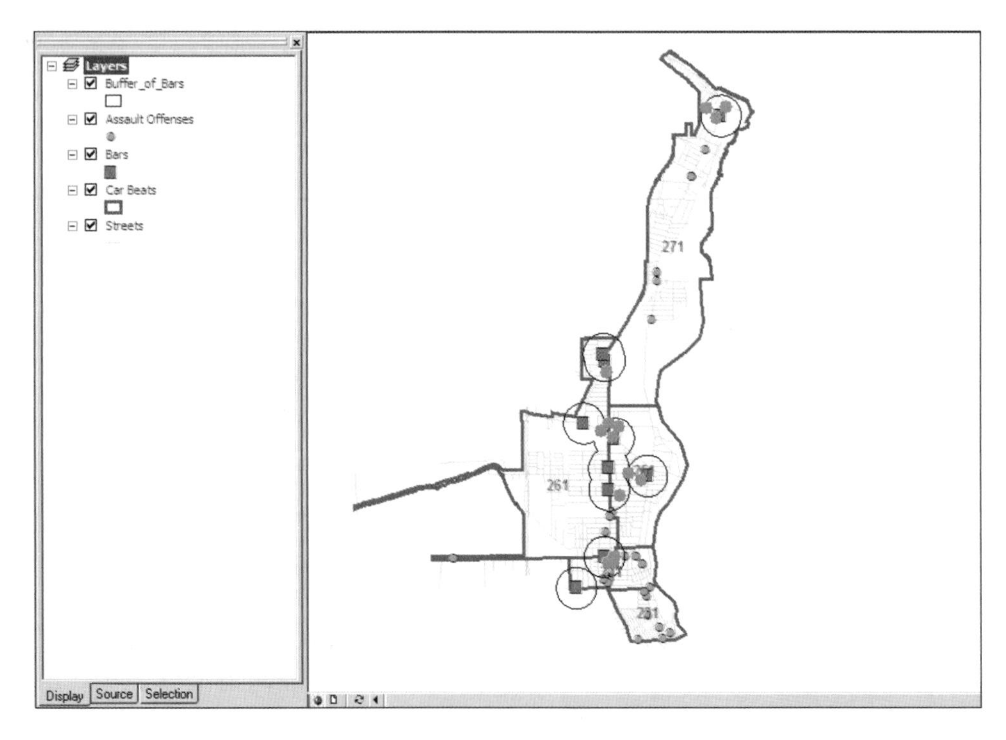

4 **In the table of contents, right-click the Assault Offenses layer and click Data and Export Data.**

5 **In the Export Data dialog box, make sure that Selected features is chosen from the Export drop-down list, name the output shapefile AssaultsInBarsBuffers.shp, and save it in the \Gistutorial\LakePrecinct folder.**

6 **Click OK, Yes.**

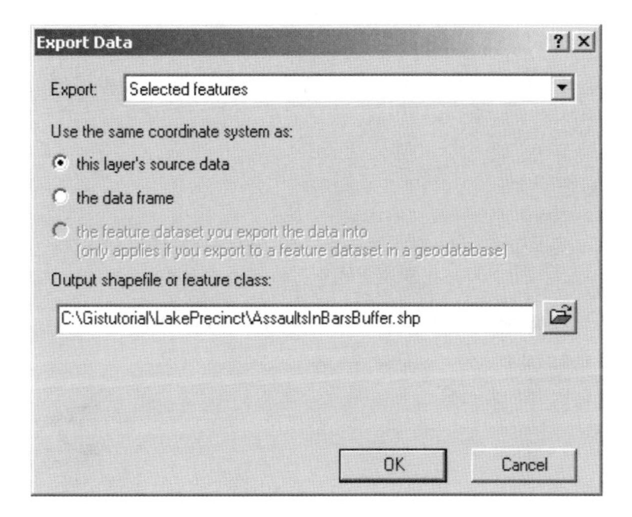

7 **Turn off the Assault Offenses layer.**

This is a map that a task force working on problem bars would want for making the case to enforce laws at bars or close bars down.

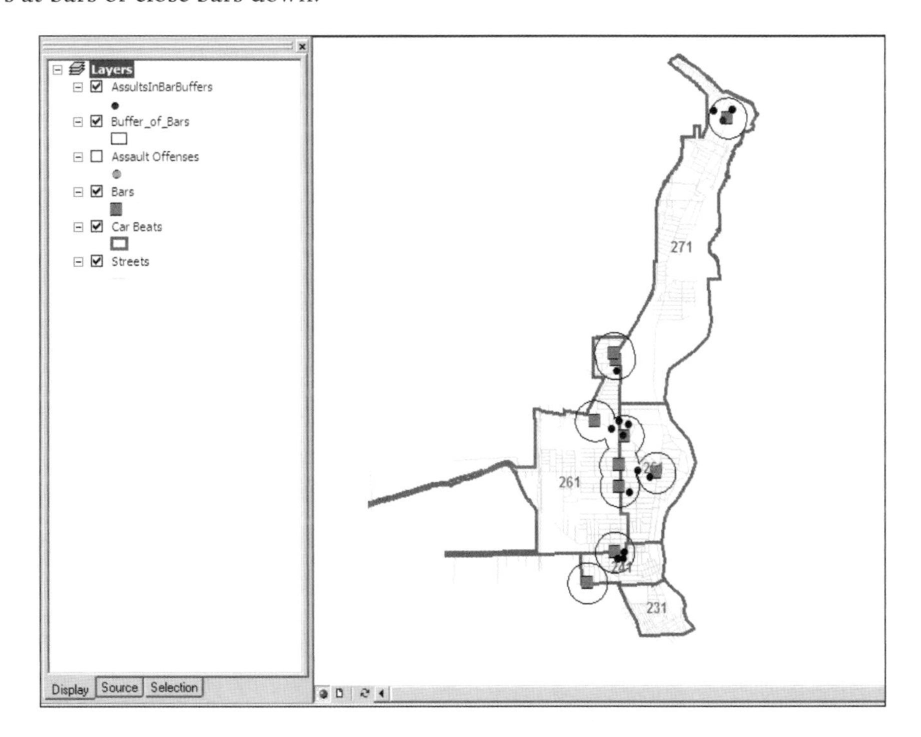

8 **From the File menu, click Save to save your map.**

Conduct a site suitability analysis

Suitability analysis is a common and classic type of GIS application. Typically, this type of analysis consists of several steps that include attribute- and location-based queries, buffers, spatial joins, and overlays. In this exercise, you will perform suitability analysis with the purpose of locating potential areas for new police satellite stations in each car beat of the Lake Precinct of Rochester, New York. Criteria for locating these stations are that the site must be centrally located in each car beat (within a 0.33-mile radius buffer of car beat centroids), and that the site must be in retail/commercial areas (be within 0.10 mile of a least one retail business and within 0.10 mile of major streets).

Open a map

1 **If necessary, start ArcMap.**

2 **In ArcMap, open the Tutorial9-2.mxd from the \Gistutorial folder.**

Tutorial9-2.mxd contains a car beat map of the Lake Precinct of the Rochester, New York, Police Department. Also shown in the map are police car beats—territories assigned to patrol units, retail business points, and street centerlines.

Add X and Y columns to car beats

1 **In the table of contents, right-click the Car Beats layer and click Open Attribute Table.**

2 **In the Attributes of Car Beats table, click Options, Add Field.**

3 **In the Add Field dialog box, name the new field X, choose Double from the Type drop-down list, then click OK.**

4 **Click Options, Add Field.**

5 **Name the new field Y, set its Type to Double, then click OK.**

Compute car beat centroids

1 **In the Attributes of Car Beats table, right-click the X column heading and click Calculate Geometry. Click Yes to calculate outside of an edit session.**

2 **In the Calculate Geometry dialog box, click the drop-down list arrow for Property and click X Coordinate of Centroid.**

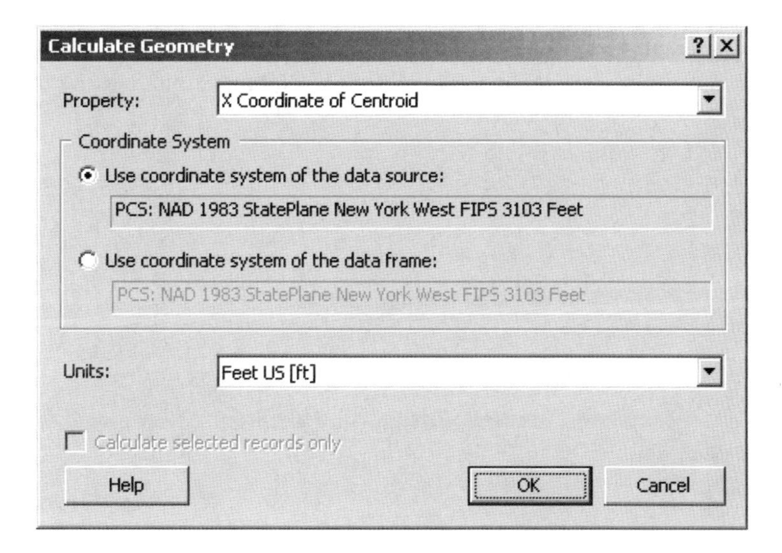

3 **Click OK.**

ArcMap calculates the x-coordinate of each car beat polygon.

YOUR TURN

Calculate the y-coordinate of car beat polygon centroids.

Each record in the Attributes of Car Beats table now contains an x- and y-coordinate value. Each x,y pair represents the centroid of a police car beat.

X	Y
1402631.8189	1156203.703772
1400522.350349	1158989.873773
1397139.549464	1165093.524371
1402264.268132	1165279.502132
1404519.587559	1179549.785764

Map car beat centroids

You can directly map x,y coordinates as points.

1 **In the Attributes of Car Beats table, click Options, Export.**

2 **In Export Data dialog box, click the browse button for the Output table field, browse to the C:\GisTutorial\LakePrecinct\ folder, double-click lake.mdb, change the Save as type to File and Personal Geodatabase tables, change the name to CarBeatCentroids, and click Save, OK, Yes.**

3 **Close the Attributes of Car Beats table.**

4 **Click Tools, Add XY Data.**

5 **In the Add XY Data dialog box, click Edit.**

6 **In the Spatial Reference Properties dialog box, click Import.**

7 **In the Browse for Dataset window, browse to \Gistutorial\LakePrecinct\Lake.mdb, click lakecarbeats, then click Add.**

8 **Click OK twice.**

CarBeatCentroids Events is added as a layer to your map. The word "Events" is used because x,y data is often collected and used to represent events that occurred at a specific time and location.

9 **Symbolize the CarBeatCentroids Events layer with a Circle 2 point marker, Mars Red, size value 10.**

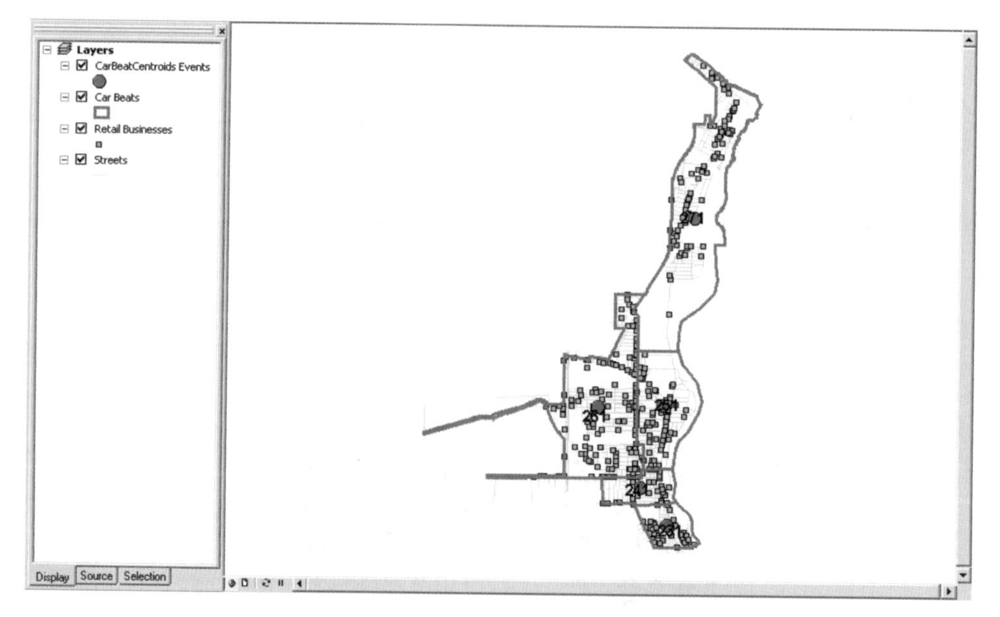

Buffer car beat centroids

1 If necessary, open ArcToolbox.

2 In ArcToolbox, expand the Analysis Tools toolbox, then expand the Proximity toolset.

3 Double-click the Buffer tool.

4 In the Buffer dialog box, click the Input Features drop-down list and choose CarBeatCentroids Events.

5 Save the Output Feature Class as **\Gistutorial\LakePrecinct\Lake.mdb\ Buffer_of_ CarBeatCentroids.**

6 Make sure the Linear Unit option is chosen, then click in the distance field directly below it and type **0.33**. From the units drop-down list, choose Miles.

7 From the Dissolve Type drop-down list, choose ALL.

8 Click OK, then click Close when the process completes.

The Buffer_of_CarBeatCentroids layer is added to the map. Now you need to find areas within the car beat buffers that meet the remaining criteria.

9 Symbolize the buffers polygons with a hollow fill, set the Outline Color to Mars Red, then set the Outline Width to 1.

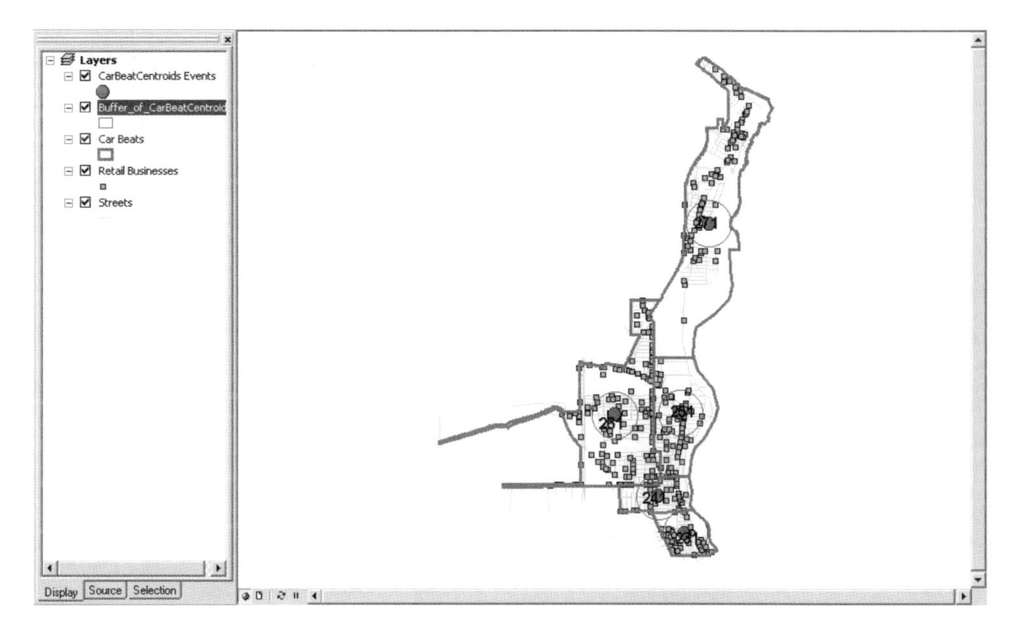

Buffer retail businesses

1 **In ArcToolbox, double-click the Buffer tool.**

2 **Select Retail Businesses from the Input Features drop-down list.**

3 **Save the Output features as \Gistutorial\LakePrecinct\Lake.mdb\Buffer_of_RetailBusinesses.**

4 **Click the Linear Unit radio button, type 0.10 for the distance, and change the distance units to Miles.**

5 **From the Dissolve Type drop-down list, choose ALL.**

6 **Click OK, then click Close when the process is complete.**

The Buffer_of_RetailBusinesses layer is automatically added to the map.

7 **Symbolize the new buffer layer with a hollow fill, set the Outline Color to Ultra Blue, and the Outline Width to 1.**

8 **In the table of contents, turn off the CarBeatCentroid Events and the Retail Businesses layers.**

The intersection of the Buffer_of_RetailBusinesses and Buffer_of_CarBeatCentroids_Events nearly satisfies the suitability criteria, but you still need to buffer the streets.

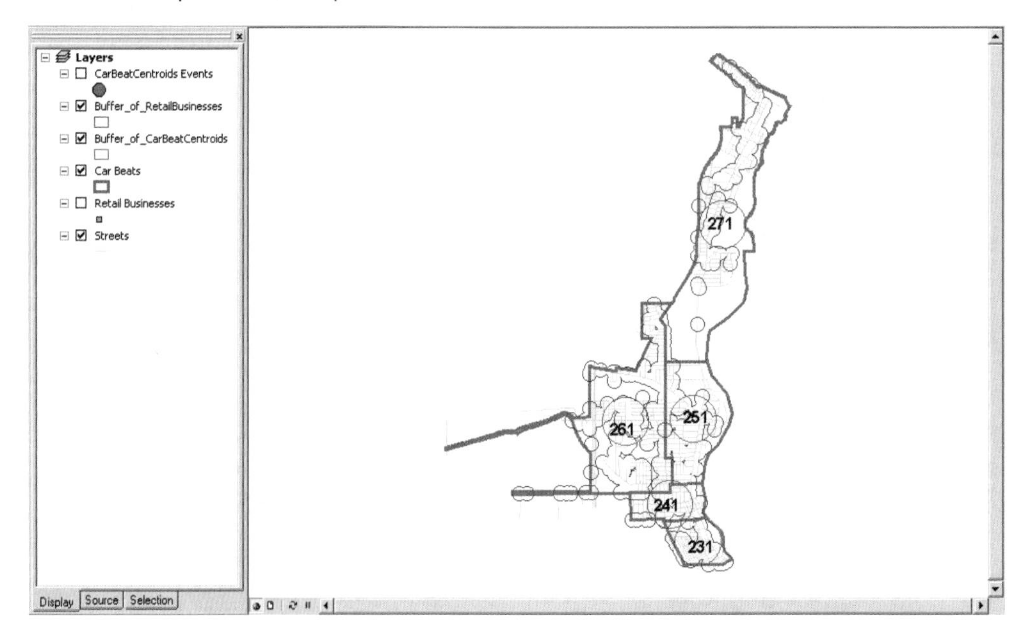

Select major streets

Major and commercial streets have FCC code values of A40 and A41 in the streets layer. You will select only those streets and then buffer them.

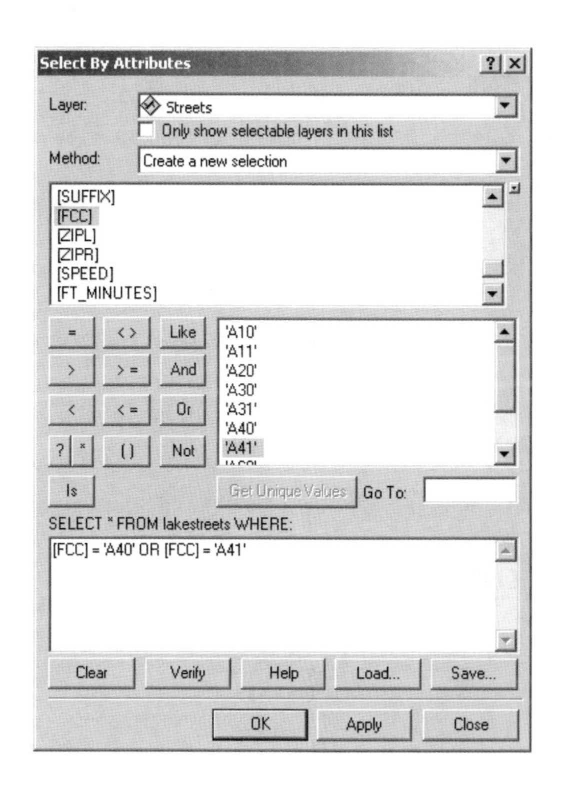

1. **In ArcMap, click Selection, Select By Attributes.**

2. **In the Select By Attributes dialog box, select Streets for the Layer and Create a new selection as the Method.**

3. **Scroll down in the Fields box and double-click [FCC], click the = button, click the Get Unique Values button, and double-click 'A40' in the Unique Values box.**

4. **Click the Or button, double-click [FCC], click the = button, and double-click 'A41'.**

5. **Click Apply, Close.**

Major and commercial streets turn the selection color.

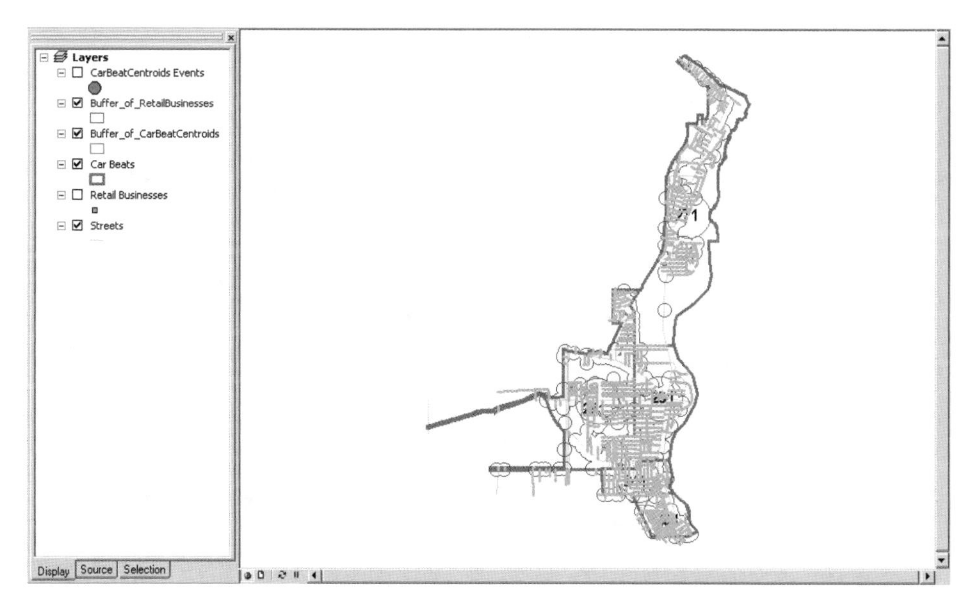

Buffer major streets

1 **In ArcToolbox, double-click the Buffer tool.**

2 **In the Buffer dialog box, set the Input Features to Streets.**

3 **Save the Output features as C:\Gistutorial\LakePrecinct\Lake.mdb\Buffer_of_Streets.**

4 **Choose the Linear Unit option, then set the distance to 0.10 Miles.**

5 **From the Dissolve Type drop-down list, choose ALL.**

6 **Click OK, then click Close when the process is complete.**

The Buffer_of_Streets polygon layer is automatically added to the map.

7 **Symbolize the new buffer layer with a hollow fill, set the Outline Color to Gray 60%, then set the Outline Width to 1.**

8 **Click Selection, Clear Selected Features, and in the table of contents, turn off the Streets layer.**

The intersection of all three sets of buffers satisfies the suitability criteria. The latest buffer is mostly redundant, but it does rule out a few retail business buffers.

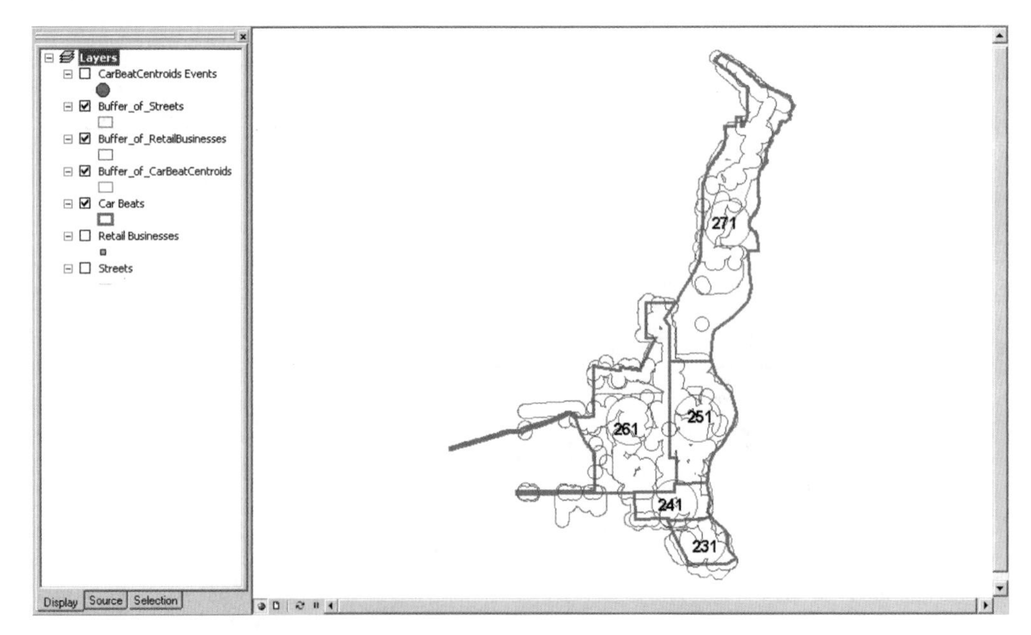

Intersect buffers

1 In ArcToolbox, expand the Analysis Tools toolbox, then expand the Overlay toolset. Double-click the Intersect tool.

2 From the Input Features drop-down list, choose Buffer_of_Streets, then choose Buffer_of_Retail_Businesses from the list.

3 Save the Output Feature Class as \Gistutorial\LakePrecinct\Lake.mdb\Streets_Retail_Intersect.

4 Click OK, then click Close when the process completes.

The resulting intersection layer contains only the areas that overlap in the two input layers.

5 From ArcToolbox, open the Intersect dialog box again.

6 In the Intersect dialog box, set the Input Features to Streets_Retail_Intersect and Buffer_of_CarBeatCentroids_Events.

7 Save the Output Feature Class as C:\Gistutorial\LakePrecinct\Lake.mdb\SuitableSites.

8 Click OK, then click Close when the process is complete.

9 Symbolize SuitableSites with a hollow fill, change the Outline Color to Leaf Green, then set the Outline Width to 1.

10 Turn off all layers except SuitableSites, Car Beats, and Streets.

11 Save the map document.

That completes the task. The green areas are suitable for satellite police stations. Next, a staff person could work with realty companies to locate specific sites within the suitable areas.

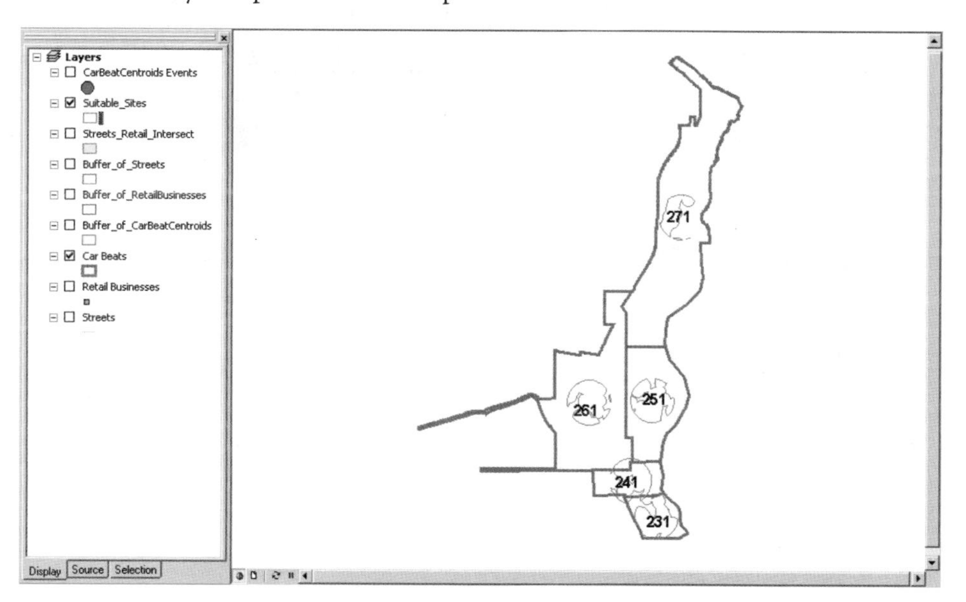

Apportion data for noncoterminous polygons

Sometimes you will not have raw point data, but only aggregate data for polygons. A good example is census data, which is tabulated for polygon layers from counties down to blocks. Nevertheless, your need may be for much different types of polygon boundaries; for example, administrative areas such as police car beats (areas served by patrol cars). The Rochester, New York, Police Department designed their administrative areas to meet police needs and, consequently, car beat boundaries do not always follow census tract boundaries. If you need census data by car beats, you have to apportion (make approximate splits of) each tract's data to two or more car beats.

1 **If necessary, start ArcMap.**

2 **In ArcMap, open the Tutorial9-3.mxd from the \Gistutorial folder.**

Tutorial9-3 contains a map of car beats and census tracts in the Lake Precinct of the Rochester Police Department. You will find several cases where car beats contain only portions of tracts.

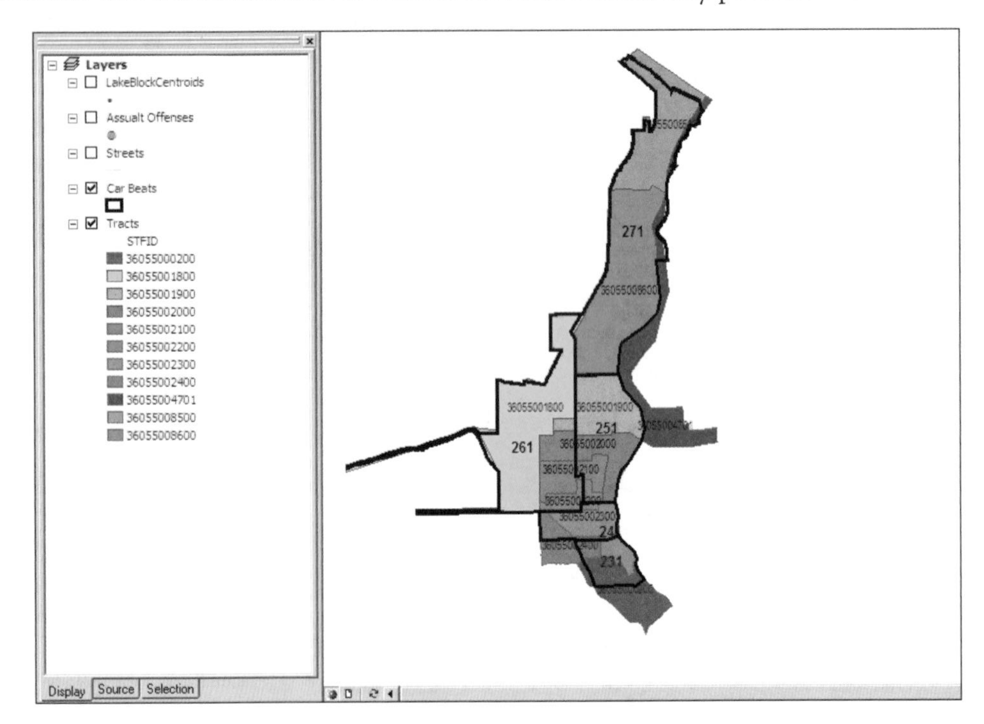

Approach to apportionment

Apportioning data from one set of polygons to another is a complex task, so much so that we thought it best to give you some background and a preview of steps. Census data from the census long form, such as data on educational attainment for the population aged 25 or older, is available at the census tract level but not at the block level. In contrast, short-form census data is available down to the city block level. You will use the block-level data for apportionment of educational attainment data.

1 **In the ArcMap table of contents, turn off Tracts layer and turn on the LakeBlockCentroids and Streets layers.**

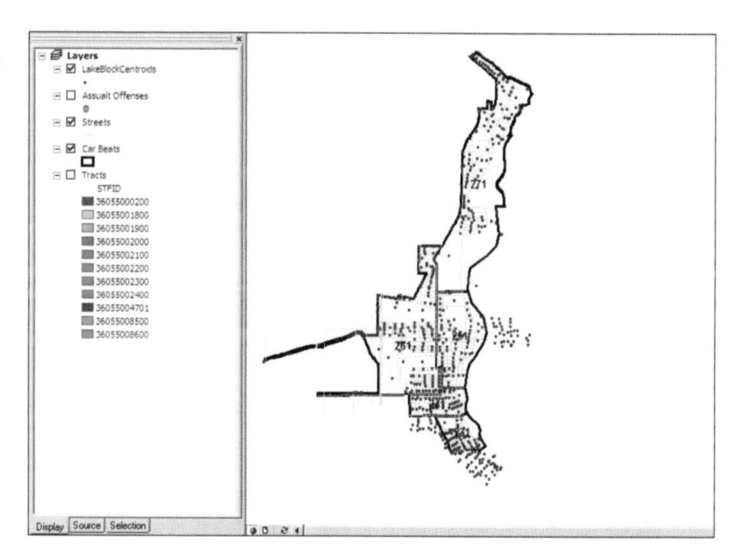

The centroids have short-form data allowing us to tabulate population for those 22 or older. The provided break points for general population do not allow us to tabulate 25 or older, but 22 or older is close enough for approximation.

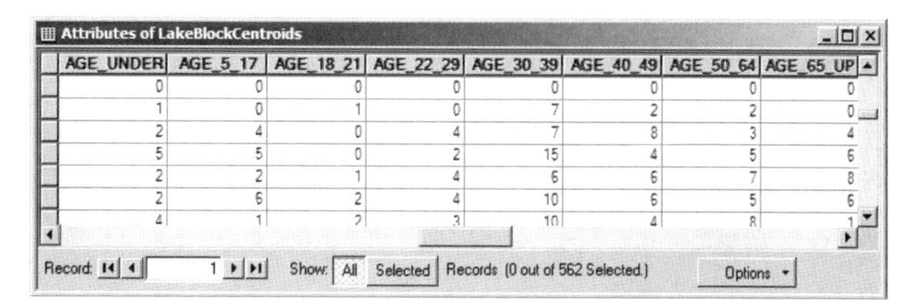
Attributes of LakeBlockCentroids

AGE_UNDER	AGE_5_17	AGE_18_21	AGE_22_29	AGE_30_39	AGE_40_49	AGE_50_64	AGE_65_UP
0	0	0	0	0	0	0	0
1	0	1	0	7	2	2	0
2	4	0	4	7	8	3	4
5	5	0	2	15	4	5	6
2	2	1	4	6	6	7	8
2	6	2	4	10	6	5	6
4	1	2	3	10	4	8	1

Record: 1 Show: All Selected Records (0 out of 562 Selected.) Options

There are several alternatives for apportioning data, for example by area, length of street network, or the block centroids population. We chose the last method because it is one of the most accurate. We *assume* that the variable being apportioned, in our case the number of people 25 or older who have less than a high school education, is spatially distributed within the census tract proportionally to general population aged 22 or older at the block level. We use block centroids instead of block polygons because the blocks are relatively small and can be safely represented by points to make the spatial data processing easier.

Note: if you merely need short-form census data, available for the block centroids, a simple spatial overlay of a polygon layer like car beats on block centroids (see tutorial 4) and the aggregation of resultant data is all that is needed instead of apportionment.

The math of apportionment

Let's take a look at one example. Below is a close-up of tract 360550002100, which is split between car beats 261 and 251.

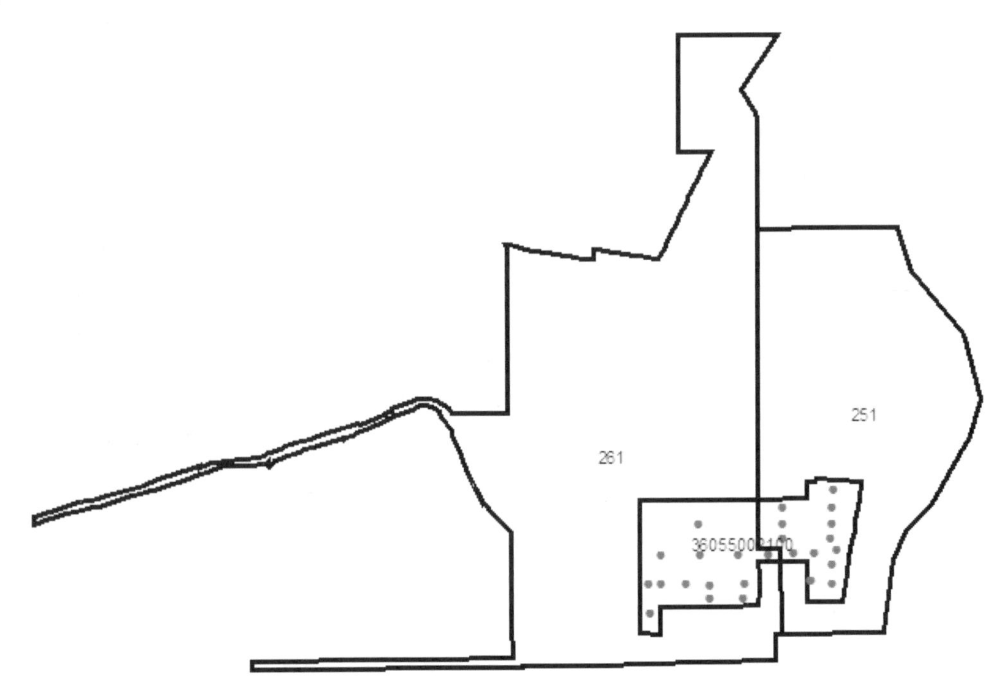

Tract 360550002100 has 205 people aged 25 or older with less than a high school education. For short, let's call this the undereducated population. How can we divide those 205 undereducated people between car beats 261 and 251?

Of the 26 blocks making up the tract, the 13 that lie in car beat 261 have 1,177 people aged 22 or older. The other 13 blocks, in car beat 251, have 1,089 such people for a total of 2,266 for the tract.

Apportionment assumes that the fraction of undereducated people aged 25 or older is the same as that for the general population aged 22 or older: 1,177 ÷ 2,266 = 0.519. For the other car beat, it is 1,089 ÷ 2,266 = 0.481.

Thus, we estimate the contribution of tract 36055002100 to car beat 261's undereducated population to be (1,177 ÷ 2,266) × 205 = 106. For car beat 251, it is (1,089 ÷ 2,266) × 205 = 99.

Eventually, by apportioning all tracts, we can sum up the total undereducated population for car beats 261 and 251.

Preview of apportionment steps

The following is a summary of the steps that you will complete in the following pages.

1 **In the attribute table for block centroids, create two new fields: the census tract ID for each block and the sum across age groups for the general population aged 22 or older in each block.**

2 **In the attribute table for block centroids, sum the field for persons aged 22 or older by tract ID to create a new table.**

3 **Spatially join the tract and car beats layers to create new polygons that each have a tract ID and car beat number.**

4 **Spatially overlay the joined layer of tracts and car beats onto the block centroids to assign all the tract attributes (including the attribute of interest, undereducated population) and car beat attributes to blocks.**

5 **Join the table from step 2 to block centroids to make the apportionment weight denominator, total population aged 22 or older by tract, available to each block centroid.**

6 **For each block centroid, create new fields to store apportionment weight and apportioned undereducated population values, then calculate these values for the new fields.**

7 **Sum the apportionment weights by tract as a check for accuracy (they should sum to 1.0 for each tract), then sum the undereducated population per car beat, storing the results in new tables.**

8 **Join the table containing undereducated population by car beat to the car beats layer, then symbolize the data for map display.**

Create Tract ID and AGE22Plus fields in the attributes of block centroids

1 In ArcMap's table of contents, right-click the LakeBlockCentroids layer and click Open Attribute Table.

2 In the Attributes of LakeBlockCentroids table, click Options, Add Field.

3 In the Add Field window, type **TractID** in the Name box, change the Type to Text, change the Length to 11, and click OK.

4 In the Attributes of LakeBlockCentroids table, click Options, Add Field.

5 Name the second new field **Age22Plus**, then click OK.

6 In the Attributes of LakeBlockCentroids table, scroll to the right end of the table, right-click the column heading for TractID, click Field Calculator, and click Yes.

7 In the Field Calculator, set the Type option to String. In the Fields box, double-click FIPSSTCO, click the & button, and in the Fields box double-click Tract2000. Verify that your settings match those shown below, then click OK.

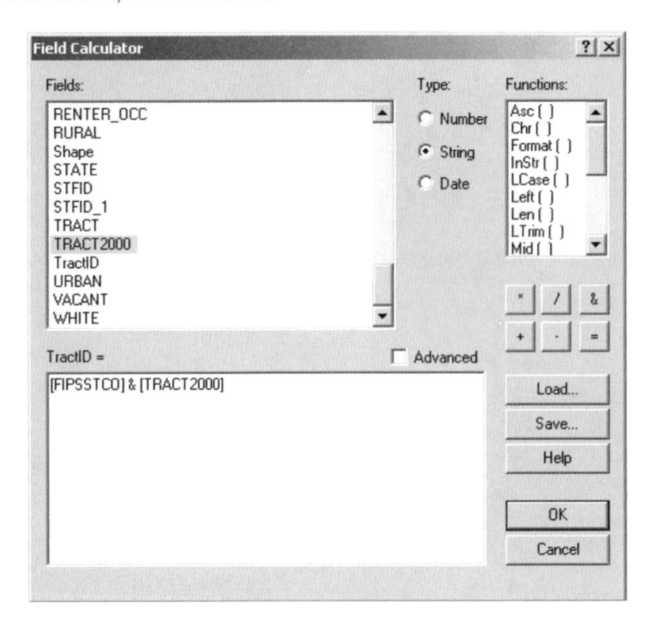

Values from the FIPSSTCO and Tract2000 field are concatenated and placed in the TractID field.

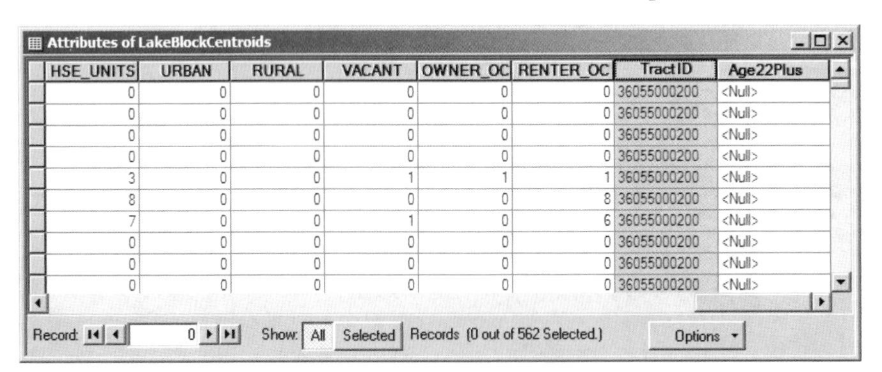

Attributes of LakeBlockCentroids

HSE_UNITS	URBAN	RURAL	VACANT	OWNER_OC	RENTER_OC	TractID	Age22Plus
0	0	0	0	0	0	36055000200	<Null>
0	0	0	0	0	0	36055000200	<Null>
0	0	0	0	0	0	36055000200	<Null>
0	0	0	0	0	0	36055000200	<Null>
3	0	0	1	1	1	36055000200	<Null>
8	0	0	0	0	8	36055000200	<Null>
7	0	0	1	0	6	36055000200	<Null>
0	0	0	0	0	0	36055000200	<Null>
0	0	0	0	0	0	36055000200	<Null>
0	0	0	0	0	0	36055000200	<Null>

Record: 0 Show: All Selected Records (0 out of 562 Selected.) Options

8 In the Attributes of LakeBlockCentroids table, scroll all the way to the right, right-click the column heading for Age22Plus, click Field Calculator, and click Yes.

9 In the Field Calculator, clear your previous expression from the expression box. In the Fields box, double-click the AGE_22_29 field, then click the + button; do the same for AGE_30_39, AGE_40_49, AGE_50_64, AGE_65_UP. Verify that your expression matches the one in the graphic below, then click OK.

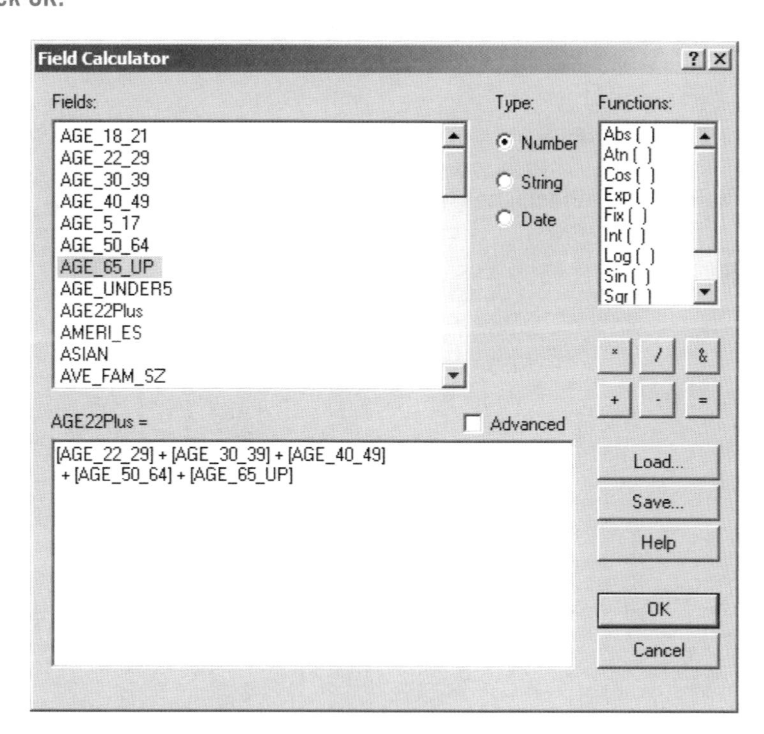

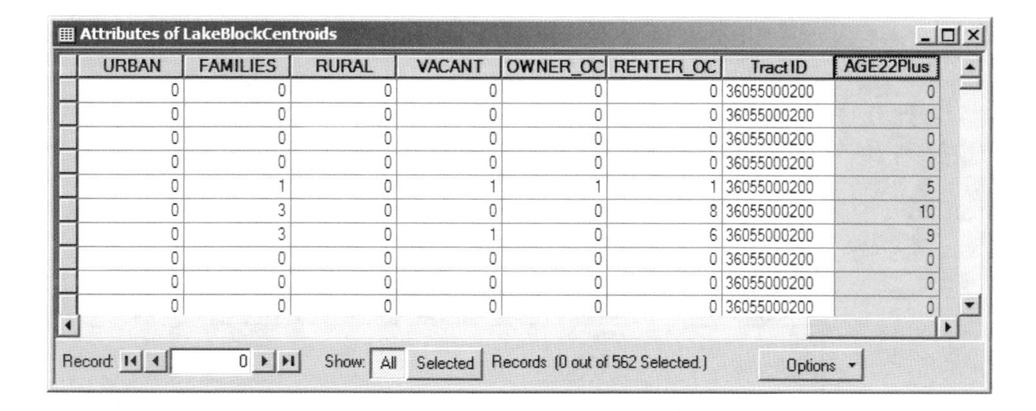

URBAN	FAMILIES	RURAL	VACANT	OWNER_OC	RENTER_OC	Tract ID	AGE22Plus
0	0	0	0	0	0	36055000200	0
0	0	0	0	0	0	36055000200	0
0	0	0	0	0	0	36055000200	0
0	0	0	0	0	0	36055000200	0
0	1	0	1	1	1	36055000200	5
0	3	0	0	0	8	36055000200	10
0	3	0	1	0	6	36055000200	9
0	0	0	0	0	0	36055000200	0
0	0	0	0	0	0	36055000200	0
0	0	0	0	0	0	36055000200	0

Sum Age22Plus by tracts

1. **In the Attributes of LakeBlockCentroids table, right-click the TractID column heading and click Summarize.**

2. **For step 1 of the Summarize dialog box, choose TractID from the drop-down list.**

3. **For step 2, scroll down the list of fields, locate and expand the Age22Plus field, then check the Sum box.**

4. **For step 3, click the browse button, browse to C:\GisTutorial\LakePrecinct\, double-click Lake.mdb, change the Save as type to File and Personal Geodatabase tables, and save the output table as \Gistutorial\LakePrecinct\Lake.mdb\Sum_Age22Plus.**

5. **Verify that your settings match those in the graphic shown at right, then click OK and Yes to add the new table to your map.**

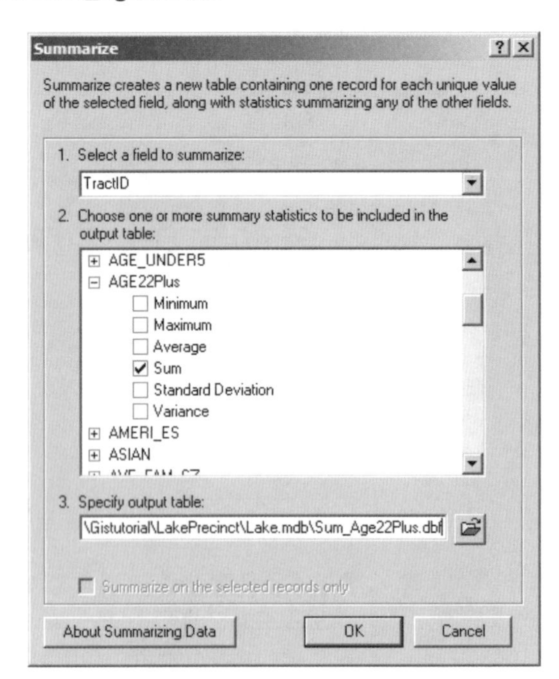

6. **Close the Attributes of LakeBlockCentroids table.**

7. **In the table of contents, right-click the Sum_Age22Plus table and click Open.**

This table contains tract-level data for the eleven tracts intersecting the Lake car beats.

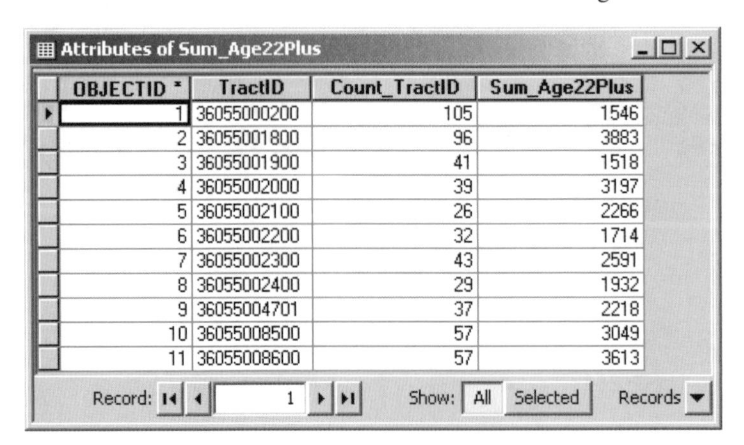

Attributes of Sum_Age22Plus

OBJECTID *	TractID	Count_TractID	Sum_Age22Plus
1	36055000200	105	1546
2	36055001800	96	3883
3	36055001900	41	1518
4	36055002000	39	3197
5	36055002100	26	2266
6	36055002200	32	1714
7	36055002300	43	2591
8	36055002400	29	1932
9	36055004701	37	2218
10	36055008500	57	3049
11	36055008600	57	3613

Record: 1 Show: All Selected Records

Intersect car beats and tracts

1. **Close any open tables.**
2. **Open ArcToolbox.**
3. **In ArcToolbox, expand the Analysis Tools toolbox, then expand the Overlay toolset.**
4. **Double-click the Intersect tool.**
5. **From the Input Features drop-down list, choose Car Beats, then click the list again and choose Tracts.**
6. **Save the Output Feature Class as \Gistutorial\LakePrecinct\Lake.mdb\Intersection_CarBeatsAndTracts.**
7. **Click OK, then click Close when the process completes.**

The resulting layer contains polygons representing the areas where the car beats and census tract polygons overlapped.

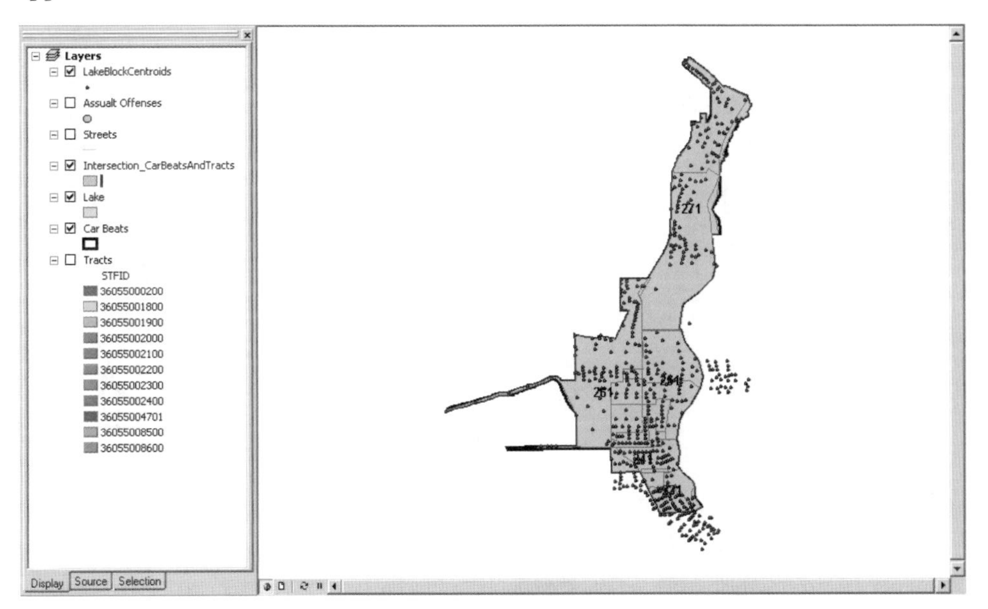

Overlay the intersection of car beats and tracts with block centroids

At this point, you will spatially overlay the joined layer of tracts and car beats on the block centroids to assign the tract and car beats attributes to the census blocks.

1 **In the table of contents, right-click the LakeBlockCentroids layer, click Joins and Relates, Join.**

2 **In the Join Data dialog box, click the What do you want to join to this layer drop-down list, and choose Join data from another layer based on spatial location.**

3 **For step 1, click the drop-down list and choose the Intersection_CarBeatsAndTracts layer.**

4 **For step 2, click the it falls inside option.**

5 **For step 3, save the output in Lake.mdb as Join_Intersection_CarBeatsAndTracts_BlockCentroids, then click OK.**

The Join_Intersection_CarBeatsAndTracts_BlockCentroids is automatically added to your map.

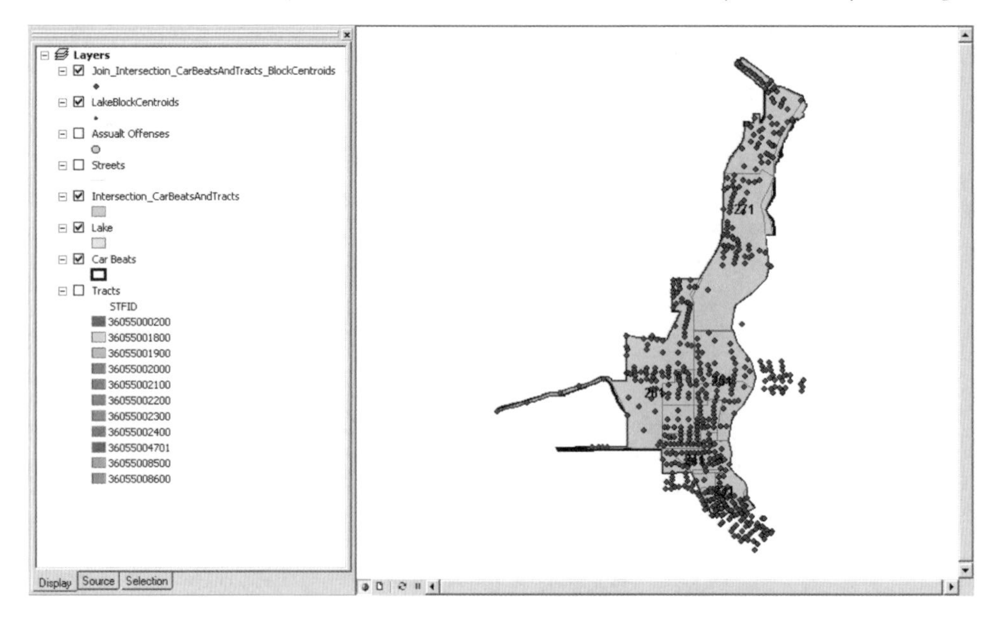

Join the summary attributes to the spatial join output

In this section, you will join the Sum_Age22Plus table to the block centroids in order to make the apportionment weight denominator—total population aged 22 or greater by tract—available to each block centroid.

1. **Click the Display tab at the bottom of the table of contents.**

2. **In the table of contents, right-click the Join_Intersection_CarBeatsAndTracts_BlockCentroids layer, and click Joins and Relates, Join.**

3. **In the Join Data dialog box, choose Join attributes from a table from the What do you want to join to this layer drop-down list.**

4. **For step 1, select TractID from the drop-down list.**

5. **For step 2, select Sum_Age22Plus from the drop-down list.**

6. **For step 3, select TractID from the drop-down list.**

7. **Verify that your setting match those in the graphic shown at right, click OK, then click Yes to create an index for the field.**

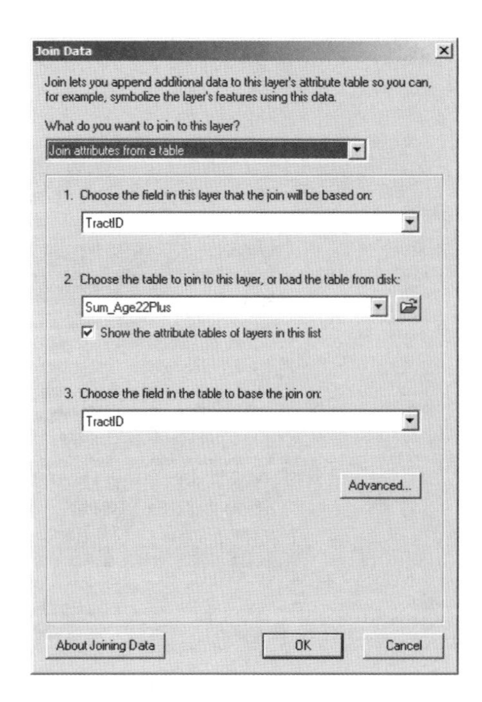

8. **In the table of contents, right-click the Join_Intersection_CarBeatsAndTracts_BlockCentroids layer, then click Open Attribute Table. In the attribute table, scroll to the right and see that Sum_AGE22Plus and other tract data have been joined to the block centroids.**

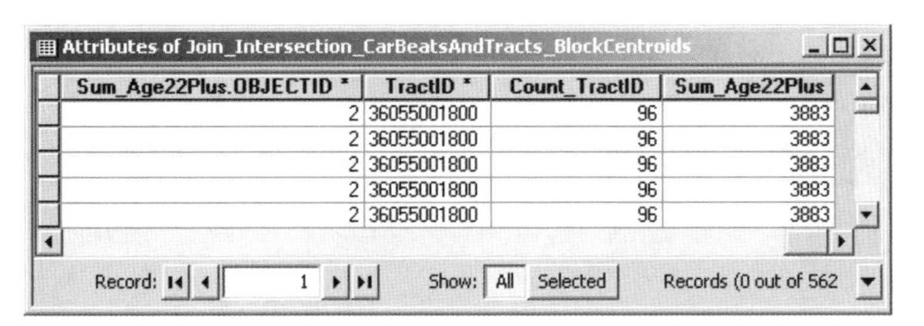

Sum_Age22Plus.OBJECTID *	TractID *	Count_TractID	Sum_Age22Plus
2	36055001800	96	3883
2	36055001800	96	3883
2	36055001800	96	3883
2	36055001800	96	3883
2	36055001800	96	3883

Compute apportionment weights

For each block centroid, you will now create and calculate new columns for the apportionment weight and apportioned undereducated population.

1 **In the Attributes of Join_Intersection_CarBeatsAndTracts_BlockCentroids table, click Options, Add Field.**

2 **Name the new field Weight, set its Type to Float, then click OK.**

3 **In the Attributes of Join_Intersection_CarBeatsAndTracts_BlockCentroids table, click Options, Add Field.**

4 **Name the new field UnderEducated, set the Type to Float, then click OK.**

5 **In the Attributes of Join_Intersection_CarBeatsAndTracts_BlockCentroids table, locate and then right-click the Join_Intersection_CarBeatsAndTracts_BlockCentroids.Weight column heading, click Field Calculator, and Yes.**

In the Field Calculator Fields panel, scroll to the right to see names of fields.

6 **In Fields box, double-click the Join_Intersection_CarBeatsAndTracts_BlockCentroids.AGE22Plus field, click the / button, scroll to the bottom of the Fields box, and double-click Sum_Age22Plus. Sum_AGE22Plus. Verify that your settings match those in the following graphic, then click OK.**

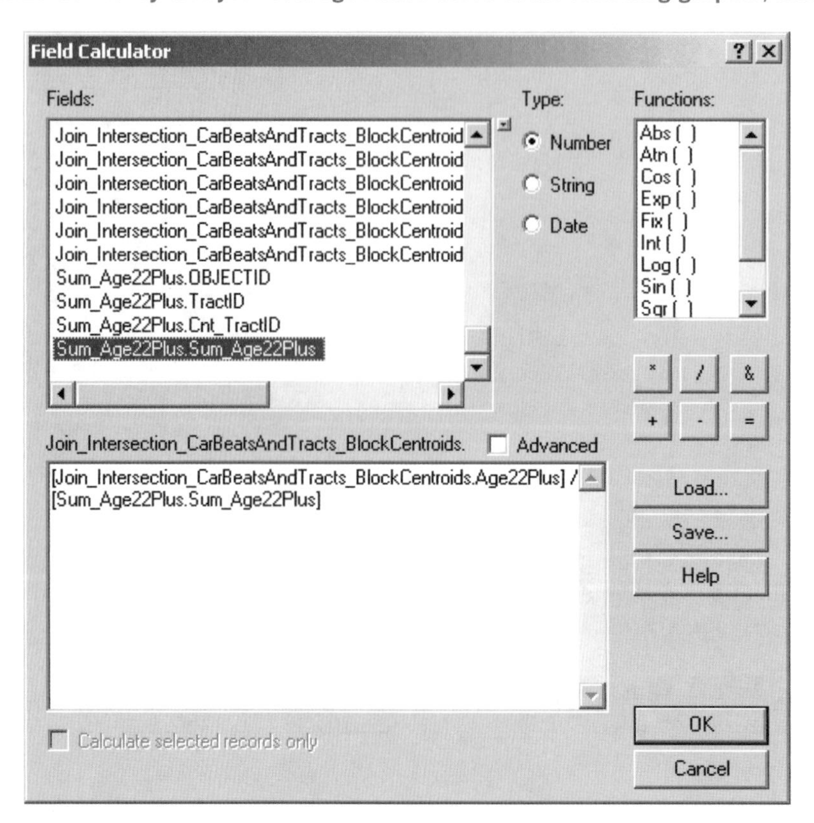

Compute apportionment values

1 **In the Attributes of Join_Intersection_CarBeatsAndTracts_BlockCentroids table, locate, then right-click the Join_Intersection_CarBeatsAndTracts_BlockCentroids.UnderEducated column heading. Click Field Calculator, then Yes.**

2 **Delete your previous expression, scroll near the bottom of the Fields list, double-click the Join_Intersection_CarBeatsAndTracts_BlockCentroids.Weight field (the sixth row up from the bottom of the list), click the * button, locate and double-click the Join_Intersection_CarBeatsAndTracts_BlockCentroids.NOHISCH field, click OK, and, when prompted with a warning message, click Yes.**

The NOHISCH field is the number of persons with no high school education.

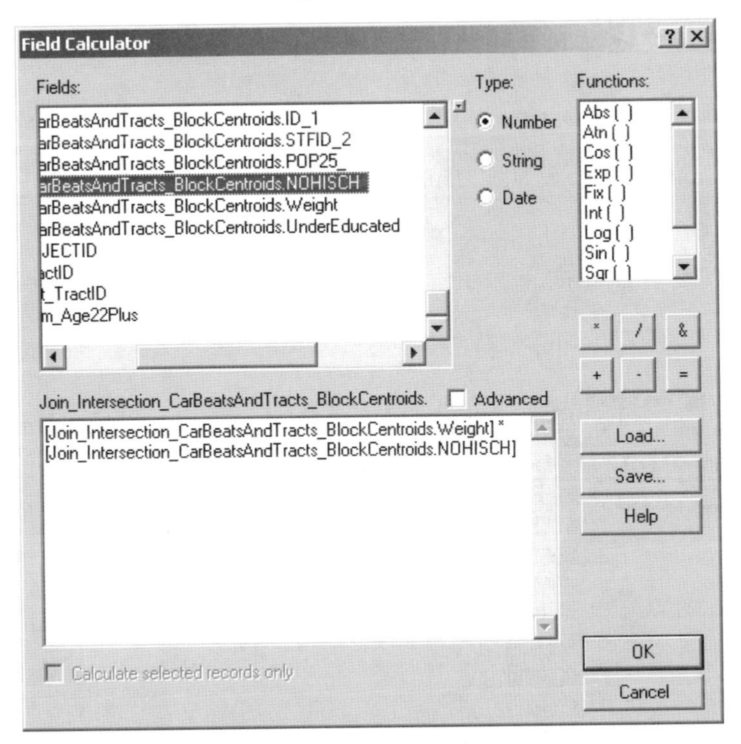

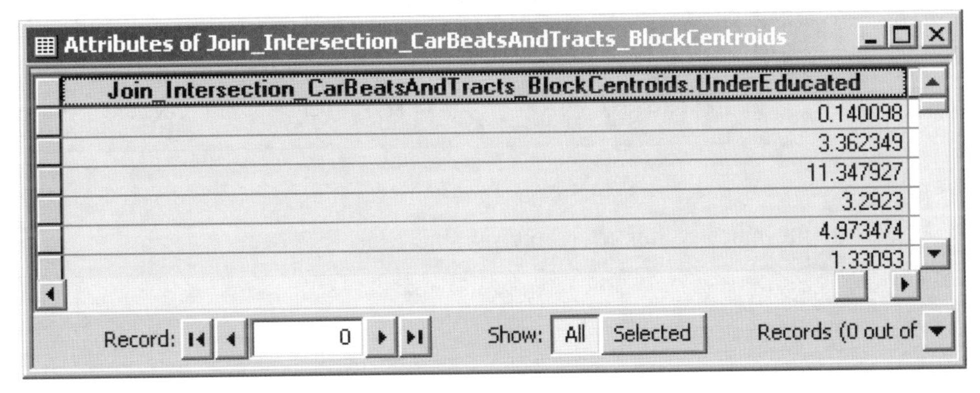

Join_Intersection_CarBeatsAndTracts_BlockCentroids.UnderEducated
0.140098
3.362349
11.347927
3.2923
4.973474
1.33093

Sum weights by tract

Sum the apportionment weights by tract; they should add up to 1.0 for each tract.

1 **Right-click the Join_Intersection_CarBeatsAndTracts_BlockCentroids layer in the table of contents, click Joins and Relates, Remove Join(s), Remove All Joins.**

You will notice that the Weight and UnderEducated columns are the last columns in the table.

2 **In the table, scroll to the left, locate and right-click the TractID column heading, and click Summarize.**

3 **For step 1, select TractID from the drop-down list.**

4 **For step 2, scroll down to and expand the Weight field, then check the Sum box.**

5 **For step 3, save the output table as \Gistutorial\LakePrecinct\Lake.mdb\Sum_WeightsByTract, click OK, then click Yes.**

6 **In the table of contents, right-click the Sum_WeightsByTract table and click Open.**

Note that the sum of weight for each tract is 1.0, which is a good check.

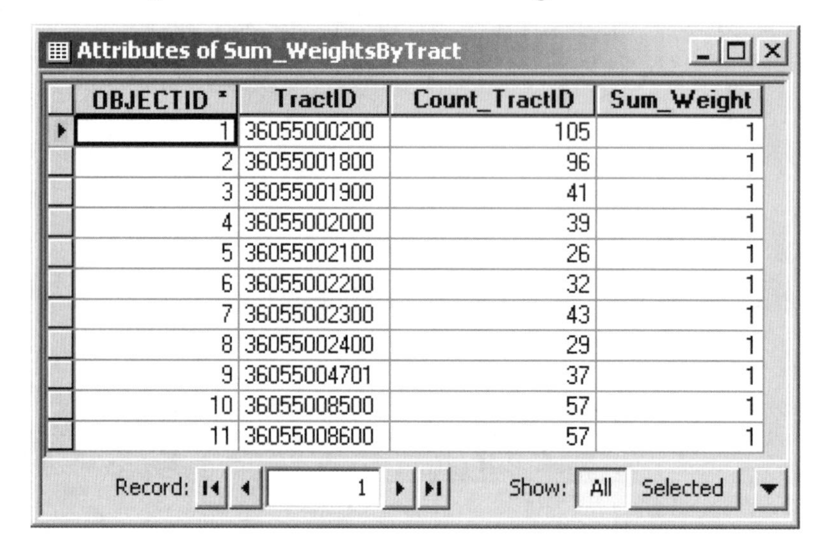

Attributes of Sum_WeightsByTract

OBJECTID *	TractID	Count_TractID	Sum_Weight
1	36055000200	105	1
2	36055001800	96	1
3	36055001900	41	1
4	36055002000	39	1
5	36055002100	26	1
6	36055002200	32	1
7	36055002300	43	1
8	36055002400	29	1
9	36055004701	37	1
10	36055008500	57	1
11	36055008600	57	1

Record: 1 Show: All Selected

7 **Close the Attributes of Sum_WeightByTract table.**

Sum undereducated by car beats

Sum the undereducated population by car beat as a new table.

1 **In the Attributes of Join_Intersection_CarBeatsAndTracts_BlockCentroids table, right-click the BEAT column heading and click Summarize.**

2 **For step 1, select BEAT.**

3 **For step 2, scroll down to and expand the UnderEducated field, then check the Sum box.**

4 **For step 3, save the output table as \Gistutorial\LakePrecinct\Lake.mdb\Sum_UnderEducated, then click OK and Yes.**

5 **Right-click the Sum_UnderEducated table in the table of contents and click Open.**

The extra row with no beat value is okay for now, because it will not join to the car beats table.

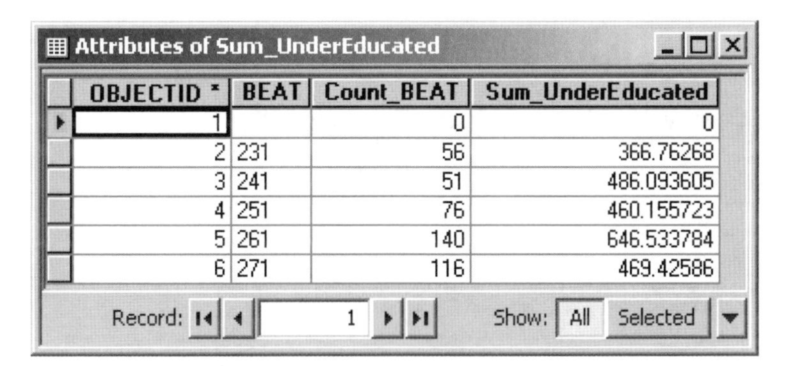
Attributes of Sum_UnderEducated

OBJECTID *	BEAT	Count_BEAT	Sum_UnderEducated
1		0	0
2	231	56	366.76268
3	241	51	486.093605
4	251	76	460.155723
5	261	140	646.533784
6	271	116	469.42586

6 **Close all open tables.**

Join Sum_UnderEducated to the car beat layer

In the following steps, you will join the table for undereducated population by car beat to the car beats layer and symbolize for display.

1 **Click the Display tab at the bottom of the table of contents.**

2 **Right-click Car Beats in the table of contents and click Joins and Relates, Join.**

3 **In the Join Data dialog box, select Join attributes from a table from the What do you want to join to this layer? drop-down list.**

4 **For step 1, select the BEAT field from the drop-down list.**

5 **For step 2, select the Sum_UnderEducated table from the drop-down list.**

6 **For step 3, select the BEAT field from the drop-down list. Make sure your settings match those in the graphic below, then click OK and Yes.**

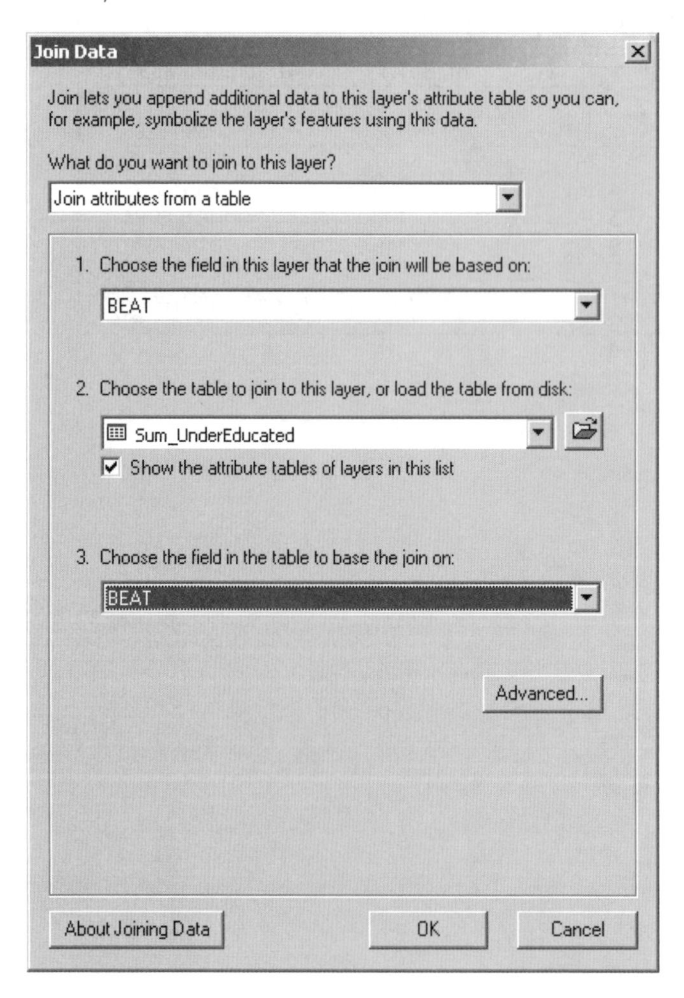

Map undereducated population by car beat

1 In the table of contents, turn all layers off except Assault Offenses and Car Beats.

2 In the table of contents, right-click the Car Beats layer, then click Properties.

3 In the Layer Properties dialog box, click the Symbology tab. In the Show box, click Quantities, Graduated Colors.

4 From the Value drop-down list choose Sum_UnderEducated, then click Classify.

5 In the Classification dialog box, change the number of Classes to 3, then select Manual from the Method drop-down list.

6 In the Break Values box, type **450**, **550**, and **99999** and click OK.

7 In the Layer Properties window, match the settings on the Symbology tab to those shown below, choose a monochromatic color ramp, and click OK.

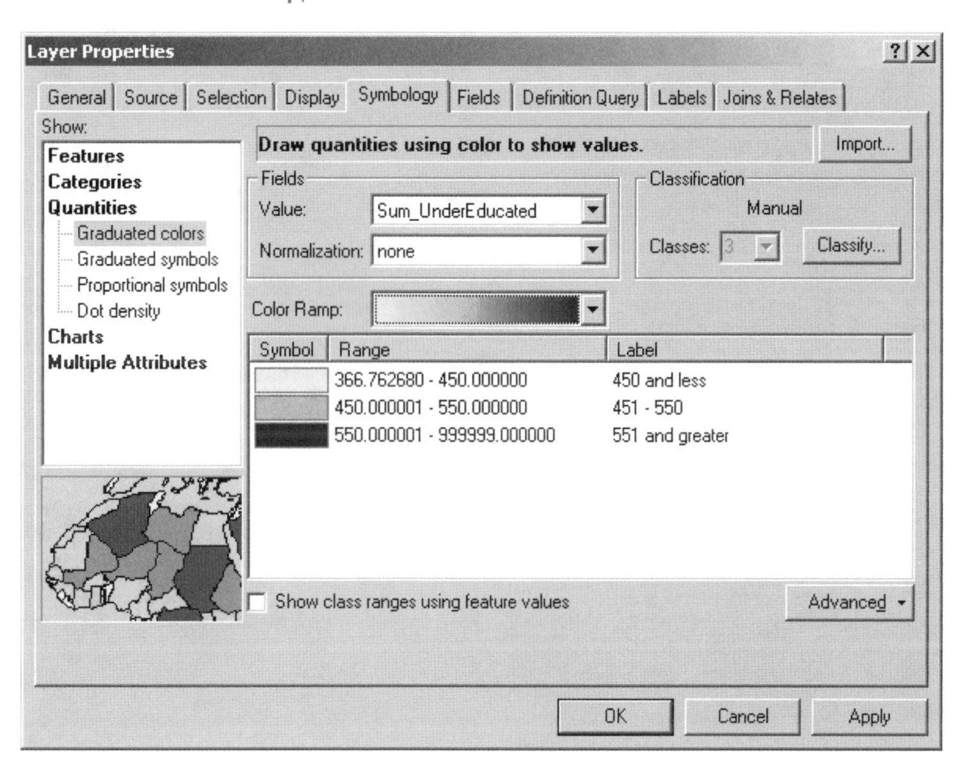

Wrap up

Now you have a census variable apportioned to car beats. The map shows that car beat 261 has the highest number of undereducated population, an indicator of crime.

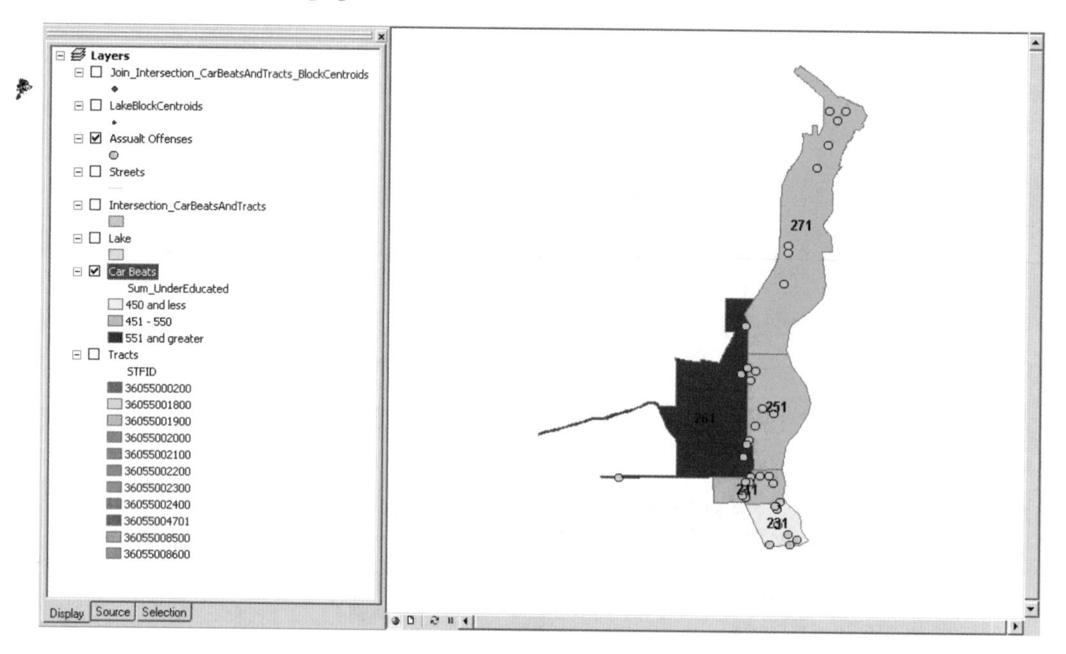

1 **Save your map.**

2 **Close ArcMap.**

Assignment 9-1

Analyze population in California cities at risk for earthquakes

When natural disasters occur, officials need to move quickly to find resources to help affected people. Earthquakes are one example of a natural disaster affecting millions of people. California is a highly populated state that is vulnerable to earthquakes. In this assignment, you will use GIS to create buffers around major earthquakes that have occurred in California and analyze how many people live within 20 miles of these events.

Start with the following:

- **C:\Gistutorial\UnitedStates\California\CACounties.shp**—polygon boundary of California counties.
- **C:\Gistutorial\UnitedStates\California\Earthquakes.dbf**—table of earthquakes in California with latitude and longitude attributes.
- **C:\Gistutorial\UnitedStates\Cities.shp**—point locations for major cities in the United States.

Create a map showing California earthquakes and population

In ArcCatalog, create a new personal geodatabase called C:\Gistutorial\Answers\Assignment9\CAEarthquakes.mdb and import any new and relevant files to the project.

Create a map document called C:\Gistutorial\Answers\Assignment9\Assignment9-1.mxd with a layout showing a 20-mile buffer around earthquakes whose magnitude is greater than 7. Use a UTM projection appropriate for California in your data frame. Include a label with the total population within buffers. See hints.

Export the map to C:\Gistutorial\Answers\Assignment9\Assignment9-1.jpg.

Assignments

Hints

It takes a couple of steps to get the desired buffers for this assignment. The map needs each separate buffer area to have a label displaying the total urban population in that area. If you were to use the ALL dissolve type when buffering earthquakes, a single polygon would result for all buffers, even though they are separate areas. Some 20-mile buffers overlap. Each set of overlapping buffers needs to be dissolved to form a single buffer polygon. Many 20-mile buffers do not overlap. These need to be separate polygons. The approach to building the needed buffers uses a dissolve field that you will create in the following steps:

- Add **Earthquakes.dbf** as an XY event file to the map document, select earthquakes with magnitude greater than 7, and export it as a feature class using the data frame's coordinate system to **C:\Gistutorial\Answers\Assignment9\CAEarthquakes.mdb\EarthquakesPlus7**. 0.20 miles in Buffer Box
- Buffer these earthquakes using a 20-mile radius, the NONE dissolve type, and saved in the feature class Buffer_NONE. Stuck Here
- Select all cities lying inside the buffers and create a new feature class for them called CAEarthquakeCities.
- Open the attribute table of EarthquakesPlus7 and add a new field called Dissolve with the Long Integer data type. Use the Field Calculator to set the values of the new field to equal ObjectID, which just has sequence numbers. You will set Dissolve equal to the same value for buffers that overlap.
- Using Buffer_NONE as a guide, select points from EarthquakesPlus7 that are in an overlapping set (there are two such sets). Then, using the Field Calculator, set the members of the selected points to the same value in the Dissolve field (use a Dissolve field value of one of the selected members). Repeat this step for the other set of overlapping buffers.
- Create a new buffer, Earthquakes_Buffer, that uses LIST as the dissolve type and then check the Dissolve field in the Dissolve Fields panel. The end result is separate polygon buffers for each nonoverlapping buffer plus two more polygons for the two sets of overlapping buffers.
- Spatially join CAEarthquakeCities with Earthquakes_Buffer to create CitiesBufferJoin. Start the join by right-clicking the polygon layer and be sure to use SUM so that city attributes are summed by buffer polygon for labeling your map.

Questions

Save the answers to your questions in a Microsoft Word document called **C:\Gistutorial\Answers\Assignment9\Assignment9-1.doc**.

1. **Which major earthquake (magnitude greater than 7) has the most detailed cities within 20 miles?**
2. **According to the detailed cities table, how many people are within 20 miles of that earthquake?**
3. **What California cities with population over 350,000 have not yet been hit by an earthquake whose magnitude is over 7?**

Assignment 9-2

Neighborhood walking distances and urban grocery store site selection

Many cities are trying to bring economic development into their downtown areas. One way to promote downtown environments is to demonstrate walking distances. Walkable catchments, sometimes referred to as "ped sheds," can be mapped to show the area within a short walking distance from attractions or amenities.

Source: *www.cnu.org/cnu_reports/CNU_Ped_Sheds.pdf*

Study area background

In this exercise, you will study the walkability of an urban area of Pittsburgh, Pennsylvania. In a recent issue of the *National Geographic Magazine,* the ZIP Code (15222) located in the heart of Pittsburgh was featured as one of the most interesting areas in the United States. You can read more about this area online at magma.nationalgeographic.com/ngm/0308/feature6.

Two neighborhoods making up the 15222 ZIP Code, the Strip District and Central Business District (CBD), are adding a number of urban living condominiums and lofts. You will create a buffer for a short walking distance to streets with major attractions such as restaurants and theaters. You can then show features within that distance (buildings, streets, and so forth).

You will also study the existing population in the 15222 ZIP Code to determine what area is most suitable for a food store. The store should be at the center of a 10-minute walking area of 2,000 population or more.

Start with the following:

- **C:\Gistutorial\PAGIS\PghZipCodes.shp**—Pittsburgh ZIP Codes.
- **C:\Gistutorial\ PAGIS\CensusBlocks.shp**—2000 Census blocks for the city of Pittsburgh.
- **C:\Gistutorial\ PAGIS\15222\15222Bldgs.shp**—building footprints near and within the 15222 ZIP Code.
- **C:\Gistutorial\ PAGIS\15222\15222Streets.shp**—street centerlines within the 15222 ZIP Code used to label street names.
- **C:\Gistutorial\AlleghenyCounty\Rivers.shp**

Map showing walkable catchment areas for neighborhood and grocery store site

Create a new map called C:\Gistutorial\Answers\Assignment9\Assignment9-2.mxd that includes a layout with two data frames:

- "15222 Walkable Catchment Area" showing a two-minute walking buffer to streets with major attractions. Label the streets with the street name. Add buildings and rivers map layers.
- "Food Store Site Suitability Study" zoomed to an ideal location for a grocery store based on the current population in the downtown area. Add buildings, curbs, and streets as background layers.

Export the map to a JPEG file called C:\Gistutorial\Answers\Assignment9\Assignment9-2.jpg.

Walking distance buffer hints

- Create and label a two-minute walking distance buffer around all Smallman Street street segments and Penn Avenue street segments up through the 1100 block as C:\Gistutorial\Answers\Assignment9\Buffer2Minute.shp. Use 528 feet (distance covered at three miles per hour in two minutes) as the buffer linear unit.
- Use the All dissolve type.

Food store site selection hints

- In the data frame called "Food Store Site Suitability Study," add the PghZIPCodes and CensusBlocks from the input layers.
- Select census blocks that have their center in the 15222 ZIP Code, then export these to a new feature class called C:\Gistutorial\Answers\Assignment9\CensusBlocks15222.shp. Add the new shapefile to the map and remove the original.
- Create two new fields to calculate the x,y centroid coordinates of the census block polygons.
- Export a table for the centroids called C:\Gistutorial\Answers\Assignment9\15222BlockCentroids.dbf, and add the new table as an x,y data map layer.
- Symbolize the census block centroids by population, showing different colors for five population classifications (0, 1–100, 101–200, 201–300, 301 and greater). Use no color for the 0 class.
- Click View, click Toolbars, and click Draw. Then draw a circle with radius 2,640 feet (yielding a 10-minute walking buffer) anywhere on the map.
- Position the circle where it appears the maximum population would be within the circle. Make your block centroids the only selectable layer, then with your circle selected click Selection and click Select by Graphics.
- Open the block centroids table and use the Statistics function to get the population of selected blocks.
- The potential grocery store owners would like to have at least 2,000 population within a 10-minute walk. Is there sufficient population yet? Place the population of the buffer area on the map, either in the title or in text.
- Draw a black dot on and label the building where you would recommend locating a grocery store.

What to turn in

If you are working in a classroom setting with an instructor, you may be required to submit the exercises you created in tutorial 9. Below are the files you are required to turn in. Be sure to use a compression program such as PKZIP or WinZIP to include all files as one .zip document for review and grading. Include your name and assignment number in the .zip document (YourNameAssn9.ZIP).

ArcMap documents

C:\Gistutorial\Answers\Assignment9\Assignment9-1.mxd
C:\Gistutorial\Answers\Assignment9\Assignment9-2.mxd

Exported maps

C:\Gistutorial\Answers\Assignment9\Assignment9-1.jpg
C:\Gistutorial\Answers\Assignment9\Assignment9-2.jpg

Personal geodatabase

C:\Gistutorial\Answers\Assignment9\CAEarthquakes.mdb (includes CAEarthquakeCities, CitiesBufferJoin, Earthquakes_Buffer, and EarthquakesPlus7)
Shapefile and table from C:\Gistutorial\Answers\Assignment9\: Buffer2Minute.shp, 15222BlockCentroids.dbf

Word document

C:\Gistutorial\Answers\Assignment9\Assignment9-1.doc

Appendix A

Data License Agreement

Important:

Read carefully before opening the sealed media package

Environmental Systems Research Institute, Inc. (ESRI), is willing to license the enclosed data and related materials to you only upon the condition that you accept all of the terms and conditions contained in this license agreement. Please read the terms and conditions carefully before opening the sealed media package. By opening the sealed media package, you are indicating your acceptance of the ESRI License Agreement. If you do not agree to the terms and conditions as stated, then ESRI is unwilling to license the data and related materials to you. In such event, you should return the media package with the seal unbroken and all other components to ESRI.

ESRI License Agreement

This is a license agreement, and not an agreement for sale, between you (Licensee) and Environmental Systems Research Institute, Inc. (ESRI). This ESRI License Agreement (Agreement) gives Licensee certain limited rights to use the data and related materials (Data and Related Materials). All rights not specifically granted in this Agreement are reserved to ESRI and its Licensors.

Reservation of Ownership and Grant of License: ESRI and its Licensors retain exclusive rights, title, and ownership to the copy of the Data and Related Materials licensed under this Agreement and, hereby, grant to Licensee a personal, nonexclusive, nontransferable, royalty-free, worldwide license to use the Data and Related Materials based on the terms and conditions of this Agreement. Licensee agrees to use reasonable effort to protect the Data and Related Materials from unauthorized use, reproduction, distribution, or publication.

Proprietary Rights and Copyright: Licensee acknowledges that the Data and Related Materials are proprietary and confidential property of ESRI and its Licensors and are protected by United States copyright laws and applicable international copyright treaties and/or conventions.

Permitted Uses:
Licensee may install the Data and Related Materials onto permanent storage device(s) for Licensee's own internal use.

Licensee may make only one (1) copy of the original Data and Related Materials for archival purposes during the term of this Agreement unless the right to make additional copies is granted to Licensee in writing by ESRI.

Licensee may internally use the Data and Related Materials provided by ESRI for the stated purpose of GIS training and education.

Uses Not Permitted:
Licensee shall not sell, rent, lease, sublicense, lend, assign, time-share, or transfer, in whole or in part, or provide unlicensed Third Parties access to the Data and Related Materials or portions of the Data and Related Materials, any updates, or Licensee's rights under this Agreement.

Licensee shall not remove or obscure any copyright or trademark notices of ESRI or its Licensors.

Term and Termination: The license granted to Licensee by this Agreement shall commence upon the acceptance of this Agreement and shall continue until such time that Licensee elects in writing to discontinue use of the Data or Related Materials and terminates this Agreement. The Agreement shall automatically terminate without notice if Licensee fails to comply with any provision of this Agreement. Licensee shall then return to ESRI the Data and Related Materials. The parties hereby agree that all provisions that operate to protect the rights of ESRI and its Licensors shall remain in force should breach occur.

Disclaimer of Warranty: THE DATA AND RELATED MATERIALS CONTAINED HEREIN ARE PROVIDED "AS-IS," WITHOUT WARRANTY OF ANY KIND, EITHER EXPRESS OR IMPLIED, INCLUDING, BUT NOT LIMITED TO, THE IMPLIED WARRANTIES OF MERCHANTABILITY, FITNESS FOR A PARTICULAR PURPOSE, OR NONINFRINGEMENT. ESRI does not warrant that the Data and Related Materials will meet Licensee's needs or expectations, that the use of the Data and Related Materials will be uninterrupted, or that all nonconformities, defects, or errors can or will be corrected. ESRI is not inviting reliance on the Data or Related Materials for commercial planning or analysis purposes, and Licensee should always check actual data.

Data Disclaimer: The Data used herein has been derived from actual spatial or tabular information. In some cases, ESRI has manipulated and applied certain assumptions, analyses, and opinions to the Data solely for educational training purposes. Assumptions, analyses, opinions applied, and actual outcomes may vary. Again, ESRI is not inviting reliance on this Data, and the Licensee should always verify actual Data and exercise their own professional judgment when interpreting any outcomes.

Limitation of Liability: ESRI shall not be liable for direct, indirect, special, incidental, or consequential damages related to Licensee's use of the Data and Related Materials, even if ESRI is advised of the possibility of such damage.

No Implied Waivers: No failure or delay by ESRI or its Licensors in enforcing any right or remedy under this Agreement shall be construed as a waiver of any future or other exercise of such right or remedy by ESRI or its Licensors.

Order for Precedence: Any conflict between the terms of this Agreement and any FAR, DFAR, purchase order, or other terms shall be resolved in favor of the terms expressed in this Agreement, subject to the government's minimum rights unless agreed otherwise.

Export Regulation: Licensee acknowledges that this Agreement and the performance thereof are subject to compliance with any and all applicable United States laws, regulations, or orders relating to the export of data thereto. Licensee agrees to comply with all laws, regulations, and orders of the United States in regard to any export of such technical data.

Severability: If any provision(s) of this Agreement shall be held to be invalid, illegal, or unenforceable by a court or other tribunal of competent jurisdiction, the validity, legality, and enforceability of the remaining provisions shall not in any way be affected or impaired thereby.

Governing Law: This Agreement, entered into in the County of San Bernardino, shall be construed and enforced in accordance with and be governed by the laws of the United States of America and the State of California without reference to conflict of laws principles. The parties hereby consent to the personal jurisdiction of the courts of this county and waive their rights to change venue.

Entire Agreement: The parties agree that this Agreement constitutes the sole and entire agreement of the parties as to the matter set forth herein and supersedes any previous agreements, understandings, and arrangements between the parties relating hereto.

Appendix B

Installing the Data and Software

GIS Tutorial includes one CD with exercise data and one DVD with ArcGIS Desktop 9.2 (ArcView license, single-use, 180-day trial) software. You will find both in the back of this book. Installation of the exercise data CD takes approximately five minutes and requires around 300 MB of hard-disk space. Installation of the ArcGIS Desktop software DVD with extensions takes approximately 25 minutes and requires at least 1.5 GB of hard-disk space. Installation times will vary with your computer's speed and available memory.

If you previously installed data for an earlier edition of *GIS Tutorial*, you cannot simply copy the current data over it. You must uninstall the previous data before you install the exercise data that comes with this book.

If you already have a licensed copy of ArcGIS Desktop 9.2 installed on your computer (or accessible through a network), do not install the software DVD. Use your licensed software to do the exercises in this book. If you have an older version of ArcGIS installed on your computer, you must uninstall it before you can install the software DVD that comes with this book.

The exercises in this book only work with ArcGIS 9.2 or higher.

Installing the exercise data

Follow the steps below to install the exercise data.

1 **Put the data CD in your computer's CD drive. A splash screen will appear.**

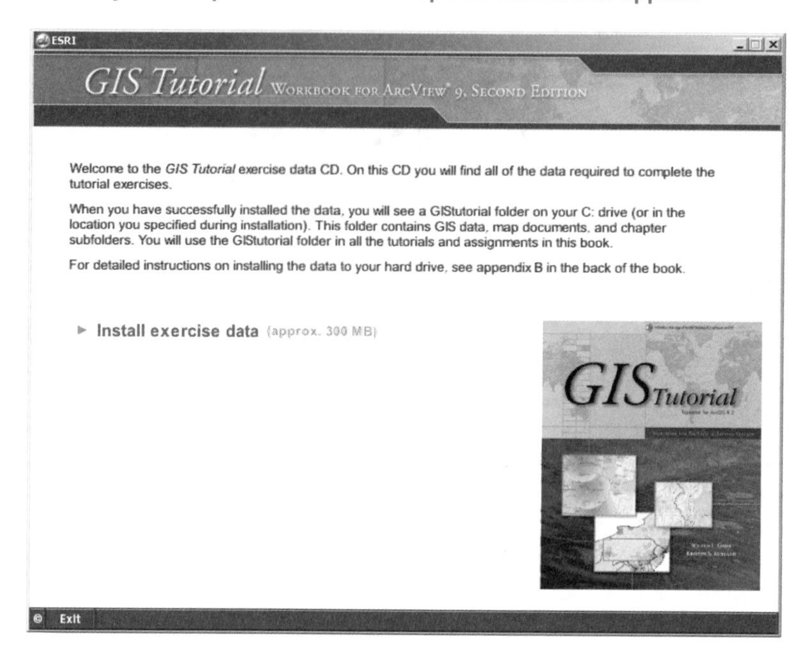

2 **Read the welcome, then click the Install exercise data link. This launches the Setup wizard.**

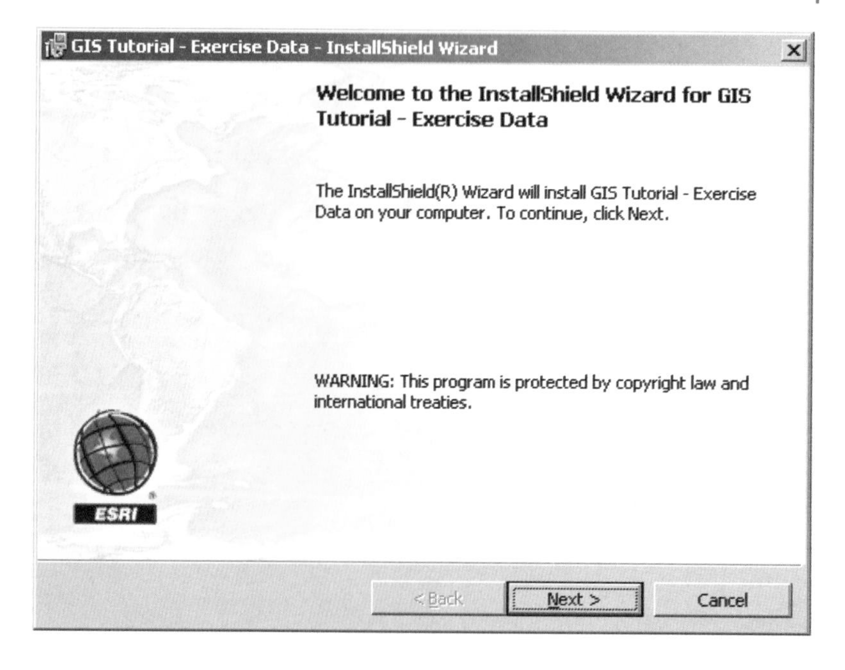

3 **Click Next. Read and accept the license agreement terms, then click Next.**

4 Accept the default installation folder. We recommend that you do not choose an alternate location.

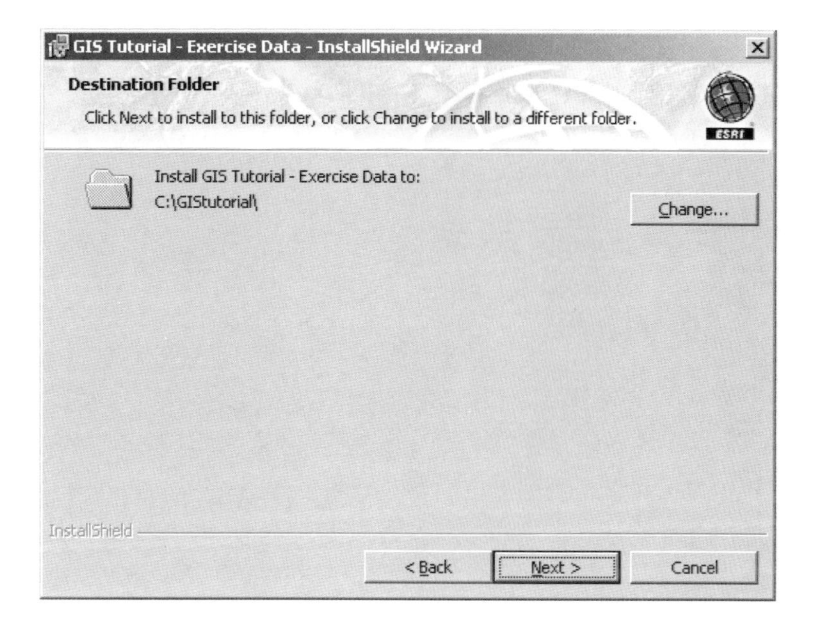

5 Click Next. The installation will take a few moments. When the installation is complete, you will see the following message:

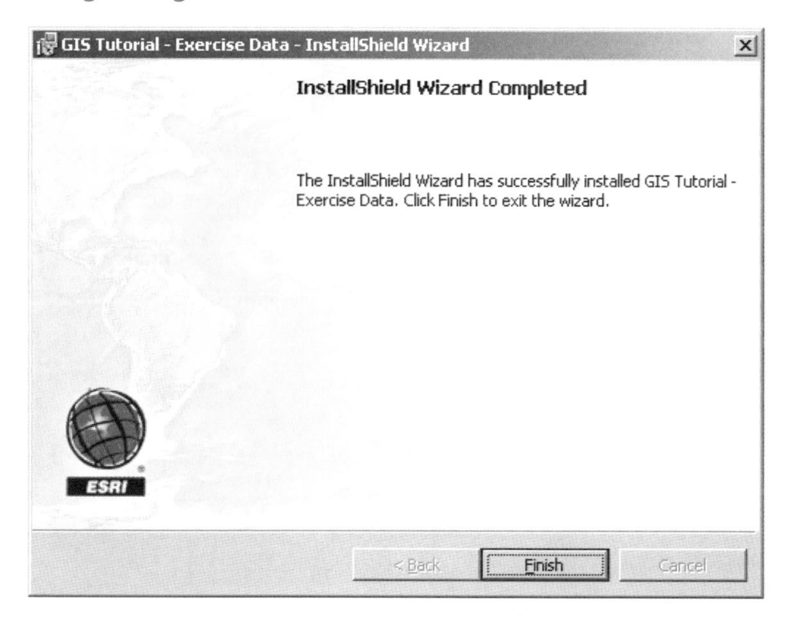

6 Click Finish. The exercise data is installed on your computer in a folder called GIStutorial.

Uninstalling the exercise data

To uninstall the exercise data from your computer, open your operating system's control panel and double-click the Add/Remove Programs icon. In the Add/Remove Programs dialog box, select the following entry and follow the prompts to remove it:

GIS Tutorial - Exercise Data

Installing the software

The ArcGIS software included on this DVD is intended for educational purposes only. Once installed and registered, the software will run for 180 days. The software cannot be reinstalled nor can the time limit be extended. It is recommended that you uninstall this software when it expires.

Follow the steps below to install the software.

1 **Put the software DVD in your computer's DVD drive. A splash screen will appear.**

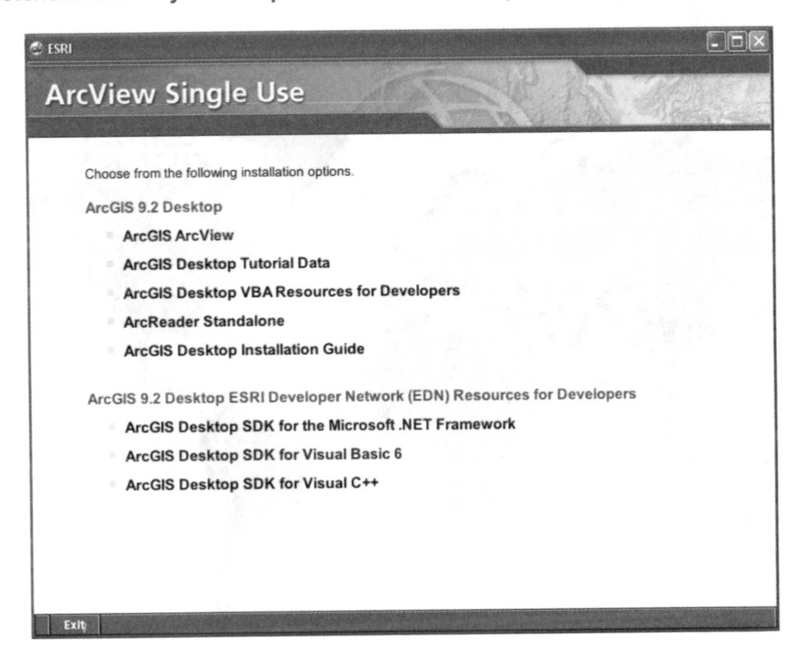

2 **Click the ArcGIS ArcView installation option. On the Startup window, click Install ArcGIS Desktop. This will launch the Setup wizard.**

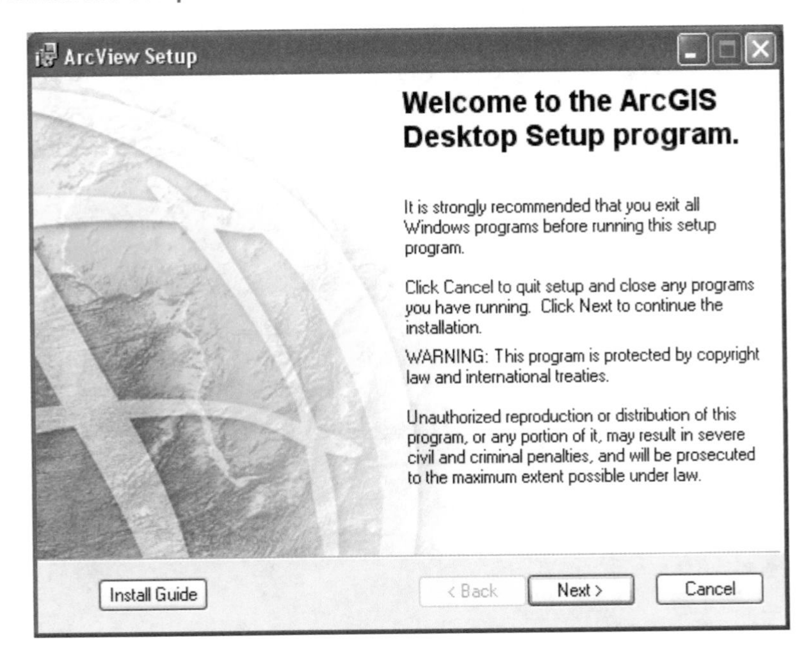

3 Read the Welcome, then click Next.

4 Read the license agreement. Click "I accept the license agreement" and click Next.

5 The default installation type is Typical, which is the one that is needed for this install. Click Next.

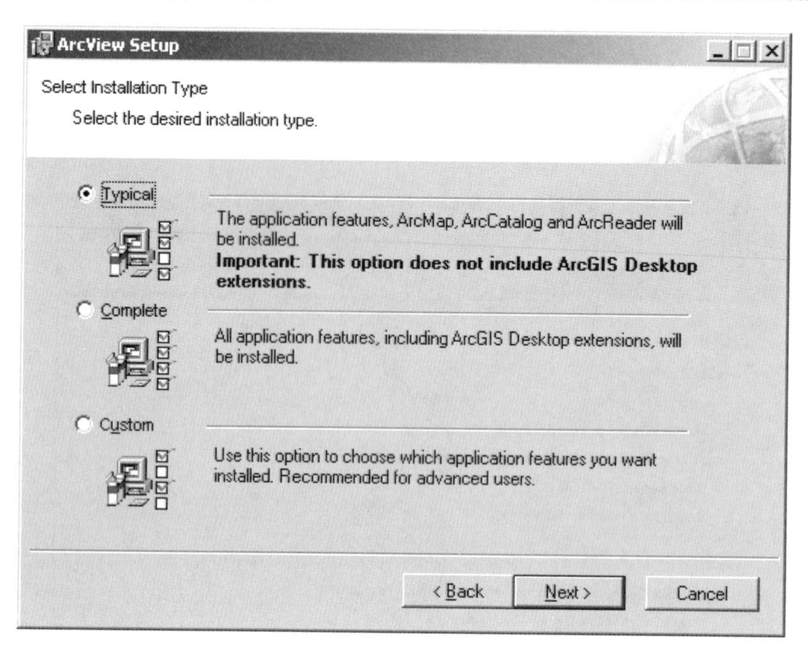

6 Click Next. Accept the default installation folder or click Browse and navigate to the drive or folder location where you want to install the software.

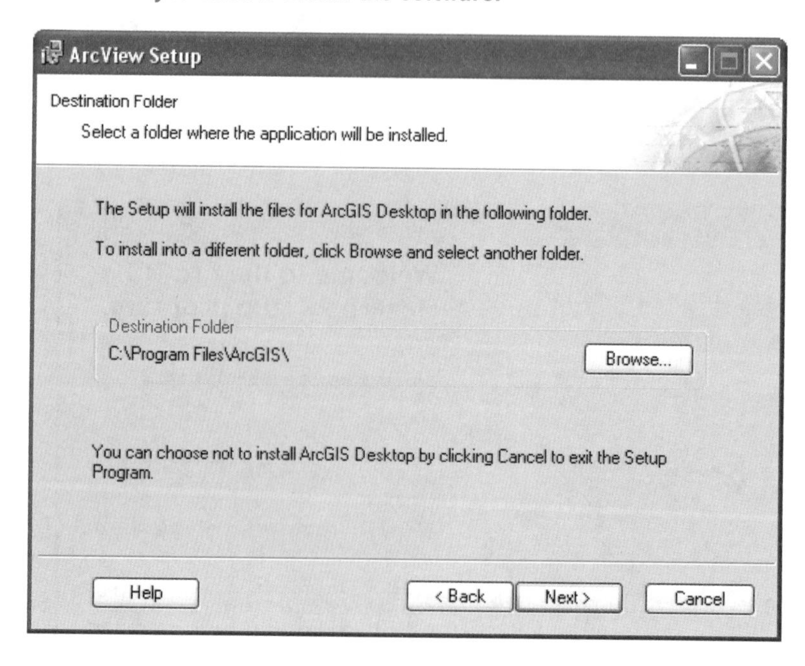

7 Click Next. Accept the default installation folder or navigate to the drive or folder where you want to install Python, a scripting language used by some ArcGIS geoprocessing functions. (You won't see this panel if you already have Python installed.) We recommend that you accept the default folder. Click Next.

8 The installation paths for ArcGIS and Python are confirmed. Click Next. The software will take several minutes to install on your computer. When the installation is finished, you see the following message:

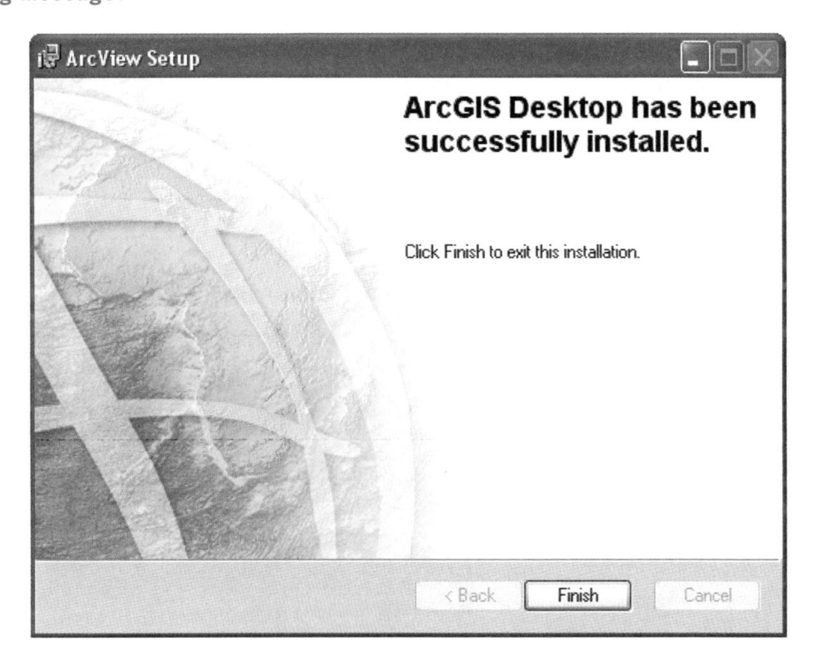

9 Click Finish.

10 On the next panel, click Register Now and follow the registration process. The registration code is located at the bottom of the software DVD jacket in the back of the book.

If you have questions or encounter problems during the installation process, or while using this book, please use the resources listed below. (The ESRI Technical Support Department does not answer questions regarding the ArcGIS 9.2 software DVD, the GIS Tutorial exercise data CD, or the contents of the book itself.)

- To resolve problems with the trial software or exercise data, or to report mistakes in the book, send an e-mail to ESRI workbook support at workbook-support@esri.com.
- To stay informed about exercise updates, FAQs, and errata, visit the book's Web page at www.esri.com/esripress/gistutorial.

Uninstalling the software

To uninstall the software from your computer, open your operating system's control panel and double-click the Add/Remove Programs icon. In the Add/Remove Programs dialog box, select the following entry and follow the prompts to remove it:

ArcGIS Desktop

Appendix C

Data Source Credits

Tutorial 1 data sources include:

\Gistutorial\UnitedStates\States.shp, from ESRI Data & Maps, 2004, courtesy of U.S. Census.

\Gistutorial\UnitedStates\Cities.shp, from ESRI Data & Maps, 2004, courtesy of U.S. Census.

\Gistutorial\UnitedStates\Colorado\Counties.shp, ESRI Data & Maps, 2004, courtesy of U.S. Census.

\Gistutorial\PAGIS\MidHill\Street, courtesy of the City of Pittsburgh, Department of City Planning.

\Gistutorial\PAGIS\MidHill\Curbs, courtesy of the City of Pittsburgh, Department of City Planning.

\Gistutorial\PAGIS\MidHill\Building, courtesy of the City of Pittsburgh, Department of City Planning.

\Gistutorial\PAGIS\MidHill\MID911, courtesy of the City of Pittsburgh, Department of City Planning.

Tutorial 2 data sources include:

\Gistutorial\UnitedStates\States.shp, from ESRI Data & Maps, 2004, courtesy of U.S. Census.

\Gistutorial\UnitedStates\Counties.shp, from ESRI Data & Maps, 2004, courtesy of U.S. Census.

\Gistutorial\UnitedStates\Utah\UtahTracts.shp, from ESRI Data & Maps, 2004, courtesy of U.S. Census.

\Gistutorial\UnitedStates\Nevada\NevadaTracts.shp, from ESRI Data & Maps, 2004, courtesy of U.S. Census.

\Gistutorial\UnitedStates\Cities.shp, from ESRI Data & Maps, 2004, courtesy of U.S. Census.

\Gistutorial\UnitedStates\Pennsylvania\PACounties.shp, from ESRI Data & Maps, 2004, courtesy of U.S. Census.

\Gistutorial\UnitedStates\Pennsylvania\PACities.shp, from ESRI Data & Maps, 2004, courtesy of U.S. Census.

\Gistutorial\PAGIS\Neighborhoods.shp, courtesy of the City of Pittsburgh, Department of City Planning.

\Gistutorial\PAGIS\Schools.shp, courtesy of the City of Pittsburgh, Department of City Planning.

\Gistutorial\UnitedStates\Pennsylvania\PATracts.shp, from ESRI Data & Maps, 2004, courtesy of U.S. Census.

Tutorial 3 data sources include:

\Gistutorial\UnitedStates\States.shp, from ESRI Data & Maps, 2004, courtesy of U.S. Census.

\Gistutorial\UnitedStates\Cities.shp, from ESRI Data & Maps, 2004, courtesy of U.S. Census.

\Gistutorial\UnitedStates\California\OrangeCountyTracts.shp, from ESRI Data & Maps, 2004, courtesy of U.S. Census.

\Gistutorial\PAGIS\CentralBusinessDistrict\CBDOutline.shp, courtesy of the City of Pittsburgh, Department of City Planning.

\Gistutorial\PAGIS\CentralBusinessDistrict\CBDBLDG.shp, courtesy of the City of Pittsburgh, Department of City Planning.

\Gistutorial\PAGIS\CentralBusinessDistrict\CBDStreets.shp, courtesy of the City of Pittsburgh, Department of City Planning.

\Gistutorial\PAGIS\Histpnts.shp, courtesy of the City of Pittsburgh, Department of City Planning.

\Gistutorial\PAGIS\Histsite.shp, courtesy of the City of Pittsburgh, Department of City Planning.

Tutorial 4 data sources include:

\Gistutorial\MaricopaCounty\tgr04013ccd00.shp, courtesy of the U.S. Census Bureau TIGER.

\Gistutorial\MaricopaCounty\tgr04013trt00.shp, courtesy of the U.S. Census Bureau TIGER.

\Gistutorial\MaricopaCounty\CensusDat.dbf, courtesy of the U.S. Census Bureau.

\Gistutorial\RochesterNY\RochesterPolice.mdb\carbeats, courtesy of the Rochester, New York Police Department.

\Gistutorial\RochesterNY\RochesterPolice.mdb\business, courtesy of the Rochester, New York Police Department.

\Gistutorial\RochesterNY\Business.shp, courtesy of the Rochester, New York Police Department.

Gistutorial\AlleghenyCounty\Munic.shp, courtesy of Southwestern Pennsylvania Commission.

\Gistutorial\PAGIS\PghTracts.shp, courtesy of the City of Pittsburgh, Department of City Planning.

\Gistutorial\PAGIS\Schools.shp, courtesy of the City of Pittsburgh, Department of City Planning.

Tutorial 5 data sources include:

Screen capture of www.esri.com home page, from ESRI Data & Maps, 2005, courtesy of the U.S. Census.

Screen capture of the United Kingdom in Data Downloader, from ESRI Data & Maps, 2005, European basemap data courtesy of Automotive Navigation Data.

Screen capture of www.esri.com/data/download/census2000_tigerline/index.html, from ESRI Data & Maps, 2000, courtesy of U.S. Census.

Screen captures of www.census.gov, courtesy of the U.S. Census. All U.S. Census Bureau materials, regardless of the media, are entirely in the public domain. There are no user fees, site licenses, or any special agreements, etc., for the public or private use, and/or reuse of any census title. As a tax funded product, it is all in the public record.

\Gistutorial\PAGIS\CentralBusinessDistrict\CBDStreets.shp, courtesy of the City of Pittsburgh, Department of City Planning.

Gistutorial\CMUCampus\CampusMap.dwg, courtesy of the Southwestern Pennsylvania Commission.

\Gistutorial\UnitedStates\States.shp, from ESRI Data & Maps, 2004, courtesy of U.S. Census.

\Gistutorial\UnitedStates\California\Earthquakes.dbf, from ESRI Data & Maps, 2004, courtesy of National Atlas of the United States, USGS.

\Gistutorial\UnitedStates\California\CACounties.shp, from ESRI Data & Maps, 2004, courtesy of U.S. Census.

\Gistutorial\World\Country.shp, from ESRI Data & Maps, 2004, courtesy of ArcWorld Supplement.

\Gistutorial\World\Ocean.shp, from ESRI Data & Maps, courtesy of ArcWorld Supplement.

\Gistutorial\AlleghenyCounty\Tracts.shp, from ESRI Data & Maps, 2004, courtesy of U.S. Census.

\Gistutorial\AlleghenyCounty\Munic.shp, from ESRI Data & Maps, 2004, courtesy of U.S. Census.

\Gistutorial\UnitedStates\Pennsylvania\2000sf1cty.dbf, courtesy of U.S. Census.

\Gistutorial\UnitedStates\Pennsylvania\2000sf1county.csv, courtesy of U.S. Census.

\Gistutorial\Flux\FLUXEvent.mdb\tAttendees, courtesy of FLUX (http://www.fluxpgh.com/).

\Gistutorial\UnitedStates\Florida\MajorCities.shp, from ESRI Data & Maps, 2004, courtesy of U.S. Census.

\Gistutorial\UnitedStates\Florida\FLCountyPopulation.dbf, from ESRI Data & Maps, 2004, courtesy of U.S. Census.

\Gistutorial\MaricopaCounty\CensusDat.dbf, courtesy of U.S. Census Bureau, TIGER.

Tutorial 6 data sources include:

\Gistutorial\PAGIS\MidHill\MIDZONE.shp, courtesy of the City of Pittsburgh, Department of City Planning.

\Gistutorial\PAGIS\MidHill\Street, courtesy of the City of Pittsburgh, Department of City Planning.

\Gistutorial\PAGIS\MidHill\Parcels, courtesy of the City of Pittsburgh, Department of City Planning.

\Gistutorial\PAGIS\MidHill\Curbs, courtesy of the City of Pittsburgh, Department of City Planning.

\Gistutorial\CMUCampus\25_45.tif, courtesy of the Southwestern Pennsylvania Commission.

\Gistutorial\CMUCampus\26_45.tif, courtesy of the Southwestern Pennsylvania Commission.

\Gistutorial\PAGIS\Zone2\streetszone2.shp, courtesy of the City of Pittsburgh, Department of City Planning.

\Gistutorial\PAGIS\Zone2\zone2.shp, courtesy of the City of Pittsburgh, Department of City Planning.

Tutorial 7 data sources include:

\Gistutorial\UnitedStates\Pennsylvania\PAZip.shp, courtesy of the Pennsylvania Resources Council.

\Gistutorial\Flux\FLUXEvent.mdb\Attendees, courtesy of FLUX (http://www.fluxpgh.com/).

\Gistutorial\PAGIS\PghStreets.shp, courtesy of the City of Pittsburgh, Department of City Planning.

\Gistutorial\PAGIS\CentralBusinessDistrict\CBDStreets.shp, courtesy of the City of Pittsburgh, Department of City Planning.

\Gistutorial\PAGIS\CentralBusinessDistrict\Clients.dbf, courtesy of the City of Pittsburgh, Department of City Planning.

\Gistutorial\UnitedStates\Pennsylvania\HHWZipCodes.dbf, courtesy of the Pennsylvania Resources Council.

\Gistutorial\UnitedStates\Pennsylvania\PAZip.shp, courtesy of the Pennsylvania Resources Council.

\Gistutorial\UnitedStates\Pennsylvania\PACounties.shp, from ESRI Data & Maps, 2004, courtesy of U.S. Census.

\Gistutorial\PAGIS\ForeignBusinesses.dbf, courtesy of the City of Pittsburgh, Department of City Planning.

\Gistutorial\PAGIS\Neighborhoods.shp, courtesy of the City of Pittsburgh, Department of City Planning.

Tutorial 8 data sources include:

\Gistutorial\UnitedStates\Cities_dtl.shp, from ESRI Data & Maps, 2004, courtesy of U.S. Census.

\Gistutorial\UnitedStates\NewYork\NYBoroughs.shp, from ESRI Data & Maps, 2004, courtesy of U.S. Census.

\Gistutorial\UnitedStates\NewYork\NYMetroZIP.shp, from ESRI Data & Maps, 2004, courtesy of U.S. Census.

\Gistutorial\UnitedStates\NewYork\ManhattanCounty.shp, from ESRI Data & Maps, 2004, courtesy of U.S. Census.

\Gistutorial\UnitedStates\NewYork\BrklynCounty.shp, from ESRI Data & Maps, 2004, courtesy of U.S. Census.

\Gistutorial\UnitedStates\NewYork\BronxCounty.shp, from ESRI Data & Maps, 2004, courtesy of U.S. Census.

\Gistutorial\UnitedStates\NewYork\QueensCounty.shp, from ESRI Data & Maps, 2004, courtesy of U.S. Census.

\Gistutorial\UnitedStates\NewYork\QueensBorough.shp, from ESRI Data & Maps, 2004, courtesy of U.S. Census.

\Gistutorial\UnitedStates\NewYork\QueensZips.shp, from ESRI Data & Maps, 2004, courtesy of U.S. Census.

\Gistutorial\UnitedStates\NewYork\StatenIslCounty.shp, from ESRI Data & Maps, 2004, courtesy of U.S. Census.

\Gistutorial\UnitedStates\Pennsylvania\PATractStatePlane.shp, from ESRI Data & Maps, 2004, courtesy of U.S. Census.

\Gistutorial\UnitedStates\Colorado\Counties.shp, from ESRI Data & Maps, 2004, courtesy of U.S. Census.

\Gistutorial\UnitedStates\Cities.shp, from ESRI Data & Maps, 2004, courtesy of U.S. Census.

\Gistutorial\PAGIS\EastLiberty\Parcel, courtesy of the City of Pittsburgh, Department of City Planning.

\Gistutorial\PAGIS\EastLiberty\EastLib, courtesy of the City of Pittsburgh, Department of City Planning.

Tutorial 9 data sources include:

\Gistutorial\LakePrecinct\Lake.mdb\lakebars, courtesy of the Rochester, New York, Police Department.

\Gistutorial\LakePrecinct\Lake.mdb\lakeassualts, courtesy of the Rochester, New York, Police Department.

\Gistutorial\LakePrecinct\Lake.mdb\LakeBlockCentroids, from ESRI Data & Maps, 2004, courtesy of U.S. Census.

Continued

Tutorial 9 data sources (continued)

\Gistutorial\LakePrecinct\Lake.mdb\LakeBusinesses, courtesy of InfoUSA.

\Gistutorial\LakePrecinct\Lake.mdb\lakecarbeats, courtesy of the Rochester, New York, Police Department.

\Gistutorial\LakePrecinct\Lake.mdb\lakeprecinct, courtesy of the Rochester, New York, Police Department.

\Gistutorial\UnitedStates\California\CACounties.shp, from ESRI Data & Maps, 2004, courtesy of U.S. Census.

\Gistutorial\UnitedStates\California\Earthquakes.dbf, from ESRI Data & Maps, 2004, courtesy of National Atlas of the United States, USGS.

\Gistutorial\UnitedStates\Cities.shp, from ESRI Data & Maps, 2004, courtesy of U.S. Census.

\Gistutorial\PAGIS\PghZipCodes.shp, courtesy of the City of Pittsburgh, Department of City Planning.

\Gistutorial\PAGIS\CensusBlocks.shp, courtesy of the City of Pittsburgh, Department of City Planning.

\Gistutorial\PAGIS\15222\15222Bldgs.shp, courtesy of the City of Pittsburgh, Department of City Planning.

\Gistutorial\PAGIS\15222\15222Streets.shp, courtesy of the City of Pittsburgh, Department of City Planning.

\Gistutorial\PAGIS\15222\15222Curbs.shp, courtesy of the City of Pittsburgh, Department of City Planning.

Related titles from ESRI Press

GIS Tutorial for Health
ISBN 978-1-58948-148-0

GIS Tutorial for Marketing
ISBN 978-1-58948-079-7

Getting to Know ArcView GIS
ISBN 978-1-879102-46-0

Extending ArcView GIS
ISBN 978-1-879102-05-7

Getting to Know ArcGIS Desktop, Second Edition
ISBN 978-1-58948-083-4

GIS for Everyone, Third Edition
ISBN 978-1-58948-056-8